Springer Water

The book series Springer Water comprises a broad portfolio of multi- and interdisciplinary scientific books, aiming at researchers, students, and everyone interested in water-related science. The series includes peer-reviewed monographs, edited volumes, textbooks, and conference proceedings. Its volumes combine all kinds of water-related research areas, such as: the movement, distribution and quality of freshwater; water resources; the quality and pollution of water and its influence on health; the water industry including drinking water, wastewater, and desalination services and technologies; water history; as well as water management and the governmental, political, developmental, and ethical aspects of water.

More information about this series at http://www.springer.com/series/13419

Steffen Mischke
Editor

Large Asian Lakes in a Changing World

Natural State and Human Impact

Springer

Editor
Steffen Mischke
Institute of Earth Sciences
University of Iceland
Reykjavík, Iceland

ISSN 2364-6934 ISSN 2364-8198 (electronic)
Springer Water
ISBN 978-3-030-42253-0 ISBN 978-3-030-42254-7 (eBook)
https://doi.org/10.1007/978-3-030-42254-7

This Springer imprint is published by the registered company Springer Nature Switzerland AG
The registered company address is: Gewerbestrasse 11, 6330 Cham, Switzerland

Preface

Lakes represent important ecosystems which are often biodiversity hotspots and important habitats for aquatic organisms and migratory birds. They represent economic resources for local fishery sectors and tourism activities, and provide invaluable ecosystem services. Concerns about biodiversity loss due to degraded water quality, shrinking lake areas and ecological consequences of species introductions inspired systematic and intensive monitoring programmes to support their sustainable use, improved management and more efficient conservation.

However, human impacts on lakes are almost omnipresent now, and there are only few regions on Earth left where lakes are not directly affected by local human activities. Even lakes in remote areas cannot be considered as pristine and in a fully natural state anymore due to anthropogenic changes of climate and biogeochemical cycles through emissions of greenhouse gases and reactive nitrogen which became apparent since the middle of the last century (Wolfe et al. 2013; Zalasiewicz et al. 2017).

Significant human impacts on lakes may have started much earlier than in the middle of the last century or even long before the Industrial Revolution in the eighteenth and nineteenth century, and it is not trivial to define a desired lake state more or less representing natural conditions as a baseline for the assessment of lake restoration measures (Kostianoy et al. 2004).

This book describes large lakes in Asia and their present state as a result of man-made alterations and conditions assumed as their natural state. Examples from mostly closed-basin lakes examine cases which are especially prone to human impacts on their catchment's water balance (Fig. 1). Past anthropogenic alterations of the lake systems may have been as dramatic as the included Aral and Dead seas. Described is also an oppositional example of a large closed-basin lake in Mongolia which experienced relatively minor impacts on the lake ecosystem so far.

The book provides a reference for the natural state of some of the largest lakes in Asia. Information on the natural states of lakes was collected from many sources including hardly accessible reports and data collections. Observational data from pre-impact periods were gathered, and proxy data from drilled or exposed lake sediment records were assessed to describe the natural state of lakes. Information

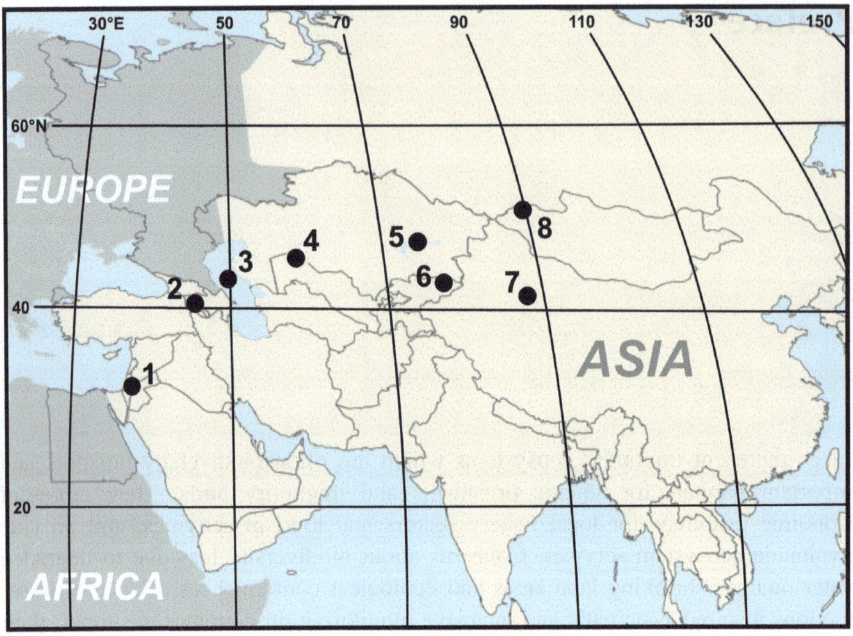

Fig. 1 Lakes included in this book: (1) the Dead Sea, (2) Lake Sevan, (3) the Caspian Sea, (4) the Aral Sea, (5) Lake Balkhash, (6) Lake Issyk-Kul, (7) Lake Lop Nur and (8) Lake Uvs Nuur

with respect to human impacts in the catchment of lakes such as water withdrawal from tributaries and concerning direct impacts on lakes (e.g. due to fishery activities and the introduction of alien species) were assembled. Both types of information, on the natural state of a lake and its alteration history as a result of human activities, will be crucial for the assessment of lake conditions in future and for endeavours of lake ecosystem restorations.

Reykjavík, Iceland Steffen Mischke

Acknowledgements The original manuscripts of the different chapters were carefully examined by the following reviewers who provided constructive assessments, detailed suggestions and corrections: Ian P. Wilkinson, Nicolas Waldmann, Siegmar W. Breckle, Philippe Sorrel, Brigitte Nixdorf, Ulrich Kamp, Hamid Alizadeh Lahijani, Suzanne Leroy, Anja Schwarz, Peter Frenzel, Tamara Yanina, Frank Neumann, Omid Beyraghdar Kashkooli and one anonymous reviewer. Their most valuable contributions to the improvements of chapter drafts are highly acknowledged.

References

Kostianoy AG, Zavialov PO, Lebedev SA (2004) What do we know about dead, dying and endangered lakes and seas? In: Nihoul JCJ, Zavialov PO, Micklin P (eds) Dying and Dead Seas. Climatic versus Anthropic Causes. NATO ARW/ASI Series. Kluwer Acad. Publ., Dordrecht, pp 1–48

Wolfe AP, Hobbs WO, Birks HH et al. (2013) Stratigraphic expressions of the Holocene-Anthropocene transition revealed in sediments from remote lakes. Earth Sci Rev 116:17–34

Zalasiewicz J, Waters CN, Summerhayes CP et al. (2017) The working group on the Anthropocene: summary of evidence and interim recommendations. Anthropocene 19:55–60

Contents

Chapter 1
The Dead Sea and Its Deviation from Natural Conditions

Revital Bookman

Abstract The Dead Sea is a hypersaline terminal lake located in a tectonic depression along the Dead Sea Transform. The regional paleogeographic setting began to take shape in the Miocene, and by the late Neogene fluvio-lacustrine sequences confined to the basin announced a new subsidence regime. Marine intrusion flooded the tectonic valley in the Pliocene, leading to the accumulation of a thick sequence of evaporites and initiated the development of the unique Ca-chloride brine of the basin. Rapid tectonic movements during the Pleistocene, or even earlier, disconnected the open sea and the valley became a deep, landlocked depression that hosted a series of lacustrine phases. The lakes deposited fine-laminated sequences during relatively wet phases and precipitated gypsum and halite during arid periods. The uppermost sedimentary fill comprises the Late Pleistocene Lisan Formation and the Holocene Ze'elim Formation. Lake Lisan existed during the last glacial period and in its highest stand extended from the Sea of Galilee to south of the Dead Sea, more than 200 m above the current level. The transition to the Holocene began with a dramatic dry-up recorded as a series of retreat strands on the margins of the basin and deposition of a thick halite unit. The Holocene Dead Sea was restricted to the deepest depression and its level fluctuated around 400 m below sea level in pace with climatic fluctuations dictated mostly by precipitation over the northern headwaters. The lake-water balance is a proxy for regional freshwater availability that influenced cultural transformations and demographic patterns during historical periods. The modern Dead Sea is being altered by intensified human activities. Water diversion and damming of freshwater for domestic and agricultural use, and brine evaporation for the potash industry have resulted in a level drop of over 1 m yr^{-1}. The negative water balance has led to erosion of the exposed margins and development of sinkholes due to subsurface evaporite dissolution. The stable stratification of the lake-water column has diminished, and seasonal halite deposition characterizes the modern lake. The level drop has also resulted in drying and migration of spring seepages, putting unique ecosystems under threat. Currently, diversion of sea water from the Red Sea to the Dead Sea is viewed as a comprehensive solution for stabilizing the level as well as for

R. Bookman (✉)
The Dr. Moses Strauss Department of Marine Geosciences, Leon H. Charney School of Marine Sciences, University of Haifa, 199 Aba Khoushy Ave., Mt. Carmel, 31905 Haifa, Israel
e-mail: rbookman@univ.haifa.ac.il

S. Mischke (ed.), *Large Asian Lakes in a Changing World*, Springer Water,
https://doi.org/10.1007/978-3-030-42254-7_1

producing hydroelectric power and desalinated water. However, this ambitious initiative must hurdle diplomatic tensions and financing difficulties, as well as intensive environmentalists' objections.

Keywords Dead Sea Basin · Lisan Formation · Ze'elim Formation · Water diversion · Lake-level drop · Red Sea–Dead Sea Water Conveyance

1.1 Introduction

The Dead Sea, the lowest place on land, is one of the saltiest lakes on the earth. The lake is a terminal basin at the border between Israel and Jordan in the Middle East (Fig. 1.1). The current lake volume is ~132 km^3, with a surface area of ~630 km^2, and a surface level elevation of ~430 meters below sea level (mbsl). The modern Dead Sea is a remnant of a series of water bodies that resided thought time in the central tectonic depression of the Dead Sea Transform, a plate boundary running from the East Anatolian Fault in southeastern Turkey to the northern end of the Red Sea (Garfunkel 1981; Garfunkel and Ben-Avraham 1996).

The tectonic inland valley, comprising lakes and freshwater environments, has been a major corridor for biological dispersal and hominin migration out of Africa (Goren-Inbar et al. 2000; Ben-Avraham et al. 2005). Despite its extreme, harsh environment during recent millennia, multiple accounts have shown that the hypersaline Dead Sea attracted humans during the course of history. The lake is mentioned in biblical stories and historical documentation and is enriched with archaeological sites that are a testimony to the wealth of the region's culture and economy (Hirschfeld 2006). However, in modern times the lake has been drastically altered by water diversion projects, industrial activities and landscape development in its watershed and along its shores. The changes are dramatic and are reflected in the lake's water balance and chemistry and the exposure of its shores. The rapid lake retreat accelerates environmental deterioration, including soft sediment erosion that results in rapid stream head-cut migration and widespread development of collapse sinkholes. These processes threaten coastal infrastructure and daily human life and have an impact on the natural environment. The lake's unique natural conditions have probably been irreversibly altered, although proposals are constantly being suggested to "save" the Dead Sea.

1.1.1 Geological Setting

The topography of the Dead Sea Basin has been shaped mostly by its tectonic nature and is bounded on the east and west by fault escarpments (Ben-Avraham and Lazar 2006; Shamir 2006). The tectonic depression is the central in a chain of grabens located at the plate boundary between the African and Arabian plates that

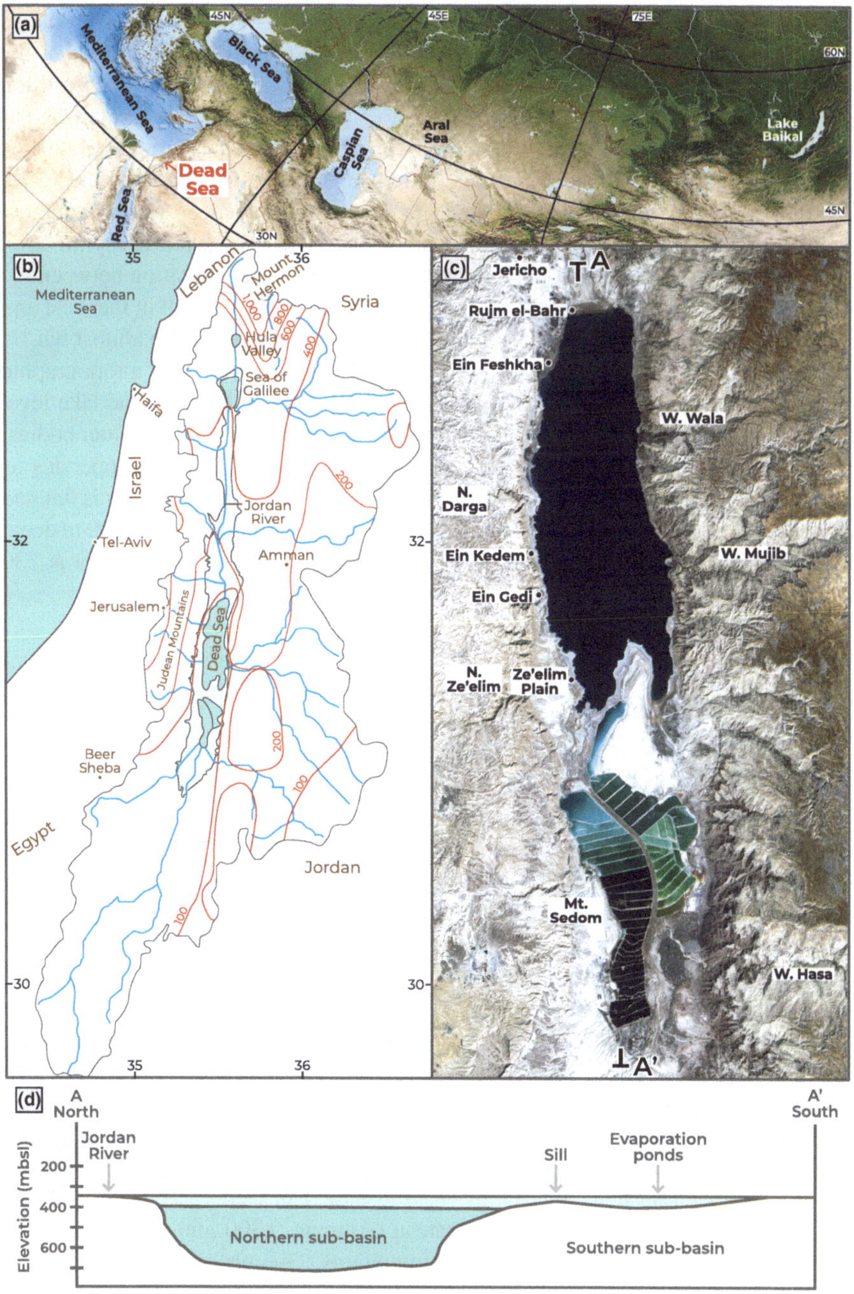

Fig. 1.1 **a** Location map of the Dead Sea. **b** The Dead Sea Basin watershed with major tributaries. In red, isohyets of average annual rainfall over the watershed. Lake Lisan at its highest stand of 165 mbsl (Bartov et al. 2003) is marked along the Jordan Valley. **c** Detailed location map of the Dead Sea. **d** North-south cross section of the Dead Sea Basin. During Holocene lowstands, the lake is restricted to the northern sub-basin (blue), while the southern shallow sub-basin is flooded during highstands (pale blue)

runs from the Red Sea to the continental collision at the Taurus-Zagros Mountains (Garfunkel 1981). Dating of fault surfaces along the transform indicates that the oceanic-to-continental plate boundary was initiated between 20.8 and 18.5 million years (Ma) ago, propagated northward around 17.1–12.7 Ma, and was established as a well-developed >500-km-long, plate-bounding fault at 3 Ma (Nuriel et al. 2017). The transform accommodated a left lateral movement of ~100 km along the Arava and Jordan valleys, which also initiated the development of deep tectonic depressions, with the Dead Sea Basin as the deepest one. The depression is asymmetric, with the mountains to the east being higher. The difference in elevation between the shoulders of the rift and the present shores of the lake is 300–500 m in the west and 400–1000 m in the east, while the southern and northern margins are almost flat.

The Dead Sea Basin is divided into two sub-basins separated by a topographic sill at the elevation of ~400 mbsl at Lynch Strait (Fig. 1.1). In 1979, the lake level dropped below the strait's elevation, dividing the lake into two separate water bodies; the northern deep sub-basin, and the southern shallow sub-basin that operates as evaporation ponds for the potash industries. While the southern sub-basin is flat and above the lake level, the northern sub-basin, the current terminal lake, is 300 m deep. This shrunken lake is ~50 km in length by ~15 km in width, with a flat floor at 730 mbsl (Hall 1996).

1.1.2 Climate and Hydrology

The Dead Sea watershed is located at the transition between the Mediterranean and desert climatic zones, reflected by sharp precipitation and temperature gradients. This climatic boundary follows the topography of the basin and stretches as a narrow tongue of semi-desert to desert conditions northward along the Lower Jordan Valley up to the Sea of Galilee. Mean winter and summer temperatures increase from the northern drainage (10–25 °C) southward to the Arava Valley (15–35 °C), and in the Dead Sea the recorded maximum temperature was above 50 °C (Israel Meteorological Service).

Average annual precipitation decreases from the northern sub-humid Mediterranean climate headwaters (100 to >1000 mm yr^{-1}) to less than 50 mm yr^{-1} over the southern hyperarid watershed. The lake surface receives less than 75 mm yr^{-1}, and its natural water balance is controlled mostly by the northern watershed. A rain shadow effect and low topography of the basin control the decrease in precipitation from west (Jerusalem, ~540 mm yr^{-1}) to east (Ein Gedi, ~75 mm yr^{-1}) and increase again towards the Jordanian Plateau (Amman, ~300 mm yr^{-1}) due to the high topography of the Edom and Moav mountains (Greenbaum et al. 2006). The wet season is from October to April, with most of the precipitation falling during the winter months (December-February). During the rainy season, cold, upper-level, low-pressure troughs migrate eastward over southern Europe and the Mediterranean Sea and drive cold and relatively dry air masses over the warm sea surface (Ziv et al. 2006; Trigo 2006). This process leads to the formation of surface low-pressure cyclones, which lift the moistened marine air to produce clouds and precipitation that feed the Jordan River and its tributaries, and subsequently the Dead Sea. During the other half of the year (May to September) the area is dry due to the strong regional atmospheric subsidence induced by the remote influence of the Indian summer monsoon system (Ziv et al. 2004).

The Dead Sea has one of the largest drainage systems in the Levant, covering ~42,000 km^2 (Fig. 1.1) and draining major tributaries to the Upper Jordan River from Mt. Hermon and the eastern Yarmouk and Zarka rivers and smaller-scale western watersheds that merge with the Jordan River south of the Sea of Galilee. The Wala, Mujib and Hasa watersheds drain directly to the Dead Sea from the east, while from the west a dense net of smaller tributaries drains the Judean Desert. From the south, the lake receives runoff occasionally from the Arava Valley watershed that spreads up to 100 km south of the Dead Sea. A water divide in the central Arava Valley separates the Dead Sea drainage basin from the Gulf of Aqaba (Greenbaum et al. 2006). This large hydrological system transfers sediment and runoff to the lake. The mean annual natural discharge at the outlet of the Jordan River to the Dead Sea was estimated at over one billion cubic meters prior to human intervention. Since 1932 when the Degania Dam was established, and particularly after the 1950s when water diversion from the Sea of Galilee and Yarmouk River intensified, the annual discharge was reduced to a few tens of millions of cubic meters (Hassan and Klein 2002; Gavrieli et al. 2011; Lensky and Dente 2015).

Freshwater to saline springs discharge to the Dead Sea Basin (Katz and Starinsky 2009). The main source of fresh groundwater that flows from the marginal aquifers is precipitation in the mountains to the west and east with groundwater residence time on a scale of tens of years (Kiro et al. 2014). Their overall contribution to the water balance is small; however, the springs have been shown to be a sensitive subsurface system that is strongly connected to changes in lake level and precipitation (Yechieli et al. 1995; Weber et al. 2018). The deep subsurface system is a density-driven flow that supports migration of brines from the Dead Sea Basin via deep aquifers (Stanislavsky and Gvirtzman 1999). Groundwater flows into the lake from the surrounding aquifers either subterraneously or through springs that discharge surficially along the margins. Submerged springs also discharge directly from the lake floor (Siebert et al. 2014); some are presently located above the water level due to the recession of the shores.

1.2 The Dead Sea Basin Lacustrine History

The Dead Sea Basin has been an effective sediment trap since the initiation of its tectonic subsidence during the Miocene and has acquired a thick sedimentary fill in the northern and southern sub-basins, where ~6–8 km and ~14 km of sediments accumulated, respectively (ten Brink and Flores 2012). The sedimentary history of the basin took place through the Oligocene and most of the Miocene when the region was a single morphological unit with low relief that drained northwestward to the

Mediterranean Sea (Garfunkel and Horowitz 1966). Fluvio-lacustrine sediments of the Hazeva Formation accumulated a >2-km thick sequence in the basin. The pre-rift paleogeographic settings were disrupted by uplifting of the rift margins in the latest Miocene; and a series of estuarine-lagoonar sequences announced a new syn-rift sedimentary regime. The Sedom Lagoon, a marine estuary that filled the tectonic depression (Starinsky 1974; Stein 2001), dated to 5–3 Ma (Belmaker et al. 2013), deposited a sequence of halite beds that are several kilometers thick (Zak 1967). The evaporitic sequence is exposed on the surface at the Sedom Diapir (Fig. 1.1) and its later partial dissolution during high lake levels had an important effect on the chemistry of Ca-chloride brines in the basin (Levy et al. 2018).

After the disconnection from the open sea and drying of the lagoon, the rift valley was occupied by a series of terminal lakes that were controlled by climatic conditions. These lakes deposited coarse to laminated fine-grained lacustrine sequences, and during arid periods evaporated and precipitated gypsum and halite (Stein 2014). Rapid tectonic subsidence began during the Pleistocene (ten Brink and Ben-Avraham 1989) and the basin became a deep, landlocked depression into which drainage of a large area was diverted. The new configuration resulted in the accumulation of the Amora Formation exposed on the eastern flanks of the Sedom Diapir and dated between ~740 and 70 thousand years (ka), covering seven glacial–interglacial cycles (Torfstein et al. 2009). The uppermost Quaternary fill in the basin comprises the Late Pleistocene Lisan Formation and the Holocene Ze'elim Formation (Bookman et al. 2006). Lake Lisan extended from the Sea of Galilee to Hazeva in the northern Arava during its maximum highstand (Neev and Emery 1967; Begin et al. 1974). The lake's existence has been dated to 70–14 ka (Schramm et al. 2000; Torfstein et al. 2013), corresponding in time to the last glacial period.

Lake Lisan operated between two distinct modes. The first was characterized by an extensive supply of freshwater that resulted in a rise in the lake's level, a density layered water column structure, and precipitation of fine laminations of alternating aragonite and clay-silt detritus. The second mode was marked by a decrease in freshwater input, resulting in mixing or complete overturn of the upper water column and precipitation of gypsum. These two modes reflect the climatic evolution of the region in the Late Pleistocene which fluctuated between drier and wetter periods (Stein et al. 1997) related to Northern Hemisphere climate (Bartov et al. 2003; Haase-Schramm et al. 2004; Torfstein et al. 2013). The transition to the Holocene was initiated with a dramatic level fall of Lake Lisan, recorded as massive deposition of gypsum (Schramm et al. 2000), a series of retreat level strands on the margins of the basin (Bookman et al. 2014) and its contraction below 465 mbsl, the lowest Late Pleistocene stands (Bartov et al. 2003). A short level rise above 400 mbsl during the Younger Dryas interval and a decline at 11–10 ka resulted in the deposition of a thick salt unit, which was recovered in cores drilled along the Dead Sea shores tens of meters below the surface (Stein et al. 2010).

1.3 The Holocene Dead Sea

1.3.1 Sediments of the Ze'elim Formation

The term "Dead Sea" relates to the water body that fills the Dead Sea Basin since the beginning of the Holocene. After the termination of the salt deposition event ca. 10 ka, the level of the Dead Sea rose and fluctuated around 400 mbsl, alternately flooding or withdrawing from the southern sub-basin (Migowski et al. 2006) in pace with climatic fluctuations mostly dictated by precipitation over the northern headwaters (Enzel et al. 2003). The northern extension of the Jordan Valley was occupied by the Sea of Galilee (Hazan et al. 2005), and the Lower Jordan River, a recent geomorphic feature, formed on the exposed Lisan lakebed (Marder et al. 2018); since then it delivers water from the wetter north to the Dead Sea.

The Holocene Ze'elim Formation, which was deposited on top of the salt unit (Yechieli et al. 1993), comprises fine-grained sequences of laminated detritus (*ld*) and alternating aragonite and detritus (*aad)* facies (Haliva-Cohen et al. 2012; Fig. 1.2). The *ld* facies appears with intercalated synchronous intervals of coarse sand, representing shallow coastal environments, or halite in the deep basin (Neugebauer et al. 2014). The facies consists of a mixture of regional dust inputs, mostly recycled from the loess surface cover of the Negev Desert, and local runoff erosion products from the catchment area. The *ld* mineralogy comprises quartz and calcite grains with minor feldspars and clays (Haliva-Cohen et al. 2012). The *ld* units are a few cm to meters thick; their thickness is not necessarily related to the humid conditions and increased runoff, but might be connected to drier conditions when the lake level was lower and more erodible material from older marginal lacustrine unconsolidated sediments were exposed to erosion, increasing the volume of suspended material in runoff (Migowski et al. 2006; Lu et al. 2017; 2020).

Although the *aad* facies is associated mainly with the Last Glacial Maximum (e.g. the upper member of the Lisan Formation; 26.2–17.7 ka; Prasad et al. 2004), it also appears in the Holocene formation, especially after 3.6 ka (Migowski et al. 2006; Waldmann et al. 2007). The detrital laminae are similar in mineralogy to the *ld* facies, however, their aeolian provenance is more distant and reflects enhanced transport of Saharan dust to the vicinity of the lake (Haliva-Cohen et al. 2012). The detrital and aragonite laminae represent discrete flashflood events that transported suspended sediments and dissolved bicarbonate in the rainy season to the lake (López-Merino et al. 2016). In the hypersaline Dead Sea, the aragonite precipitation is an inorganic process resulting from mixing of freshwater runoff and brine in the surface layer of the lake during the rainy season (Barkan et al. 2001). Aragonite deposition is also enhanced during dust storms that are often associated with Mediterranean rain fronts. The carbonate fraction in the dust dissolves in the runoff and increases the availability of bicarbonate for aragonite deposition (Belmaker et al. 2019). In parts of the Ze'elim Formation, a primary gypsum lamina formed instead of or additionally to the aragonite (Heim et al. 1997; Migowski et al. 2006; Torfstein et al. 2008) and

Fig. 1.2 Environments and sediments of the Dead Sea Basin. **a** The Dead Sea modern shore. View along the western shore towards the south. Notice the abandoned shorelines on the mud plain that used to be the lake floor. **b** Gully development on the retreating mud flats. The rapid lake retreat accelerates environmental deterioration, including soft sediment erosion that results in rapid stream head-cut migration. **c** Salt deposition on the Dead Sea modern shore. **d** The Ein Feshkha spring system, the largest on the western Dead Sea coast. The Dead Sea level drop resulted in drying and migration of spring seepages, development of gullies that drain the spring outflow and incise the exposed mud flats. **e** Flashflood plume of runoff water laden with suspended detrital sediments entering the Dead Sea. These suspended sediments will later spread and settle on the lake floor. **f** Southward view of the Ze'elim Plain on the western coast of the Dead Sea. Shore terraces abandoned by the shrinking lake run parallel to the shore. Sinkholes can be seen near the shore and on the mud plain. Notice the Dead Sea works pumping station on the top left corner. Picture by Rotem Shachal

represents drier years with lower water input and enhanced evaporation (Stein et al. 1997).

In the Holocene, prior to the modern human-induced limnological conditions, halite deposits occurred only in the deep environment of the northern sub-basin (Heim et al. 1997; Neugebauer et al. 2015), or in the shallow southern sub-basin when it became disconnected from the main lake (Neev and Emery 1967, 1995). These deposits are a proxy for extreme dry conditions and a negative water balance and were used to determine the periods when the Dead Sea level shrank below the sill elevation (Bookman (Ken-Tor) et al. 2004).

1.3.2 The Holocene Dead Sea Level: Climatic and Human Implications

The Dead Sea level curve (Fig. 1.3) was reconstructed from deep-cut exposures in the Ein Qedem steep shores (Stern 2010), the Darga fan delta (Kadan 1997; Bartov 2004), David fan delta in the proximity of Ein Gedi, and Ze'elim Plain (Bookman (Ken-Tor) et al. 2004; Kagan et al. 2015). These outcrops were exposed due to the anthropogenic retreat of the Dead Sea in recent decades. The early Holocene record was recovered from sediment cores from the lake margins and the deep basin floor (Yechieli et al. 1993; Heim et al. 1997; Migowski et al. 2006; Neugebauer et al. 2014).

Near-shore sections on the margins of the Dead Sea include shore ridges, evaporitic crusts, lagoonar deposits, and ripple marks that are all indicators for lake level elevation (Fig. 1.4). A detailed discussion of their sedimentological use as absolute level indicators is available in Machlus et al. (2000), Bookman (Ken-Tor) et al. (2004) and Bartov et al. (2006). The record from sediment cores could only yield a relative lake-level reconstruction, as the level was above their location most of the time. Nevertheless, the level reconstruction is based on stratigraphic correlation between cores and lithological comparison with the exposed sections in near-shore environments (Migowski et al. 2006). This comparison allows integration of information from the deeper-lacustrine environment with that of the near-shore to give the fullest and most detailed reconstruction.

The Dead Sea level is a sensitive proxy for changes in Holocene climate that can be used to explore the connection between environmental-hydrological conditions and human social changes. As freshwater availability and agriculture are highly connected to climate, high dependency of human activities upon climatic factors is expected, and climatic conditions were discussed in association with transformations in settlement and demographic patterns (Migowski et al. 2006; Neumann et al. 2007a; Litt et al. 2012; Langgut et al. 2013, 2015; Roberts et al. 2019).

The deposits overlying the late Pleistocene–Holocene transition salt unit were dated roughly between 10 and 8 ka. A marginal shore ridge from the beginning of

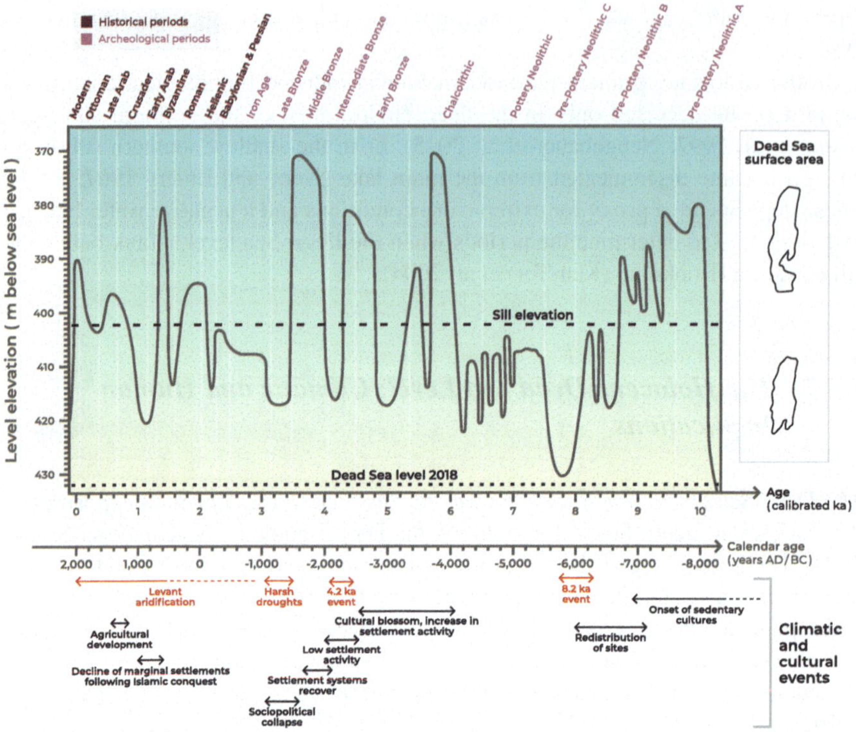

Fig. 1.3 The composite Holocene Dead Sea lake-level curve reconstructed from deep-cut exposures in the Ein Qedem steep shores, the Darga and David fan delta, and the Ze'elim Plain (Bookman (Ken-Tor) et al. 2004; Bartov 2004; Stern 2010; Kagan et al. 2015). The chronology is based on over 100 radiocarbon ages of organic debris. Main climatic and cultural shifts are presented below the curve. The southern sub-basin of the Dead Sea dries at a lake level below 402 mbsl

the Holocene marked an elevation of 380 mbsl meaning the shallow southern sub-basin was flooded, after the dramatic desiccation (Migowski et al. 2006). The early Holocene climate amelioration reflected in the Dead Sea highstands can be linked to the onset of sedentary cultures, belonging to the Pre-Pottery Neolithic culture in the Near East (Bar-Yosef 2000), such as Jericho in the Lower Jordan River Valley and Ain Ghazal in the Jordanian highlands (Migowski et al. 2006). This period represented a major transformation in prehistoric societies from small bands of mobile hunter–gatherers to large settled farming and herding villages in the Mediterranean zone (Barker and Goucher 2015).

After 9 ka the lake level dropped below the sill elevation separating the Dead Sea sub-basins, which resulted in halite deposition at the southern sub-basin (Neev and Emery 1995; Charrach 2018). Evaporite crusts from on-shore exposures and a hiatus in sediment cores suggest a rapid drop of lake level ca. 8.1 ka to below 430 mbsl (Migowski et al. 2006; Stern 2010). This level drop, which lasted only a few hundred years, was the lowest stand of the Dead Sea level during the Holocene. Within the

Fig. 1.4 Sedimentary lake-level indicators. **a** Shore ridges that build-up as a respond to wave action during the winter are abandoned with the level retreat (>1 m yr^{-1}), leaving annual steps on the lake margins. They are composed of well-rounded, clast-supported gravels commonly with platy pebbles. The shore ridge is ~1 m high. **b** Holocene lacustrine sediments composed of alternating aragonite and silt- to clay-size detrital laminae (*aad* facies). Laminations are less than a mm to a couple of cm thick. **c** An outcrop that consists of shore-ridge deposits interlayered within fine laminated lacustrine sediments in a Holocene fan-delta. The shore deposit is composed of rounded cobbles some covered with aragonite crusts, and represent deposition at the waterfront. Scale bar is 0.5 m. **d** Symmetric ripple marks at the exposed lake floor. The ripples are covered by coarse salt crystals that characterize the shore environment. Scale bar is 0.1 m

dating uncertainties, this lowstand correlates with the 8.2 ka cold event (Rohling and Pälike 2005), the globally most extreme cooling event during the Holocene (Alley and Ágústsdóttir 2005). The sudden cooling and simultaneous decrease in rainfall intensity was also inferred from a southeastern Mediterranean Sea pollen record (Langgut et al. 2011) and the Soreq Cave speleothem record (Bar-Matthews et al. 1999). In the following two millennia the lake fluctuated around 410 mbsl and was restricted to the northern sub-basin.

The arid phase that began after 9 ka is synchronous with a major environmental and societal change of the first farmer and last hunter-gatherer communities over the entire Mediterranean area. Frequent cultural gaps as redistribution of sites and changes in material culture are observed in the northern Mediterranean, and a socio-economic rupture is associated with a spatial redistribution of sites in the Near East

(Berger and Guilaine 2009). In the Dead Sea area, the degeneration of the first city-like settlement of Jericho into a village during the Pre-Pottery Neolithic B (Bar-Yosef 2000) was temporarily associated with the climatic shift within the dating limits. With the continuation of climatic deterioration, the Ain Ghazal population and its activity dropped sharply within only a few generations ca. 8 ka. This change has been observed at many Neolithic sites throughout the Levant, where once thriving Pre-Pottery Neolithic B villages were abandoned and often replaced by new, but less sophisticated, settlements. Ain Ghazal stands out as it was never reoccupied, probably because it is a climatically and environmentally sensitive location at the steppe-forest boundary close to the 250 mm isohyet which is considered the minimum amount of precipitation required for non-irrigation farming (Simmons et al. 1988). This explanation is supported by evidence from the sites located along the Jordan Valley with access to freshwater even during droughts (Bar-Yosef and Kra 1995).

In the Middle Holocene, after 6 ka, the level rose again above the sill elevation and fluctuated with large amplitudes between 370 and 415 mbsl (Migowski et al. 2006; Bartov 2004), flooding and receding from the southern sub-basin, respectively. This climate amelioration peaked in the Early Bronze Age and was associated with cultural blossoming in the entire region. This was a humid phase (Langgut et al. 2015) with an increase in settlement activity (Finkelstein and Gophna 1993; Neev and Emery 1995; Amiran and Ilan 1996; Finkelstein et al. 2006) and a developed olive orchard economy (Neumann et al. 2007b; Langgut et al. 2013). Bitumen exploitation from the lake was an important driving force behind the emergence of maritime activity including the use of watercraft on the Dead Sea; lumps of archaeological bitumen found in excavations of Canaan and Egypt have been geochemically linked to natural asphalts of the Dead Sea area (Oron et al. 2015). One of them, the Early Bronze I Tel Irani bitumen, has been shown to be identical to floating asphalt blocks on the Dead Sea (Connan et al. 1992).

A lowstand dated between 4.4 and 4.1 ka (Kagan et al. 2015), probably corresponds to the broader framework of the 4.2 ka event, a phase of environmental stress characterized by severe and prolonged drought in the Levant (Kaniewski et al. 2018). This dry period is consistent with the onset of sudden aridification that led to a cultural collapse of the Akkadian civilization in Mesopotamia (Cullen et al. 2000; DeMenocal 2001). In the Dead Sea region, pollen-based reconstructions also suggest drier climate conditions during that period (Neumann et al. 2007a; Langgut et al. 2014, 2016). This climatic shift corresponds to archaeological finds of low settlement activity at the end of the Intermediate Bronze Age that continued in the Judean Highlands into the Middle Bronze Age I (Ofer 1994; Finkelstein 1995). The dry phase in the very late Intermediate Bronze Age and the Middle Bronze Age I had a significant impact on settlement patterns in the entire Levantine region. During that time, the 400-mm rainfall isohyet, marking the boundary between the Mediterranean and Irano–Turanian vegetation zones, probably shifted to the north and west (Langgut et al. 2015). As a result, permanent settlements withdrew from the southern margins of Canaan, while populations in northeastern semi-arid zones, such as the Beq'a of Lebanon and the Jezirah in Syria, shrank in size, and communities moved to "greener" parts of the Levant. Wetter climatic conditions in the Middle Bronze

Age II–III caused the settlement system to recover and re-expand in the south in areas such as the Beer Sheva Valley in the northern Negev (Finkelstein and Langgut 2014).

However, this re-expansion due to improved climatic conditions did not last long, and the Dead Sea level record soon points toward broader climatic deterioration. This climatic shift, reflected by a large lake-level drop of ~40 m to ~415 mbsl in less than 200 years, was dated to ca. 3.5–3.3 ka. The most pronounced evidence for this lowstand is an anomalously thick (>1 m) beach ridge made of shore deposits and aragonite crusts in the Ze'eim outcrops (Bookman (Ken-Tor) et al. 2004; Kagan et al. 2015). Pollen records from that time, the end of the Late Bronze Age from the thirteenth century BCE to the end of the twelfth century BCE, show extremely sparse arboreal vegetation and evidence for the low settlement activity. This suggests that the shrinkage of Mediterranean forests was most probably not the result of human pressure, but due to a dry spell that took place across a vast geographic area (Litt et al. 2012; Langgut et al. 2015). By the end of the Bronze Age, harsh, long-term droughts had a major effect on the sociopolitical collapse in the Eastern Mediterranean (Weiss 1982; Neumann and Parpola 1987; Alpert and Neumann 1989; Ward and Joukowsky 1992; Issar 1998). The crisis years are represented by destruction of urban centers, shrinkage of major sites, and an overall change in the settlement pattern. Textual evidence from several places in the ancient Near East attests to drought and famine beginning in the mid-thirteenth century BCE and continuing until the second half of the twelfth century BCE (Astour 1965; Na'aman 1994; Zaccagnini 1995; Singer 1999, 2000).

After this major level drop, the Dead Sea never rose again to the Middle Holocene highstands and an increase in the *aad* facies is evident (Migowski et al. 2006). It has been suggested that the aragonite-laminae deposition became dominant due to enhanced precipitation and penetration of moist air into the Judean Mountains (Waldmann et al. 2007). However, this suggestion is in conflict with the moderate Dead Sea levels that did not rise above 390 mbsl in the Late Holocene (Ken-Tor (Bookman) et al. 2004). Results from modern flashflood sampling show that aragonite precipitation is probably not controlled solely by the input of freshwater but also requires high dust fluxes that increase carbonate saturation in the lake-surface water (Belmaker et al. 2019). It appears that the Late Holocene Dead Sea recorded the Levant aridification, which is evident as a decrease in the overall water balance of the lake, accompanied by an increase in dust storms (Frumkin and Stein 2004).

The Late Holocene landscape is anthropogenised by increased agricultural and herd-based communities. It is also when the demographic trends appear to decouple from the climatic shifts given the advancements in technology and the extensive social networks of empires that geared the capability of local communities to deal with environmental stress (Palmisano et al. 2019). Along the Late Holocene period, the Dead Sea level fluctuated around 400 mbsl, rising above the sill and flooding the southern sub-basin for short periods (Bookman (Ken-Tor) et al. 2004). Highstands occurred in the second and first centuries BCE and the fourth century CE during the Roman and early Byzantine periods, respectively. Although the age resolution of the reconstructed lake level is not as high as archaeological chronologies from that time, it is clear that the wet climate occurred during a period of cultural prosperity in the region. These periods are characterized by an increase in population, and settlement of desert margins (Issar et al. 1992; Hirschfeld 2004; Palmisano et al. 2019). Isotopic analysis of archaeological wood remains from the Roman siege rampart at Masada supports the evidence for wetter climatic conditions for the region (Yakir et al. 1994), and annual rainfall in Jerusalem was calculated to be above the modern mean annual rainfall (Orland et al. 2009). The longest Dead Sea lowstand during historical times occurred after the Byzantine period and continued at least until the ninth century CE. Aridity during early Medieval times and droughts in the southern Levant and other regions, notably Anatolia, may be one of the main factors behind the gradual long-term decline of settlement on the marginal lands in the Levant following the Islamic conquest (Izdebski et al. 2016). Although proximate causes such as invasions by the Roman Empire and altered imperial fiscal policies probably had a significant impact as well, an economic downturn evidenced by numismatic trends was shown to coincide with palaeoclimatic evidence for drought in the fifth century CE (Fuks et al. 2017).

A short-term highstand occurred in the 11th and 12th centuries CE during the Crusader period. Comprehensive archaeological evidence from this period suggests widespread agricultural development in the region (Ellenblum 1991), which could reflect favorable hydrologic conditions, as indicated by the reconstructed lake-level curve. The enhanced rainfall during this period did not directly lead to cultural flourishing, but it certainly could have reinforced and stabilized communities which were highly dependent on water resources and local agriculture.

The youngest highstand, which is also the last time the *aad* facies was deposited in the Dead Sea Basin (the upper laminated unit; Bookman (Ken-Tor) et al. 2004), lasted from the late 19th century CE to the first two decades of the 20th century. This highstand was instrumentally measured at 390 mbsl (Klein 1986) and marks a significant change in the annual rainfall in the region that exceeded the modern average. During this highstand the southern sub-basin was flooded for a few decades, giving the impression for the first modern expeditions and European Jewish immigrants that this was the natural surface area of the lake, although the lake level was lower and the lake restricted to the northern sub-basin during most of the historical period.

1.4 The Modern Dead Sea

1.4.1 Lake Level Records Since the Commencement of Scientific Research

Although historical references for Dead Sea levels are available from travelers' books and pilgrims' notes, instrumental lake-level measurements began only at the end of the 19th century (Klein 1986; Fig. 1.5) with the arrival of scientific expeditions to the area (Goren 2002). In 1865, on behalf of the Palestine Exploration

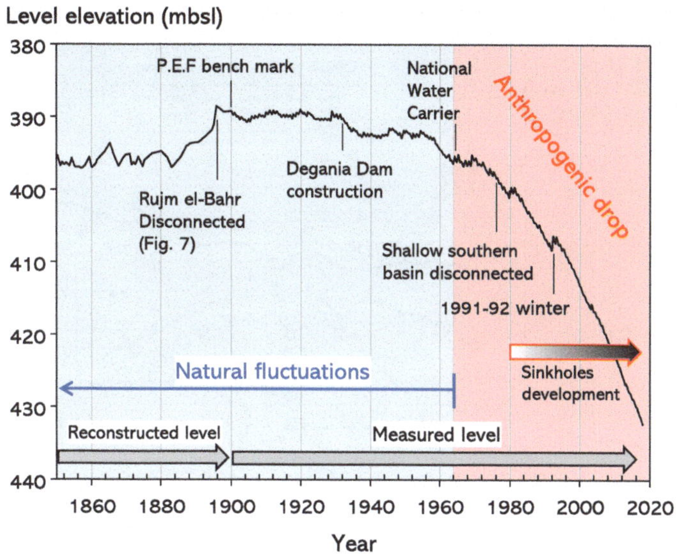

Fig. 1.5 Dead Sea level since 1850. Evidence for the natural level was recorded in historical documentations describing Rujm el-Bahr, a mound of rocks that functioned as an ancient mooring site at the northern shore of the Dead Sea in the second and first century BCE. Later, the level was recorded by a benchmark engraved on a rock used by the Palestine Exploration Fund in the beginning of the twentieth century. Since the completion of the Israel National Water Carrier water is diverted from the Sea of Galilee, resulting in much-reduced flows to the Dead Sea. By 1976–1977, the Dead Sea level fell to 402 mbsl, exposing the sill in the Lynch Strait and separating the lake into its two sub-basins. The rate of level drop that intensified after the 1980s was interrupted twice after the exceptionally rainy winters of 1979–1980 and 1991–1992. Sinkholes were first described in the 1980s, however, their rate of occurrence has increased dramatically since 2000

Fund, founded that year, the absolute Dead Sea level was determined to be ~400 mbsl (Klein 1961; Underhill 1967). During the latter part of that century, the level started to rise after a lowstand that exposed the southern basin floor (Klein 1961, 1986) as is also evident in historical maps (Fig. 1.6). By the 1880s, Rujm el-Bahr (Fig. 1.7), an ancient mooring site (Bar-Adon 1989), was disconnected from the northern shore, and by the 1890s, the lake level covered the site reaching ~390 mbsl. In 1900, a horizontal mark with the initials “P.E.F.” (Palestine Exploration Fund) was engraved on a rock at Ein Feshkha (also known as Einot Tzukim) marking this highstand, which prevailed with minor fluctuations between 1896 and 1929 (Klein 1982). In 1932, Degania Dam was built on the Jordan River immediately downstream its outlet from the Sea of Galilee. Its influence on the Dead Sea level was negligible, and during the 1930s, the 4–5 m level drop resulted mostly from severe droughts. Between 1937 and 1954, the lake level stabilized around 395 mbsl, and started dropping again in the late 1950s due to several years of drought. In 1964, the Israel National Water Carrier was completed, and water was diverted from the Sea of Galilee, resulting in much-reduced flows to the Dead Sea. Since the late 1960s–early 1970s, Jordan has diverted a large volume of flow from the second largest water

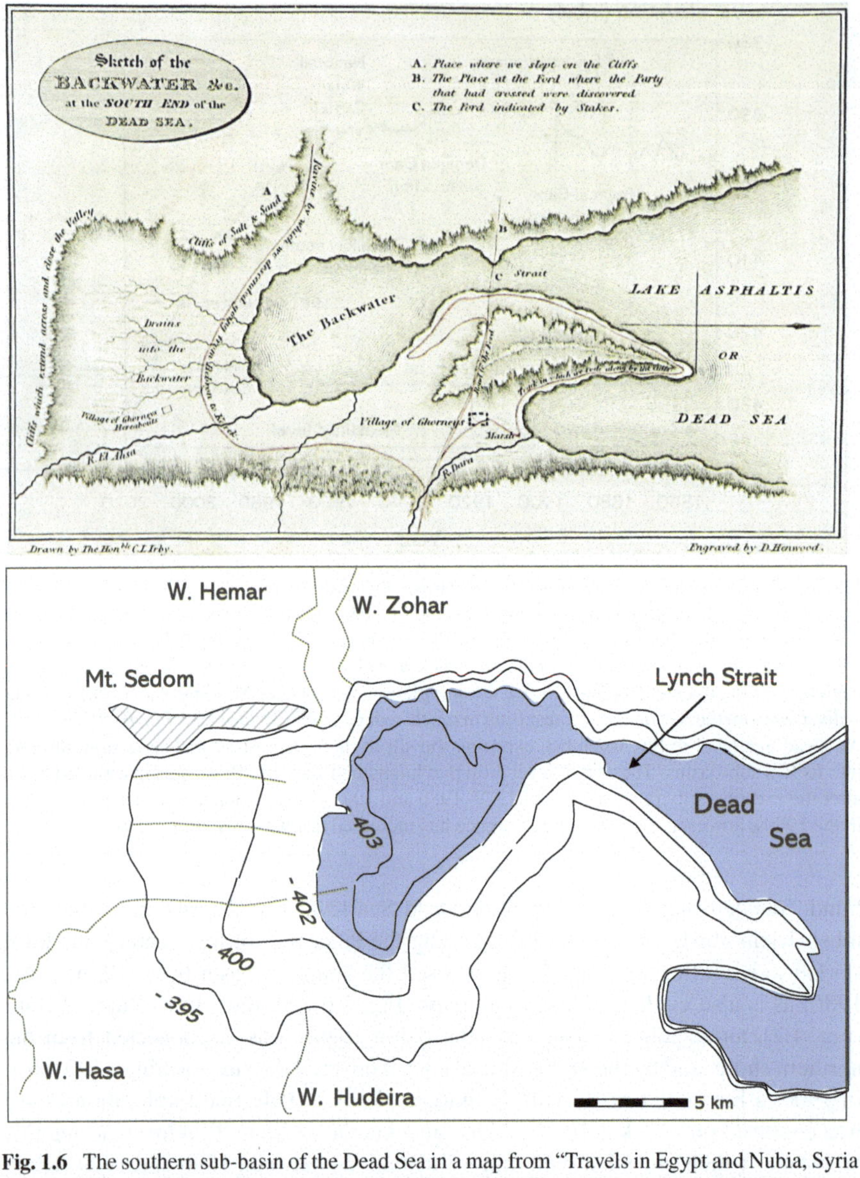

Fig. 1.6 The southern sub-basin of the Dead Sea in a map from "Travels in Egypt and Nubia, Syria and Asia Minor: during the years 1817 and 1818" by Irby, Charles Leonard and Mangles James. The historical map is compared with the Dead Sea bathymetry for a level at around 402 mbsl during the map drawing. The water depth at the strait was very shallow and a ford was connecting between the western and eastern shores

Fig. 1.7 The northern shore of the Dead Sea with the island of Rujm el-Bahr, ca. 1867–96 by Bonfils Studio PEF Catalogue PEF-P-1898

source to the Dead Sea, the Yarmouk River (Klein 1990), withdrawing additional water from the Dead Sea catchment. By 1976–1977, the Dead Sea level fell to 402 mbsl, exposing the sill in the Lynch Strait and separating the lake into its two sub-basins for the first time in more than a century. Since then, the Dead Sea has been confined to the northern sub-basin, reducing its surface area by >30%.

The level-drop rate increased with time, from 0.7 m yr^{-1} in the 1970–80s to 1.2 m yr^{-1} in 2015, although the rate of evaporation was supposed to decrease by more than 20% due to the reduction in the lake's surface and increase in its salinity (Lensky and Dente 2015). A seasonal examination of the level-drop found that during the summers no change in the level-drop rate was detected. Thus, the deficit cannot be attributed to the brine pumping and evaporation by the potash industries. However, the level-drop rate increased during the winter months, and since no significant change was observed in the annual precipitation, it is attributed to increase in effective utilization of runoff in reservoirs (Lensky and Dente 2015).

1.4.2 Limnology of a Lake with a Negative Water Balance

The 300-m-deep Dead Sea is unique in its lacustrine configuration, as all the other known, modern, hypersaline waterbodies are shallow continental basins or coastal lagoons (Sirota et al. 2016). The Ca-chloridic nature of the saline waters in the Dead Sea evolved from the relicts of the Sedom Lagoon waters and evaporites, and the subsequent ancient lacustrine phases (Katz and Starinsky 2009). The Dead Sea brine is presently found within deep boreholes, in saline springs along the Dead Sea shores, and in the Dead Sea itself, reaching a salinity of ~340 g l^{-1}. Extreme salinities of up to 550 g l^{-1} have been measured in sinkholes along the Dead Sea coast (Zilberman-Kron 2008).

Indigenous microbial communities, adapted to the extremely harsh conditions and unique ionic composition of the Dead Sea, have been documented since the early surveys of the lake. These organisms live at or near the upper limit of their tolerance (Wilkansky 1936; Elazari-Volcani 1943). Enrichment cultures and quantitative assessments of the microbial community densities in the Dead Sea water and sediment samples have revealed a variety of photosynthetic microorganisms, including eukaryotic flagellate algae and different types of cyanobacteria (Volcani 1944; Kaplan and Friedmann 1970). The diversity and abundance of biota is extremely poor and dominated by the unicellular green alga *Dunaliella parva*, as the primary producer, and various species of halophilic Archaea from the Halobacteriaceae family as the main consumers, appearing only during exceptionally rainy years when extensive rain and floods cause a significant dilution of the upper layer of the water column by at least 10% (Oren 2003).

The first in-depth studies of the properties of the Dead Sea water column showed that the lake was stratified (meromictic) and the shallow southern sub-basin was flooded at least from the late 19th century until the 1960s. A less saline upper water mass (epilimnium) floated over a denser water mass of brine (hypolimnium; Neev and Emery 1967). In 1959, the pycnocline was at a depth of about 40 m. The average density of the surface layer was 1.205 g cm^{-3}, compared with an in-situ density of 1.233 g cm^{-3} in the bottom layer. The long-term stability of the water column weakened with the drop in lake level, and successive deepening of the pycnocline from 70 m to beyond 200 m was recorded between 1975 and 1978. Complete overturn was observed during the winter of 1978–1979 (Steinhorn et al. 1979), and the lower water mass was estimated to have been disconnected from the atmosphere for about 300 years (Stiller and Chung 1984). This decline was accompanied by shrinkage of the shallow southern basin, which was finally disconnected in 1976. The complete overturn resulted in homogenization and oxidation of the entire water column (Steinhorn and Gat 1983). The Dead Sea waters were also continuously pumped by the Dead Sea Potash Works into the southern sub-basin that serves as evaporation ponds above the declining level until today (Fig. 1.1). The highly concentrated industrial end-brines, spilled back into the northern sub-basin, added to the gradual increase in the density of the mixolimnion that resulted in the end of the stratified water-column period (Gavrieli and Oren 2004). Currently, a thin upper layer with relatively low salinity develops in the winter and is separated by a halocline from an anaerobic hypolimnion. However, at least once a year, the density of the upper layer equals that of the deep water, causing destruction of the stratification by vertical mixing (Anati and Stiller 1991). As a result, the deep layers of the lake become exposed to the atmosphere and change from anaerobic to aerobic.

The holomictic state has been interrupted twice since the overturn, after the exceptionally rainy winters of 1979–1980 and 1991–1992. The runoff inflows led to dilution of the surface water and development of a stabilized halocline that maintained the stratification for 3–4 years (Anati and Stiller 1991). These events, and in particular, the anomalous winter that followed the volcanic eruption of Mt Pinatubo in 1991 (Bookman et al. 2014), served as a modern analogue for discussing the Dead Sea Basin limnological system during the last glacial highstands. Thermodynamic calculations based on the changes in the carbonate system in the surface-mixed-layer during this anomalous winter showed that 10% Dead Sea brine with a high Ca content mixed with 90% freshwater runoff with high bicarbonate content triggered aragonite deposition (Barkan et al. 2001).

The modern thermohaline stratification of the Dead Sea results in seasonal and depth variations in the chemistry of the water column. Monthly observations of the thermohaline epilimnion and hypolimnion evolution along with the anthropogenic level drop have shown that during the winter, the water column cools to a uniform temperature of ~23 °C, and with the onset of summer, the epilimnion warms, reaching a maximum temperature of ~35 °C in August. The salinity peak of the epilimnion lags a few weeks behind its temperature maximum and then declines toward fall and winter (Sirota et al. 2016). This seasonal variation controls the characteristics of halite deposition (Sirota et al. 2017). During summer, the epilimnion is undersaturated and halite is dissolved due to the high temperature, whereas the entire water column is supersaturated and crystallizes halite during winter. Although the variations in the water balance suggest the opposite since summer is associated with a higher loss of water by evaporation from the lake compared to the winter, the thermal effect overcomes the hydrological balance effect and thus governs the seasonal saturation cycle. In addition, precipitation of halite on the hypolimnetic lake floor at the expense of dissolution of halite from the epilimnetic floor, results in lateral focusing and thickening of halite deposits in the deeper part of the basin (Sirota et al. 2018).

1.4.3 Responses to the Anthropogenic Level Drop

Following the level drop and shrinkage of the Dead Sea surface area, a rapid widening of the lake margin occurred, particularly on the western coasts (Fig. 1.2). Coastal flats composed of coarse sediments developed at the proximal parts of the alluvial fans, and fine-grained muds at their distal parts, in places reaching up to 3 km from the coast of the last natural highstand at the beginning of the last century. On these flats, beach ridges that accumulated in response to wave action during the winter (Bartov 2004) are deserted annually, leaving steps that follow the dropping lake level (Fig. 1.4). The newly exposed areas that were the lake floor just a few decades ago are prone to erosion processes. Gullies begin to develop as soon as the previous lake floor is exposed (Fig. 1.2), particularly after exposure of the distal parts of the alluvial fans that are characterized by steep slopes (Filin et al. 2014).

The level drop forced the fluvial system to adjust. Decades of water diversion reduced the flows of the Jordan River to the Dead Sea, while the base level continued dropping. The greatest changes in the channel width near the mouth of the Jordan River were recorded between 1930 and 1980 as the base-level drop was relatively slow (Hassan and Klein 2002). However, after the 1980s, the increase in level drop resulted in upstream incision, bank collapses and a decline in channel width. Near the river mouth, large floods exert high shear stress within the confined channel increasing sinuosity (Dente et al. 2019). Channels draining across the western escarpment also present intensive incision with mean vertical entrenchment rates of 0.4–0.8 m y^{-1} and much higher widening rates of 2.1–24 m y^{-1} (Bowman et al. 2010). The measured and modeled annual sediment yield in these alluvial channels in response to the rapid, ongoing base-level drop has been estimated between 4.5 and 152 tons y^{-1} km^{-2}, also depending on the stream watershed characteristics and rainfall regime (Laronne et al. 2003; Ben Moshe et al. 2008).

As a result of the Dead Sea coastline withdrawal, the shallow groundwater system was altered, and spring migration was detected. In the Ein Feshkha spring system (Fig. 1.1), the largest on the western Dead Sea coast, the level drop resulted in drying and migration of spring seepages, development of gullies that drain the spring outflows and incise the exposed mud flats, and an enlargement of the wet, swamp-like zone (Fig. 1.2). Although the overall water balance of the lake is negative, the spring system maintained the groundwater discharge volume during the last century (Galili 2011), with only a slight change of mostly climate-induced variability (Burg et al. 2016). The hydrological and morphological changes also altered the aquatic habitats in Ein Feshkha and threaten its ecological systems and inhabitants, such as the fish *Aphafnius richarsoni*, which is endemic to the Dead Sea (Goren and Ortal 1999). An integrated intervention, using artificial habitat development, ensured the preservation of these unique fish communities that thrived in the spring system. This intervention was an unusual action; but it was needed due to the deviation in natural conditions resulting from the high rate of level drop (Millstein et al. 2017).

The most pronounced geomorphic response of the Dead Sea level retreat is the development of thousands of collapse sinkholes (Yechieli et al. 2006). The sinkholes developed along a narrow strip of several meters to over one km width along the coasts and result from dissolution of a thick salt layer buried below the surface (Abelson et al. 2003, 2006; Yechieli et al. 2003; Shalev et al. 2006; Ezersky 2008; Frumkin et al. 2011). This evaporite unit comprises salt that was deposited at the transition of the last glacial to the Holocene, during a dry spell dated to 11–10 ka (Yechieli et al. 1993; Stein et al. 2010). The elevation of the unit top in the subsurface varies between 415 and 443 mbsl and it probably remained submerged below the lake water during its deposition. During the Holocene, the adjacent shallow groundwater system was sufficiently saline to inhibit salt dissolution, and as the salt unit was covered by the Holocene sediment sequence it was not exposed to freshwater runoff prior to the modern level drop. The sinkholes started to appear in the 1980s (Frumkin and Raz 2001). However their rate of occurrence has increased dramatically since 2000 with the Dead Sea level-drop-rate increase (Abelson et al. 2017).

Seaward migration of the Dead Sea shoreline also drives subsurface brine migration. Topography-driven fresh groundwater that flows from the surrounding highlands (Stanislavsky and Gvirtzman 1999) penetrates the coastal shallow aquifer and replaces the retreating brine. Fault lines that ruptured the underlying silt and clay aquitard serve as conduits for the unsaturated groundwater, enabling access across the aquiclude layers (Abelson et al. 2003, 2006; Closson and Karaki 2009) that dissolve the salt layer that was previously immersed in the Dead Sea brine (Yechieli et al. 2006). The development of sinkholes is further enhanced in tributaries of alluvial fans, after flash-flood waters drain into existing sinkholes and enhance sub-surface salt dissolution (Shviro 2015; Avni et al. 2016).

Since the sinkhole phenomenon has hazardous consequences for the local industrial and tourist infrastructure, and for agriculture and daily human life (Fig. 1.8), it has been monitored intensively (Nof et al. 2013). The collapse of a sinkhole is often associated with gradual land subsidence that can precede the collapse by periods that range from days to a few years. Recently, systematic interferometric synthetic aperture radar observations, combined with detailed light detection and ranging measurements, are being used to sense minute precursory subsidence and alert before the catastrophic collapse (Nof et al. 2019). In addition, seismic monitoring by borehole geophones, which has been used to identify subsurface instabilities preceding sinkhole collapse, has shown that the hypocenter depths range from the base of the salt layer upwards. Mapping of the plan-view distribution provides evidence for the propagation of the salt-unit-front dissolution in the subsurface even in areas where no land subsidence was detected (Abelson et al. 2018). Examination of sinkhole evolution along the western shore of the Dead Sea in recent decades has revealed that although the main trigger for their development is the human-induced level drop, natural variations of precipitation in the recharge-source region, the Judean Mountains, ~25 km to the west, dominate their evolution rate with a lag period of 5–6 years (Abelson et al. 2017).

Fig. 1.8 Recent sinkhole development at Mineral Beach, a touristic site on the western shore of the Dead Sea. Picture by Rotem Shachal

1.4.4 The Dead Sea Water Projects — A Future Perspective

Geopolitical issues, scarce water resources, and the Jordan Valley Rift topography set the background for a variety of projects that have been suggested to connect the Dead Sea with the Mediterranean and Red seas for human uses. The incentives changed during the last 150 years from the unrealistic proposal to dig a navigable waterway that connects the Red Sea and the Mediterranean Sea (Vardi 1990; Al-Rubaiy 2000), to suggestions for the use of the topographic differences to generate an alternative hydroelectric energy source (Beyth 2007).

In the early 20th century, the Zionist movement advanced the Dead Sea Canal project. Various proposals were presented including the proposal for hydroelectric use by Theodore Herzl, first president of the World Zionist Organization (Powell 2005). Walter Clay Lowdermilk, an American soil conservation expert, outlined a vision for the economic rejuvenation of Palestine after a visit in the region in 1939. His proposal primarily concentrated on developing the Jordan River Valley and diverting the upper Jordan water to irrigate the Jezreel Valley, the coastal plain and the northern Negev, together with diversion of the Yarmouk and other tributaries into valley-side canals to facilitate irrigation of the surrounding areas. Mediterranean water would then be led by a tunnel and canal system from the vicinity of Haifa to hydroelectricity stations at the eastern edge of the valley, and then south to maintain the levels of the Dead Sea for its loss of the freshwater diverted from the Jordan River (Miller 2003). The outline became known as the "Lowdermilk plan" and although it was not implemented, the vision held importance for the later development of the National Water Carrier of Israel that was established in 1964.

The 1970s energy crisis revived the idea of the canal for generating hydroelectric power using the topographic gradient between the Mediterranean Sea and the Dead Sea, and led the Israeli government to establish the Mediterranean–Dead Sea Company Ltd. However, the project failed since Jordan, which did not have any formal relations with Israel at that time, filed complaints to the U.N. (Rauschning et al. 1997). It was recognized that the construction of the canal that will introduce sea water to the Dead Sea would violate the principals of international law and cause direct, irreparable damage to Jordanian potash production.

Following the Israel–Jordan peace treaty in 1994, both sides agreed to cooperate in activities related to the Dead Sea environment. This agreement together with an interim agreement between the Palestinians and Israelis lead to a pre-feasibility study on integrated development in the Jordan Rift Valley, which considered the Red Sea–Dead Sea Canal as the prominent alternative (Harza JRV Group 1996) and turned the main focus from energy production to desalination and halting of the chronic Dead Sea level drop (Beyth 2007). Negotiations between the three parties finally led to an agreement in 2005 on the terms of reference for conducting a feasibility study on one particular route of the proposed Red Sea–Dead Sea Canal (Fischhendler et al. 2015).

Despite the bureaucratic hurdles, financing difficulties, and intensive environmentalists' objections, as well as diplomatic tensions between the countries, this project is considered to be a comprehensive solution for stabilizing the Dead Sea level and producing desalinated water (Hussein 2017). The proposal calls for a transfer of up to 2000 million cubic meters of seawater a year from the Red Sea through the Arava Valley to the Dead Sea (Husein and Tsur 2016). The project would incorporate a hydroelectric power plant exploiting the elevation difference of 425 m between the Red and Dead seas, and a desalination plant with an output capacity of up to 800 million cubic meters a year. Jordan, the Palestinian Authority, and Israel, together with the World Bank, which guides the project, drafted terms of reference for a feasibility study and an environmental evaluation to assess whether the project was feasible and determine the environmental implications (Markel et al. 2013).

The expected physical, chemical and biological changes in the Dead Sea with different inflow volumes and brine concentrations (seawater and rejected brine from desalination) were delineated in a report submitted to the World Bank (Gavrieli et al. 2011). A main conclusion of the study program was that inflow of over 400 million cubic meters is needed to stabilize and raise the Dead Sea level and that over 1500 million cubic meters will result in flooding of the evaporation ponds in the southern sub-basin (Fig. 1.9). Meromictic conditions that will develop in the event of high inflow volumes have the potential to affect groundwater flow patterns adjacent to the impending stratified lake. Numerical simulations showed that intrusion of Dead Sea waters into the freshwater aquifer are possible if a thick mixolimnion (>30 m) with a relatively large density difference between the Dead Sea and regional groundwater develops (Oz et al. 2011). The diluted water may flow into the aquifer, dissolve the subsurface salt unit and increase sinkhole formation. Groundwater discharge may temporarily decrease due to a rise in lake level and landward migration of the shore before reaching a new controlled surface-lake elevation (Gavrieli et al. 2011) (Fig. 1.9).

Stratification can also affect biological blooming of the unicellular green alga *Dunaliella* and halophilic Archaea (Oren 2003). Mesocosm simulations showed that development of *Dunaliella* was possible only when Dead Sea water (340 g l^{-1} total dissolved salts) was diluted with a minimum of 10% of Red Sea water (40 g l^{-1} total dissolved salts). However, the availability of PO_4^{3-} is essential for the algae to grow, and growth rates and yields increased with increasing phosphate concentration and decreasing salinity. Field simulation showed that the development of algae was rapidly followed by development of dense blooms of red halophilic Archaea, which remained high for over two years in some scenarios, raising the concern that the biological properties may have important environmental implications for the lake (Oren et al. 2009).

An additional environmental concern related to the seawater or rejected brine inflow and stratification is massive nucleation and growth of gypsum crystals due to mixing of the SO_4^{2-}—rich Red Sea water and Ca^{2+}—rich Dead Sea brine (Reznik et al. 2009a). Mixing of Dead Sea and Red Sea waters resulted in precipitation of minute gypsum crystals that could lead to "whitening" of the surface waters of the Dead Sea (Reznik et al. 2009b). Gypsum precipitation may also remove essential nutrients, which will be no longer available for biological interactions and thus prevent algae blooms (Gavrieli et al. 2011).

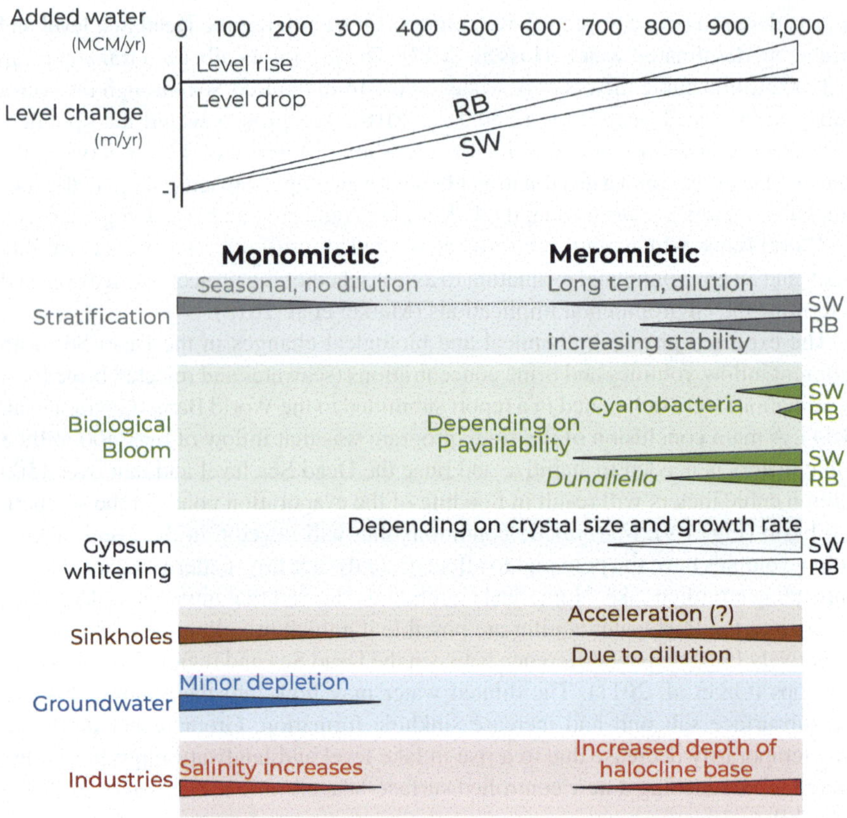

Fig. 1.9 Schematic illustration presenting the expected physical, chemical and biological changes in the Dead Sea with different inflow volumes and brine concentrations of seawater (SW) and rejected brine (RB) from desalination (Modified from Gavrieli et al. 2011)

Alternatives to the Red Sea–Dead Sea Water Conveyance have also been suggested, such as the Mediterranean Sea–Dead Sea southern and northern gravity tunnels, although the economic costs and risks are larger than for the Red Sea–Dead Sea project (Allan et al. 2014). Transfer of desalinated Mediterranean Sea water through the Jordan Valley for the restoration of the Jordan River flow was rejected (Markel 2017), however inflow of treated wastewater into the Jordan River from the wider region around the Jordan Valley was suggested within a regional NGO master plan for the year of 2050 (EcoPeace 2015). A "No Action Alternative" was also discussed (et Bellier 2014) and it was concluded that this scenario involves substantial, adverse changes to the Dead Sea and its surrounding environment. Under this alternative, the chemical industries would also eventually close and the Dead Sea would stabilize.

The future of the Dead Sea under a negative water balance and present climatic conditions was also studied by examining the extremely hypersaline brines in sinkholes along the western coast and in evaporation experiments. It was concluded that the level drop may continue down to 516–537 mbsl, corresponding to a water activity range of 0.46–0.39 in its brine, in equilibrium with the overlying median air humidity. This suggests that the lake level will drop more than 100 m from its present level in about 400 yrs before a new equilibrium is likely to be reached. The increase in salinity and the associated decrease in water activity will decrease the rate of evaporation (Yechieli et al. 1998; Zilberman et al. 2017). Even though, the Dead Sea will still be relatively deep (~200 m) and its surface area will not change dramatically as most of its previous shallow lake floor is already exposed. However, the topographic gradient will probably enhance marginal erosional processes that can compete with the tectonic subsidence and eventually can fill the basin with sediments.

Acknowledgements The writing of this paper was made possible through the support of the Israel Science Fund grant (ISF, Grant # 1093/10) to R. Bookman.

References

Abelson M, Yechieli Y, Baer G, Lapid G, Behar N, Calvo R, Rosensaft M (2017) Natural versus human control on subsurface salt dissolution and development of thousands of sinkholes along the Dead Sea coast. J Geophys Res-Earth 122(6):1262–1277

Abelson M, Aksinenko T, Kurzon I, Pinsky V, Baer G, Nof R, Yechieli Y (2018) Nanoseismicity forecasts sinkhole collapse in the Dead Sea coast years in advance. Geology 46(1):83–86

Abelson M, Yechieli Y, Crouvi O, Baer G, Wachs D, Bein A, Shtivelman V (2006) Evolution of the Dead Sea sinkholes. Geol S Am Spec Pap 401:241–254

Abelson M, Baer G, Shtivelman V, Wachs D, Raz E, Crouvi O, Kurzon I, Yechieli Y (2003) Collapse-sinkholes and radar interferometry reveal neotectonics concealed within the Dead Sea Basin. Geophys Res Lett 30(10) https://doi.org/10.1029/2003gl017103

Alley RB, Ágústsdóttir AM (2005) The 8k event: cause and consequences of a major Holocene abrupt climate change. Quat Sci Rev 24(10–11):1123–1149

Allan JA, Malkawi AH, Tsur Y (2014) Red Sea–Dead Sea water conveyance study program, study of alternatives. Final Report Executive Summary and Main Report. The World Bank, 186pp

Alpert P, Neumann J (1989) An ancient "correlation" between streamflow and distant rainfall in the Near East. J Near Eastern Stud 48:313–314

Al-Rubaiy S (2000) Water security, and the concept of sovereignty and peace in Jordan River Valley States. Dar al Hasad, Damascus, Syria

Amiran R, Ilan O (1996) Early Arad II. The chalcolithic and Early Bronze IB settlements and the Early Bronze II city—architecture and town planning. Israel Museum and Israel Exploration Society, Jerusalem

Anati DA, Stiller M (1991) The post-1979 thermohaline structure of the Dead Sea and the role of double-diffusive mixing. Limnol Oceanogr 36(2):342–353

Astour MC (1965) New evidence on the last days of Ugarit. Am J Archaeol 69(3):253–258

Avni Y, Lensky N, Dente E, Shviro M, Arav R, Gavrieli I, Yechieli Y, Abelson M, Lutzky H, Filin S, Haviv I (2016) Self-accelerated development of salt karst during flash floods along the Dead Sea Coast, Israel. J Geophys Res-Earth 121(1):17–38

Bar-Adon P (1989) Excavations in the Judean Desert. In: Zusman A, Straws D (eds) Antiquities. Israel Antiquities Authority, Jerusalem (in Hebrew)

Barker G, Goucher C (2015) The Cambridge world history Volume II: a world with agriculture, 12,000 BCE–500 CE. Cambridge University Press, Cambridge

Bar-Matthews M, Ayalon A, Kaufman A, Wasserburg GJ (1999) The Eastern Mediterranean paleoclimate as a reflection of regional events: Soreq cave, Israel. Earth Planet Sci Lett 166(1–2):85–95

Barkan E, Luz B, Lazar B (2001) Dynamics of the carbon dioxide system in the Dead Sea. Geochim Cosmochim Ac 65(3):355–368

Bartov Y (2004) Sedimentary fill analysis of a continental basin - the Late Pleistocene Dead Sea. Ph.D. thesis, Hebrew University of Jerusalem, Israel (in Hebrew with English abstract)

Bartov Y, Goldstein SL, Stein M, Enzel Y (2003) Catastrophic arid episodes in the Eastern Mediterranean linked with the North Atlantic Heinrich events. Geology 31(5):439–442

Bartov Y, Bookman R, Enzel Y (2006) Current depositional environments at the Dead Sea margins as indicators of past lake levels. Geol S Am Spec Pap 401:127–140

Bar-Yosef O (2000) The impact of radiocarbon dating on old world archaeology: past achievements and future expectations. Radiocarbon 42(1):23–39

Bar-Yosef O, Kra RS (eds) (1995) Late Quaternary chronology and paleoclimates of the Eastern Mediterranean. University of Arizona, Tucson

Belmaker R, Lazar B, Beer J, Christl M, Tepelyakov N, Stein M (2013) ^{10}Be dating of Neogene halite. Geochim Cosmochim Ac 122:418–429

Belmaker R, Lazar B, Stein M, Taha N, Bookman R (2019) Constraints on aragonite precipitation in the Dead Sea from geochemical measurements of flood plumes. Quat Sci Rev 221:105876

Ben Moshe LB, Haviv I, Enzel Y, Zilberman E, Matmon A (2008) Incision of alluvial channels in response to a continuous base level fall: field characterization, modeling, and validation along the Dead Sea. Geomorphology 93(3–4):524–536

Ben-Avraham Z, Lazar M (2006) The structure and development of the Dead Sea Basin: recent studies. Geol S Am Spec Pap 401:1–13

Ben-Avraham Z, Lazar M, Schattner U, Marco S (2005) The Dead Sea fault and its effect on civilization. In: Wenzel F (ed) Perspectives in modern seismology. Springer, Heidelberg

Berger JF, Guilaine J (2009) The 8200 cal BP abrupt environmental change and the Neolithic transition: a Mediterranean perspective. Quat Int 200(1–2):31–49

Begin ZB, Erlich A, Nathan I (1974) Lisan lake, the representative of the Pleistocene age in the dead sea. Geol Surv Isr Bull 63:1–30

Beyth M (2007) The Red Sea and the Mediterranean-Dead Sea canal project. Desalination 214(1–3):365–371

Bookman (Ken-Tor) R, Enzel Y, Agnon A, Stein M (2004) Late Holocene lake levels of the Dead Sea. Geol Soc Am Bull 116(5–6):555–571

Bookman R, Bartov Y, Enzel Y, Stein M (2006) Quaternary lake levels in the Dead Sea Basin: two centuries of research. Geol Soc Am Spec Pap 401:155–170

Bookman R, Filin S, Avni Y, Rosenfeld D, Marco S (2014) Possible connection between large volcanic eruptions and level rise episodes in the Dead Sea Basin. Quat Sci Rev 89:123–128

Bowman D, Svoray T, Devora S, Shapira I, Laronne JB (2010) Extreme rates of channel incision and shape evolution in response to a continuous, rapid base-level fall, the Dead Sea, Israel. Geomorphology 114(3):227–237

Burg A, Yechieli Y, Galili U (2016) Response of a coastal hydrogeological system to a rapid decline in sea level; the case of Zuqim springs–the largest discharge area along the Dead Sea coast. J Hydrol 536:222–235

Charrach J (2018) Investigations into the Holocene geology of the Dead Sea Basin. Carbonate Evaporite 34:1415–1442

Closson D, Karaki NA (2009) Human-induced geological hazards along the Dead Sea coast. Environ Geol 58(2):371–380

Connan J, Nissenbaum A, Dessort D (1992) Molecular archaeology: export of Dead Sea asphalt to Canaan and Egypt in the Chalcolithic-Early Bronze Age (4th-3rd millennium BC). Geochim Cosmochim Ac 56(7):2743–2759

Cullen HM, deMenocal PB, Hemming S, Hemming G, Brown FH, Guilderson T, Sirocko F (2000) Climate change and the collapse of the Akkadian empire: evidence from the Deep Sea. Geology 28(4):379–382
DeMenocal PB (2001) Cultural responses to climate change during the late Holocene. Science 292(5517):667–673
Dente E, Lensky NG, Morin E, Dunne T, Enzel Y (2019) Sinuosity evolution along an incising channel: new insights from the Jordan River response to the Dead Sea level fall. Earth Surf Proc Landforms 44(3):781–795
EcoPeace (2015) Regional NGO master plan for sustainable development in the Jordan valley, 181pp
Elazari-Volcani B (1943) Bacteria in the bottom sediments of the Dead Sea. Nature 152(3853):274
Ellenblum R (1991) Frankish rural settlement in crusader Palestine. Ph.D. thesis, Hebrew University of Jerusalem, Israel (in Hebrew)
Enzel Y, Bookman R, Sharon D, Gvirtzman H, Dayan U, Ziv B, Stein M (2003) Late Holocene climates of the Near East deduced from Dead Sea level variations and modern regional winter rainfall. Quat Res 60(3):263–273
et Bellier C (2014) Red Sea–Dead Sea water conveyance study program, feasibility study. Final Feasibility Study Report - Summary of Final FS Report. Accessed 14 Jan 2015
Ezersky M (2008) Geoelectric structure of the Ein Gedi sinkhole occurrence site at the Dead Sea shore in Israel. J Appl Geophys 64(3–4):56–69
Filin S, Avni Y, Baruch A, Morik S, Arav R, Marco S (2014) Characterization of land degradation along the receding Dead Sea coastal zone using airborne laser scanning. Geomorphology 206:403–420
Finkelstein I (1995) The great transformation: the "conquest" of the highlands frontiers and the rise of the territorial states. In: Levy TE (ed) The Archaeology of society in the Holy Land. Leicester University Press, London
Finkelstein I, Halpern B, Lehmann G, Niemann HM (2006) The Megiddo hinterland project. In: Finkelstein I, Ussishkin D, Halpern B (eds) Megiddo IV: the 1998–2002 seasons. Institute of Archaeology, Tel Aviv, pp 1998–2002
Finkelstein I, Gophna R (1993) Settlement, demographic, and economic patterns in the highlands of Palestine in the Chalcolithic and Early Bronze periods and the beginning of urbanism. B Am Sch Orient Res 289:1–22
Finkelstein I, Langgut D (2014) Dry climate in the middle Bronze I and its impact on settlement patterns in the Levant and beyond: new pollen evidence. J Near Eastern Stud 73(2):219–234
Fischhendler I, Cohen-Blankshtain G, Shuali Y, Boykoff M (2015) Communicating mega-projects in the face of uncertainties: Israeli mass media treatment of the Dead Sea Water Canal. Public Underst Sci 24(7):794–810
Frumkin A, Stein M (2004) The Sahara–East Mediterranean dust and climate connection revealed by strontium and uranium isotopes in a Jerusalem speleothem. Earth Planet Sci Lett 217(3–4):451–464
Frumkin A, Ezersky M, Al-Zoubi A, Akkawi E, Abueladas AR (2011) The Dead Sea sinkhole hazard: geophysical assessment of salt dissolution and collapse. Geomorphology 134(1–2):102–117
Frumkin A, Raz E (2001) Collapse and subsidence associated with salt karstification along the Dead Sea. Carbonate Evaporite 16(2):117–130
Fuks D, Ackermann O, Ayalon A, Bar-Matthews M, Bar-Oz G, Levi Y, Maeir AM, Weiss E, Zilberman T, Safrai Z (2017) Dust clouds, climate change and coins: consiliences of palaeoclimate and economy in the Late Antique southern Levant. Levant 49(2):205–223
Galili E (2011) Summary of hydrometric measurements in Einot Zukim, 2003–2011. Hydro Report: 1/2012, Gilat, Israel
Garfunkel Z (1981) Internal structure of the Dead Sea leaky transform (rift) in relation to plate kinematics. Tectonophysics 80(1–4):81–108

Garfunkel Z, Ben-Avraham Z (1996) The structure of the Dead Sea Basin. Tectonophysics 266(1–4):155–176
Garfunkel Z, Horowitz A (1966) The upper Tertiary and Quaternary morphology of the Negev, Israel. Isr J Earth Sci 15(3):101–117
Gavrieli I, Lensky N, Abelson M, Ganor J, Oren A, Brenner S, Lensky I, Shalev E, Yechieli Y, Dvorkin Y, Gertman I, Wells S, Simon E, Rosentraub Z, Reznik I (2011) Dead Sea study. Final Report, August 2011. Isr Geol Surv Rep. GSI/10/2011. Tel Aviv: TAHAL, Il-201280-R11-218 (Red Sea - Dead Sea Water Conveyance Study Program)
Gavrieli I, Oren A (2004) The Dead Sea as a dying lake. In: Nihoul JCJ, Zavialov PO, Micklin PP (eds) Dying and Dead Seas: climatic versus anthropogenic causes. Springer, New York
Goren-Inbar N, Feibel CS, Verosub KL, Melamed Y, Kislev ME, Tchernov E, Saragusti I (2000) Pleistocene milestones on the out-of-Africa corridor at Gesher Benot Ya'aqov, Israel. Science 289(5481):944–947
Goren H (2002) Sacred, but not surveyed: nineteenth-century surveys of Palestine. Imago Mundi 54(1):87–110
Goren M, Ortal R (1999) Biogeography, diversity and conservation of the inland water fish communities in Israel. Biol Conserv 89(1):1–9
Greenbaum N, Ben-Zvi A, Haviv I, Enzel Y (2006) The hydrology and paleohydrology of the Dead Sea tributaries. Geol Soc Am Spec Pap 401:63–93
Haase-Schramm A, Goldstein SL, Stein M (2004) U-Th dating of Lake Lisan (late Pleistocene Dead Sea) aragonite and implications for glacial East Mediterranean climate change. Geochim Cosmochim Ac 68(5):985–1005
Haliva-Cohen A, Stein M, Goldstein SL, Sandler A, Starinsky A (2012) Sources and transport routes of fine detritus material to the Late Quaternary Dead Sea Basin. Quat Sci Rev 50:55–70
Hall JK (1996) Digital topography and bathymetry of the area of the Dead Sea depression. Tectonophysics 266(1–4):177–185
Hassan MA, Klein M (2002) Fluvial adjustment of the Lower Jordan River to a drop in the Dead Sea level. Geomorphology 45(1–2):21–33
Harza JRV Group (1996) Red Sea–Dead Sea Canal Project: draft prefeasibility report. Jordan Rift Valley Integrated Development Study
Hazan N, Stein M, Agnon A, Marco S, Nadel D, Negendank JFW, Schwab MJ, Neev D (2005) The late Quaternary limnological history of Lake Kinneret (Sea of Galilee), Israel. Quat Res 63(1):60–77
Heim C, Nowaczyk NR, Negendank JF, Leroy SA, Ben-Avraham Z (1997) Near East desertification: evidence from the Dead Sea. Naturwissenschaften 84(9):398–401
Hirschfeld Y (2006) The archaeology of the Dead Sea valley in the Late Hellenistic and Early Roman periods. Geol S Am Spec Pap 401:215–229
Hirschfeld Y (2004) A climatic change in the early Byzantine period? Some archaeological evidence. Palest Explor Q 136(2):133–149
Hussein H (2017) Politics of the Dead Sea Canal: a historical review of the evolving discourses, interests, and plans. Water Int 42(5):527–542
Issar AS, Govrin Y, Geyh MA, Wakshal E, Wolf M (1992) Climate changes during the Upper Holocene in Israel. Isr J Earth Sci 40:219–223
Issar A (1998) Climate change and history during the Holocene in the eastern Mediterranean region. In: Issar A, Brown N (eds) Water, environment and society in times of climate change. Kluwer Academic, Dordrecht, pp 113-128
Izdebski A, Pickett J, Roberts N, Waliszewski T (2016) The environmental, archaeological and historical evidence for regional climatic changes and their societal impacts in the Eastern Mediterranean in Late Antiquity. Quat Sc Rev 136:189–208
Kadan G (1997) Evidence for Dead Sea Lake-level fluctuations and recent tectonism from the Holocene Fan-Delta of Nahal Darga, Israel. M.Sc. thesis, Ben Gurion University of the Negev, Israel (in Hebrew with English abstract)

Kagan EJ, Langgut D, Boaretto E, Neumann FH, Stein M (2015) Dead Sea levels during the Bronze and Iron ages. Radiocarbon 57(2):237–252
Kaniewski D, Marriner N, Cheddadi R, Guiot J, Van Campo E (2018) The 4.2 ka BP event in the Levant. Clim Past 14(10):1529–1542
Kaplan IR, Friedmann A (1970) Biological productivity in the Dead Sea Part I. Microorganisms in the water column. Isr J Chem 8(3):513–528
Katz A, Starinsky A (2009) Geochemical history of the Dead Sea. Aquat Geochem 15(1–2):159–194
Kiro Y, Weinstein Y, Starinsky A, Yechieli Y (2014) The extent of seawater circulation in the aquifer and its role in elemental mass balances: a lesson from the Dead Sea. Earth Planet Sci Lett 394:146–158
Klein C (1961) On the fluctuations of the level of the Dead Sea since the beginning of the 19th century. Hydrological Service, Jerusalem, Hydrological Paper 7, revised edition
Klein C (1986) Fluctuations of the level of the Dead Sea and climatic fluctuations in Israel during historical times. Ph.D. thesis, Hebrew University of Jerusalem, Israel (in Hebrew with English abstract)
Klein C (1982) Morphological evidence of lake level changes, western shore of the Dead-Sea. Isr J Earth Sci 31(2–4):67–94
Klein M (1990) Dead Sea level changes. Isr J Earth Sci 39:49–50
Langgut D, Almogi-Labin A, Bar-Matthews M, Weinstein-Evron M (2011) Vegetation and climate changes in the South Eastern Mediterranean during the Last Glacial-Interglacial cycle (86 ka): new marine pollen record. Quat Sci Rev 30(27–28):3960–3972
Langgut D, Finkelstein I, Litt T (2013) Climate and the Late Bronze Collapse: new evidence from the Southern Levant. J Inst Archaeol Tel Aviv 40(2):149–175
Langgut D, Neumann FH, Stein M, Wagner A, Kagan EJ, Boaretto E, Finkelstein I (2014) Dead Sea pollen record and history of human activity in the Judean Highlands (Israel) from the Intermediate Bronze into the Iron Ages (~2500–500 BCE). Palynol 38(2):280–302
Langgut D, Adams MJ, Finkelstein I (2016) Climate, settlement patterns and olive horticulture in the southern Levant during the Early Bronze and Intermediate Bronze Ages (c. 3600–1950 BC). Levant 48(2):117–134
Langgut D, Finkelstein I, Litt T, Neumann FH, Stein M (2015) Vegetation and climate changes during the Bronze and Iron Ages (~3600–600 BCE) in the southern Levant based on palynological records. Radiocarbon 57(2):217–235
Laronne J, Lekach J, Cohen H, Alexandrov Y (2003) Experimental drainage basins in Israel: rainfall, runoff, suspended sediment and bedload monitoring. Proceedings of the First Interagency Conference on Research in the Watersheds, October 2003, Benson, Arizona
Lensky NG, Dente E (2015) The hydrological processes driving the accelerated Dead Sea level decline in the past decades. Geol Surv Isr Rep, GSI/16/2015, Jerusalem
Levy EJ, Yechieli Y, Gavrieli I, Lazar B, Kiro Y, Stein M, Sivan O (2018) Salt precipitation and dissolution in the late Quaternary Dead Sea: evidence from chemical and δ^{37}Cl composition of pore fluids and halites. Earth Planet Sci Lett 487:127–137
Litt T, Ohlwein C, Neumann FH, Hense A, Stein M (2012) Holocene climate variability in the Levant from the Dead Sea pollen record. Quat Sci Rev 49:95–105
López-Merino L, Leroy SA, Eshel A, Epshtein V, Belmaker R, Bookman R (2016) Using palynology to re-assess the Dead Sea laminated sediments – indeed varves? Quat Sci Rev 140:49–66
Lu Y, Waldmann N, Nadel D, Marco S (2017) Increased sedimentation following the Neolithic Revolution in the Southern Levant. Glob Planet Change 152:199–208
Lu Y, Bookman R, Waldmann N, Marco S (2020) A 45 kyr laminae record from the Dead Sea: Implications for basin erosion and floods recurrence. Quat Sci Rev 229:106143
Machlus M, Enzel Y, Goldstein SL, Marco S, Stein M (2000) Reconstructing low levels of Lake Lisan by correlating fan-delta and lacustrine deposits. Quat Int 73:137–144
Malkawi AIH, Tsur Y (2016) Reclaiming the Dead Sea: alternatives for action. Society-Water-Technology. Springer, Cham, pp 205–225

Marder E, Bookman R, Filin S (2018) Geomorphological response of the lower Jordan River basin to active tectonics of the Dead Sea transform. Geomorphology 317:75–90
Markel D, Alster J, Beyth M (2013) The Red Sea-Dead Sea conveyance feasibility study. In: Becker N (ed) Water policy in Israel. Springer, Dordrecht, pp 2008–2012
Markel D, Sagiv M, Gavrieli I, Goldstein N, Cohen G, Arieli N (2011) Discussion on the Dead Sea rehabilitation. Ecol Environ 1:55–67 (in Hebrew)
Migowski C, Stein M, Prasad S, Negendank JF, Agnon A (2006) Holocene climate variability and cultural evolution in the Near East from the Dead Sea sedimentary record. Quat Res 66(3):421–431
Miller R (2003) Bible and Soil: Walter Clay Lowdermilk, the Jordan valley project and the Palestine debate. Middle Eastern Stud 39(2):55–81
Millstein D, Uzon A, Hazan E, Lidar N, Sabah A, Nesher R, Keshet N (2017) Development of wet habitats in Zukim nature reserve as a tool for nature preservation values. Ecol Environ 8(1):354–360 (in Hebrew)
Na'aman N (1994) The 'Conquest of Canaan' in the Book of Joshua and in history. In: Finkelstein I, Na'aman N (eds) From nomadism to monarchy: archaeological and historical aspects of early Israel. Israel Exploration Society, Jerusalem
Neev D, Emery KO (1967) The Dead Sea: depositional processes and environments of evaporates. Isr Geol Surv B 41:1–147
Neev D, Emery KO (1995) The destruction of Sodom, Gomorrah, and Jericho: geological, climatological, and archaeological background. Oxford University Press, Oxford
Neugebauer I, Brauer A, Schwab MJ, Dulski P, Frank U, Hadzhiivanova E, Kitagawa H, Litt T, Schiebel V, Taha N, Waldmann ND (2015) Evidences for centennial dry periods at ~3300 and ~2800 cal. yr BP from micro-facies analyses of the Dead Sea sediments. Holocene 25(8):1358–1371
Neugebauer I, Brauer A, Schwab MJ, Waldmann ND, Enzel Y, Kitagawa H, Torfstein A, Frank U, Dulski P, Agnon A, Ariztegui D, Ben-Avraham Z, Goldstein SL, Stein M, DSDDP Scientific Party (2014) Lithology of the long sediment record recovered by the ICDP Dead Sea Deep Drilling Project (DSDDP). Quat Sci Rev 102:149–165
Neumann FH, Kagan EJ, Schwab MJ, Stein M (2007a) Palynology, sedimentology and palaeoecology of the Late Holocene Dead Sea. Quat Sci Rev 26(11–12):1476–1498
Neumann F, Schölzel C, Litt T, Hense A, Stein M (2007b) Holocene vegetation and climate history of the northern Golan Heights (Near East). Veg Hist Archaeobot 16(4):329–346
Neumann J, Parpola S (1987) Climatic change and the eleventh-tenth-century eclipse of Assyria and Babylonia. J Near Eastern Stud 46(3):161–182
Nof RN, Baer G, Ziv A, Raz E, Atzori S, Salvi S (2013) Sinkhole precursors along the Dead Sea, Israel, revealed by SAR interferometry. Geology 41(9):1019–1022
Nof R, Abelson M, Raz E, Magen Y, Atzori S, Salvi S, Baer GSAR (2019) Interferometry for sinkhole early warning and susceptibility assessment along the Dead Sea, Israel. Remote Sens-Basel 11(1):89
Nuriel P, Weinberger R, Kylander-Clark AR, Hacker BR, Craddock JP (2017) The onset of the Dead Sea transform based on calcite age-strain analyses. Geology 45(7):587–590
Ofer A (1994) 'All the hill country of Judah': from a settlement fringe to a prosperous monarchy. In: Finkelstein I, Na'aman N (eds) From nomadism to monarchy: archaeological and historical aspects of early Israel. Israel Exploration Society, Jerusalem
Oren A (2003) Biodiversity and community dynamics in the Dead Sea: archaea, bacteria and eucaryotic algae. In: Nevo E, Oren A, Wasser SP (eds) Fungal life in the Dead Sea Ruggell. ARG Gantner Verlag
Oren A, Gavrieli J, Kohen M, Lati J, Aharoni M, Gavrieli I (2009) Long-term mesocosm simulation of algal and archaeal blooms in the Dead Sea following dilution with Red Sea water. Nat Resour Env Iss 15(1):27
Orland IJ, Bar-Matthews M, Kita NT, Ayalon A, Matthews A, Valley JW (2009) Climate deterioration in the Eastern Mediterranean as revealed by ion microprobe analysis of a speleothem that grew from 2.2 to 0.9 ka in Soreq Cave, Israel. Quat Res 71(1):27–35

Oron A, Galili E, Hadas G, Klein M (2015) Early maritime activity on the Dead Sea: bitumen harvesting and the possible use of reed watercraft. J Marit Archaeol 10(1):65–88
Oz I, Shalev E, Gvirtzman H, Yechieli Y, Gavrieli I (2011) Groundwater flow patterns adjacent to a long-term stratified (meromictic) lake. Water Resour Res 47(8):W08528
Palmisano A, Woodbridge J, Roberts CN, Bevan A, Fyfe R, Shennan S, Cheddadi R, Greenberg R, Kaniewski D, Langgut D, Leroy SAG, Litt T, Miebach A (2019) Holocene landscape dynamics and long-term populations trends in the Levant. Holocene 29(5):708–727
Powell JM (2005) The empire meets the new deal. Geograph Res 43(4):337–360
Prasad S, Vos H, Negendank JFW, Waldmann N, Goldstein SL, Stein M (2004) Evidence from Lake Lisan of solar influence on decadal-to centennial-scale climate variability during marine oxygen isotope stage 2. Geology 32(7):581–584
Rauschning D, Wiesbrock Katja, Lailach M (eds) (1997) Key resolutions of the United Nations General Assembly 1946–1996. CUP Archive
Reznik IJ, Ganor J, Gal A, Gavrieli I (2009a) Gypsum saturation degrees and precipitation potentials from Dead Sea–seawater mixtures. Environ Chem 6(5):416–423
Reznik IJ, Gavrieli I, Ganor J (2009b) Kinetics of gypsum nucleation and crystal growth from Dead Sea brine. Geochim Cosmochim Ac 73(20):6218–6230
Roberts N, Woodbridge J, Palmisano A, Bevan A, Fyfe R, Shennan S (2019) Mediterranean landscape change during the Holocene: synthesis, comparison and regional trends in population, land cover and climate. Holocene 29(5):923–937
Rohling EJ, Pälike H (2005) Centennial-scale climate cooling with a sudden cold event around 8,200 years ago. Nature 434(7036):975–979
Schramm A, Stein M, Goldstein SL (2000) Calibration of the ^{14}C time scale to >40 ka by ^{234}U–^{230}Th dating of Lake Lisan sediments (last glacial Dead Sea). Earth Planet Sci Lett 175(1–2):27–40
Shalev E, Lyakhovsky V, Yechieli Y (2006) Salt dissolution and sinkhole formation along the Dead Sea shore. J Geophys Res-Sol Earth 111(B3)
Simmons AH, Köhler-Rollefson I, Rollefson GO, Mandel R, Kafafi Z (1988) 'Ain Ghazal: a major Neolithic settlement in central Jordan. Science 240(4848):35–39
Siebert C, Rödiger T, Mallast U, Gräbe A, Guttman J, Laronne JB, Storz-Peretz Y, Greenman A, Salameh E, Al-Raggad M, Vachtman D (2014) Challenges to estimate surface-and groundwater flow in arid regions: the Dead Sea catchment. Sci Total Environ 485:828–841
Singer I (1999) A political history of Ugarit. In: Watson GEW, Wyatt N (eds) Handbook of Ugaritic Studies. Brill, Leiden
Singer I (2000) New evidence on the end of the Hittite empire. In: Oren ED (ed) The Sea peoples and their world: a reassessment. University of Pennsylvania Press, Philadelphia
Sirota I, Arnon A, Lensky NG (2016) Seasonal variations of halite saturation in the Dead Sea. Water Resour Res 52(9):7151–7162
Sirota I, Enzel Y, Lensky NG (2017) Temperature seasonality control on modern halite layers in the Dead Sea: in situ observations. GSA Bull 129(9–10):1181–1194
Sirota I, Enzel Y, Lensky NG (2018) Halite focusing and amplification of salt layer thickness: from the Dead Sea to deep hypersaline basins. Geology 46(10):851–854
Shamir G (2006) The active structure of the Dead Sea depression. Geol S Am Spec Pap 401:15–32
Shviro M (2015) The influence of flash-floods on subsidence rates in sinkholes sites along the Dead Sea: insights from high-resolution InSAR. Ph.D. thesis, Ben-Gurion University of the Negev, Israel
Starinsky A (1974) Relationship between Ca chloride brines and sedimentary rocks in Israel. Ph.D. thesis, The Hebrew University of Jerusalem, Israel (in Hebrew)
Stanislavsky E, Gvirtzman H (1999) Basin-scale migration of continental-rift brines: paleohydrologic modeling of the Dead Sea Basin. Geology 27(9):791–794
Stein M (2001) The sedimentary and geochemical record of Neogene-Quaternary water bodies in the Dead Sea Basin-inferences for the regional paleoclimatic history. J Paleolimnol 26(3):271–282

Stein M, Starinsky A, Katz A, Goldstein SL, Machlus M, Schramm A (1997) Strontium isotopic, chemical, and sedimentological evidence for the evolution of Lake Lisan and the Dead Sea. Geochim Cosmochim Ac 61(18):3975–3992
Stein M, Torfstein A, Gavrieli I, Yechieli Y (2010) Abrupt aridities and salt deposition in the post-glacial Dead Sea and their North Atlantic connection. Quat Sci Rev 29(3–4):567–575
Stein M (2014) The evolution of Neogene-Quaternary water-bodies in the Dead Sea rift valley. In: Garfunkel Z, Ben-Avraham Z, Kagan E (eds) Dead Sea Transform Fault System: Reviews. Springer, Dordrecht, pp 279–316
Steinhorn I, Assaf G, Gat JR, Nishry A, Nissenbaum A, Stiller M, Beyth MT, Neev D, Garber R, Friedman GM, Weiss W (1979) The Dead Sea: deepening of the mixolimnion signifies the overture to overturn of the water column. Science 206(4414):55–57
Steinhorn I, Gat JR (1983) The Dead Sea. Sci Am 249(4):102–111, C1–C8
Stern O (2010) Geochemistry, hydrology and paleo-hydrology of Ein Qedem spring system. Geol Surv Isr Rep, GSI/17/2010, Jerusalem
Stiller M, Chung YC (1984) Radium in the Dead Sea: a possible tracer for the duration of meromixis. Limnol Oceanogr 29(3):574–586
ten Brink US, Ben-Avraham Z (1989) The anatomy of a pull-apart basin: seismic reflection observations of the Dead Sea Basin. Tectonics 2:333–350
ten Brink US, Flores CH (2012) Geometry and subsidence history of the Dead Sea Basin: a case for fluid-induced mid-crustal shear zone? J Geophys Res-Sol Earth 117(B1)
Torfstein A, Gavrieli I, Katz A, Kolodny Y, Stein M (2008) Gypsum as a monitor of the paleo-limnological–hydrological conditions in Lake Lisan and the Dead Sea. Geochim Cosmochim Ac 72(10):2491–2509
Torfstein A, Haase-Schramm A, Waldmann N, Kolodny Y, Stein M (2009) U-series and oxygen isotope chronology of the mid-Pleistocene Lake Amora (Dead Sea Basin). Geochim Cosmochim Ac 73(9):2603–2630
Torfstein A, Goldstein SL, Kagan EJ, Stein M (2013) Integrated multi-site U-Th chronology of the last glacial Lake Lisan. Geochim Cosmochim Ac 104:210–231
Trigo IF (2006) Climatology and interannual variability of torm-tracks in the Euro-Atlantic sector: a comparison between ERA-40 and NCEP/NCAR reanalyses. Clim Dynam 26 (2–3):127–143
Underhill HW (1967) Dead Sea levels and the P.E.F. mark. Palest Explor Quarterly 99(1):45–53
Vardi J (1990) Mediterranean–Dead Sea Project – historical review. In: Arad V, Beyth M, Vardi J (eds) Geol Surv Isr Rep, GSI/9/90, Jerusalem
Volcani B (1944) Studies on the microflora of the Dead Sea. Ph.D. thesis, The Hebrew University of Jerusalem (in Hebrew)
Waldmann N, Starinsky A, Stein M (2007) Primary carbonates and Ca-chloride brines as monitors of a paleo-hydrological regime in the Dead Sea Basin. Quat Sci Rev 26(17–18):2219–2228
Ward WA, Joukowsky M (1992) The crisis years: the 12th century BC: from beyond the Danube to the Tigris. Kendall Hunt, Dubuque
Weber N, Yechieli Y, Stein M, Yokochi R, Gavrieli I, Zappala J, Mueller P, Lazar B (2018) The circulation of the Dead Sea brine in the regional aquifer. Earth Planet Sci Lett 493:242–261
Weiss B (1982) The decline of Late Bronze Age civilization as a possible response to climatic change. Clim Change 4(2):173–198
Wilkansky B (1936) Life in the Dead Sea. Nature 138(3489):467
Yakir D, Issar A, Gat J, Adar E, Trimborn P, Lipp J (1994) ^{13}C and ^{18}O of wood from the Roman siege rampart in Masada, Israel (AD 70–73): evidence for a less arid climate for the region. Geochim Cosmochim Ac 58(16):3535–3539
Yechieli Y, Magaritz M, Levy Y, Weber U, Kafri U, Woelfli W, Bonani G (1993) Late Quaternary geological history of the Dead Sea area, Israel. Quat Res 39(1):59–67
Yechieli Y, Ronen D, Berkowitz B, Dershowitz WS, Hadad A (1995) Aquifer characteristics derived from the interaction between water levels of a terminal lake (Dead Sea) and an adjacent aquifer. Water Resour Res 31(4):893–902
Yechieli Y, Gavrieli I, Berkowitz B, Ronen D (1998) Will the Dead Sea die? Geology 26(8):755–758

Yechieli Y, Abelson M, Bein A, Crouvi O, Shtivelman V (2006) Sinkhole "swarms" along the Dead Sea coast: reflection of disturbance of lake and adjacent groundwater systems. Geol Soc Am Bull 118(9–10):1075–1087

Yechieli Y, Abelson M, Wachs D, Shtivelman V, Crouvi O, Baer G (2003) Formation of sinkholes along the shore of the Dead Sea - preliminary investigation. In: Beck BF (ed) Sinkholes and the Engineering and Environmental Impacts of Karst. Americal Society of Civil Engineers, San Antonio, Texas

Zaccagnini C (1995) War and famine at Emar. Orientalia 64:92–109

Zak I (1967) The geology of Mount Sedom. Ph.D. thesis, The Hebrew University of Jerusalem, Israel (in Hebrew with English abstract)

Zilberman T, Gavrieli I, Yechieli Y, Gertman I, Katz A (2017) Constraints on evaporation and dilution of terminal, hypersaline lakes under negative water balance: the Dead Sea, Israel. Geochim Cosmochim Ac 217:384–398

Zilberman-Kron T (2008) The source and geochemical evolution of the brines in sinkholes along the western shore of the Dead Sea. M.Sc. thesis, The Hebrew University of Jerusalem, Israel

Ziv B, Dayan U, Kushnir Y, Roth C, Enzel Y (2006) Regional and global atmospheric patterns governing rainfall in the southern Levant. Int J Climatol 26:55–73

Ziv B, Saaroni H, Alpert P (2004) The factors governing the summer regime of the eastern Mediterranean. Int J Climatol 24(14):1859–1871

Chapter 2
Lake Sevan: Evolution, Biotic Variability and Ecological Degradation

Ian P. Wilkinson

Abstract Lacustrine conditions in the Sevan Basin, Armenia, have a history that extends back to the late Miocene (Pontian), although the modern morphology of Lake Sevan was established after volcanic activity during the Pleistocene and Holocene. A diverse fauna and flora was established in this high-altitude lake during the Holocene, but profound anthropogenic pressures commenced in the mid-twentieth century. The model of ecological degradation predicts a phase of inception, followed by a phase of accelerated degradation which reaches an apogee. In some circumstances, this is followed by a phase of recovery. In Lake Sevan, the phase of inception was relatively short lived, confined to the 1930s, with the phase of acceleration commencing in the 1940s and continuing through the 1950s and 1960s resulting in significant changes in water depth, temperature, sedimentation, pellucidity and water quality, causing profound changes in the trophic structure and diversity. The phase of recovery was initiated in 2002 and it remains to be seen whether this will return the lake to its former condition and the biota to its former diversity.

Keywords Armenia · Holocene · Anthropocene · Biotic variability · Anthropogenic change

2.1 Introduction

Lake Sevan is situated in Gegharkounik Province, in the eastern part of Armenia, at an altitude of 1900 m above sea level (asl) and surrounded by peaks that rise up to 3598 m asl (Figs. 2.1 and 2.2). It is one of the largest high-mountain lakes with a maximum length of about 70 km and a maximum width of about 50 km. The Artanish and Noraduz peninsulas reduce the width of Lake Sevan at one point to only 7 km and the two parts of the lake have been named Great and Little Sevan. The lake is fed by 28 rivers and two large springs and is drained by the River Hrazdan, which

I. P. Wilkinson (✉)
British Geological Survey, Nicker Hill, Keyworth, Nottinghamshire NG12 5GG, UK
e-mail: ipw@bgs.ac.uk

School of Geography, Geology and the Environment, University of Leicester, University Road, Leicester LE1 7RH, UK

S. Mischke (ed.), *Large Asian Lakes in a Changing World*, Springer Water,
https://doi.org/10.1007/978-3-030-42254-7_2

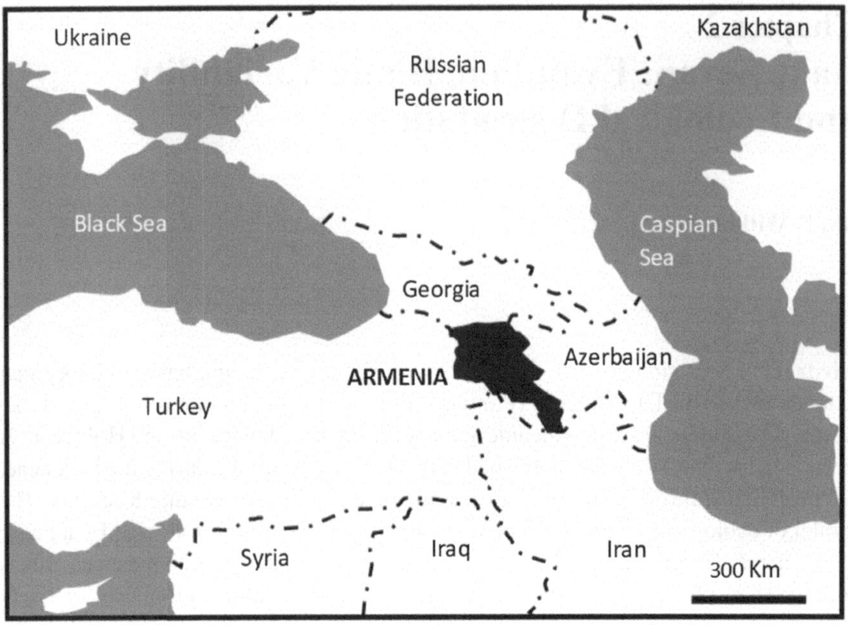

Fig. 2.1 The regional position of Armenia (coloured black)

leaves at the northern end, joins the River Arax on the Armenian/Turkish border and flows south and east into the Caspian Sea.

Lake Sevan is of vital importance to Armenia in providing water for hydroelectricity generation, irrigation and human consumption, as well as providing a source of food (90% of the country's fish and 80% of the crayfish). Water quality and pollution in the lake and its tributaries are, therefore, of great concern.

The ecology of Lake Sevan underwent continuous natural evolution throughout the Neogene and into the early twentieth century (Hakobyan and Djomin 1982; Filatov 1983; Gulakyan and Wilkinson 2002). The pristine ecology of Lake Sevan is seen in the preserved Holocene archive through to about 1927–1933 AD, but the inception of anthropogenic ecological degradation then began, accelerating during the 1940s and 1950s (Table 2.1). It became a significant problem during the late twentieth century, coinciding with the global phenomenon which is regarded by some as defining the Anthropocene Epoch (e.g. Waters et al. 2014, 2016; Wilkinson et al. 2014; Zalasiewicz et al. 2018, 2019). It seems likely, therefore, that the profound and accelerating environmental degradation, which began in approximately 1950, marks the start of the Anthropocene in the Sevan Basin.

The so-called "Sevan Problem", was caused predominantly by mistaken political decisions during the Soviet period together with over-exploitation and pollution of the lake and its catchment area (Babayan et al. 2006).

Fig. 2.2 Sketch map of Lake Sevan and localities mentioned in the text. Shaded areas indicate the extent of Lake Sevan prior to the fall in water level. SP: Sevan Peninsula

i. Aridity and limited energy resources in Armenia led to intensive use of the lake for human and livestock consumption, irrigation, and hydroelectric generation in the catchment area. Water loss was at a faster rate than replenishment resulting in a major fall in water level and changes in temperature, substrate, light penetration, etc. (Fig. 2.3).
ii. Population growth and tourism together with inadequate waste removal schemes caused problems associated with sewage, including hypoxia.
iii. Domestic and industrial solid waste were washed into the lake via rivers and streams.
iv. Agricultural pollutants entered the lake, particularly during the late twentieth century, although since the end of the Soviet period, the use of fertilizers and the number of cattle has decreased so that agricultural pollution has also reduced. However, agriculture contributes approximately 53% of Lake Sevan phosphorus load and c. 70% of its nitrogen load.
v. Heavy metal and chemical pollutants resulting from mining for gold at Sotq, chrome at Shorzha and lime at Artanish were washed into the lake via streams and rivers.

The phase of degradation continued into the early years of the twenty-first century, however, between 1996 and 1998, the Program of Reconstruction of the Ecological Balance of Lake Sevan was instigated to increase the lake's water level, reduce pollution in the catchment area, improve waste management, protect fish stocks and increase biodiversity. This was given judicial backing in 2001 when the National

Table 2.1 A timeline of key Anthropogenic environmental events in Lake Sevan resulting severe ecological degradation

Epoch	Date (AD)	Key event
Anthropocene	2001–2015	Renewal of water diversion into the lake resulted in a rise in lake level from 1896.46 m in 2001 to 1900.12 m in 2015. The Arpa-Voratan Tunnel extension, to divert water to Lake Sevan, was completed in 2003
	1997–2001	Continued reduction in lake level to 1896.46 m above sea level
	1996	Water level reached 20 m below the 1927 level
	1988–1994	Increased hydroelectricity generation, predominantly during the 1992–1994 and renewed drop in water level
	1983	Mass death of fish (cause unknown). Continued periods of anoxia
	1981–1991	The Arpa-Sevan water diversion scheme resulted in a slight rise in water level (+1.5 m) by 1991
	1970–73	First occurrence anoxia near the lake bed with presence of H_2S and CH_4. Nitrate and primary production were at maximum levels. Major decline of fish stocks. Eutrophication
	1964–1972	Blooms of blue-green cyanobacteria *Anabaena flos aquae* and *A. lemermanii*
	1963–1976	Water level of Lake Sevan stabilized 18 m below the 1927 level
	1950–1978	Reduction of macrophyte biomass from c. 900,000 to 8000 tons/year
Holocene	1949	Start of significant fall in water levels of Lake Sevan associated with power generation, irrigation and post-war population growth
	1933	The start of water removal for the generation of hydroelectricity
	1927	Water depth at a maximum (c. 100 m in Little Sevan). Water surface at approximately 1916.20 m above sea level

Parliament of the Republic of Armenia passed environmental legislation to protect Lake Sevan. The commencement of the Lake Sevan Action Programme initiated a phase of recovery which accelerated after about 2002.

2.2 Geological Evolution of the Sevan Basin

Although the origin of Lake Sevan, in its current form, dates from only about 2000 years before present (BP), lacustrine conditions have been centred on the Sevan region for a considerable period of time; certainly since the Pliocene, and probably from the late Miocene. In this sense it can be thought of as an ancient lacustrine depocentre.

2.2.1 Eocene-Oligocene

During the Palaeogene, deposition in Armenia was controlled by the evolution of the Paratethys Sea in which foraminifera and molluscs were indicative of the Mediterranean palaeogeographic province. Calcareous siltstones and mudstones, which accumulated in open marine conditions, are rich in nannofossils and planktonic foraminifera, whereas the shallow marine larger benthonic foraminifera are found mainly in the interbedded limestones. Temporal variability can also be observed with several internationally recognised biostratigraphical zones being represented in the Eocene to Oligocene successions.

Eocene planktonic foraminifera include common species of *Subbotina, Turborotalia, Globigerinatheka, Dentoglobigerina* and *Pseudohastigerina* (Krasheninnikov et al. 1986; Cotton et al. 2017). A number of biostratigraphical zones have been recognised including the Eocene *Truncorotaloides rohri, Globigerinatheka semiinvoluta* and *Globigerinatheka index* zones through to the Lower Oligocene *Globigerina tapuriensis* and *Globigerina sellii* zones. Larger benthonic foraminifera are also present, notably the Eocene *Nummulites millecaput*-group, *Nummulites fabianii* and *Heterostegina reticulata* lineages and, in the Oligocene, *Nummulites intermedius*. Calcareous nannofossil assemblages are dominated by the placoliths *Reticulofenestra, Cyclicargolithus floridanus, Coccolithus* and *Dictyococcites,* but nannoliths such as species of *Lanternithus* and *Zygrablithus* are common. The presence of *Chiasmolithus oamaruensis, Isthmolithus recurvus* and *Cribrocentrum isabellae, Discoaster barbadiensis* and *Ericsonia subdisticha* occur in the late Eocene (Cotton et al. 2017; Shcherbinina et al. 2017).

2.2.2 Miocene

The evolution of the Sevan region during the Neogene was outlined by Aslanian and Vehuni (1984) and Aslanian and Sayadian (1984). It was during the Miocene that the environment of deposition began to change from essentially marine to non-marine. The Eastern Paratethys, which covered the Sevan Depression during the Late Sarmatian (c. 12 Ma), regressed during the Meotian (c. 8 Ma) ultimately resulting in the formation of a large lake during the Pontian (c. 5.5 Ma; Nevesskaya et al. 2003). This is demonstrated by the change from the globigerine- and nummulites-dominated foraminiferal faunas of the Eocene and Oligocene to shallow marine and lagoonal faunas, including smaller foraminifera and ostracods dating to the Mid-Miocene. Aslanian and Vehuni (1984) recorded foraminifera (*Nonion subgranosus, Streblus beccarii*) and ostracods (*Cyprideis sarmatica, C. torosa* and *Candonella schubinae*) from the Sarmatian; molluscs (*Mactra* (*Sarmatimactra*) *caspia, M.* (*S.*) *podolica, M.* (*S.*) *timida, M.* (*S.*) *bulgarica*) and plant remains in the Upper Sarmatian; and bryozoa (*Membranipora* and *Nitcheina* (*Membranipora*) *kirschenevensis*) in the Meotian reefal deposits. These are in turn overlain by

Pontian freshwater clays characterised by molluscs *Dreissena*, *Limnaea* and *Planorbis* together with diatomaceous oozes including species of *Cyclotella* and *Stephanodiscus* (Poretzky 1953; Milanovskii 1968; Golovenkina 1977; Aleshinskaya and Pirumova 1982; Sayadyan 2006). Bubikyan (1984) recorded several non-marine ostracods from the Pontian of the Sevan region, including *Darwinula stevensoni, Ilyocypris bradyi, Ilyocypris gibba, Neglecandona neglecta, Pseudocandona albicans* and *Cyprideis torosa,* suggesting the development of fluviatile and lacustrine environments (Fig. 2.4).

2.2.3 Pliocene

By the Pliocene there was a complete absence of marine conditions in the Sevan region, and lignite-bearing volcanogenic successions are present (Aslanian and Sayadian 1984). Tectonic uplift in the Lesser Caucasus and Armenian Highland (Neill et al. 2015) was accompanied by episodes of volcanism, resulting in the damming of the River Hrazdan and the establishment of Lake Sevan in the intermontane depression. During the Early Pliocene 45 lacustrine and fluviatile diatom species (including *Cyclotella* and *Stephanodiscus)* inhabited the basin and, in the Late Pliocene, the lacustrine bivalves *Mactra subcaspia, Avicardium nikitini* and *Cerastoderma dombra* were present (Sayadyan 2006). In addition to those ostracods recovered from the late Miocene (listed above), Bubikyan (1984) recorded common *Neglecandona neglecta, Candona candida* (as *C. kirgizica)* and *Limnocythere inopinata* in the Pliocene lacustrine deposits of the Sevan intermontane basin (Fig. 2.4).

2.2.4 Pleistocene

Pleistocene through to Holocene sedimentation within the Sevan intermontaine basin was dominated by volcanogenic sediments and lavas which dammed the River Hrazdan resulting in the modern lacustrine environment (Aslanian and Sayadian 1984). Middle Pleistocene lacustrine sediments yield molluscs such as *Dreissena diluvia* and *Micromelania caspia* and oligotrophic diatoms including species of *Melosira, Fragilaria, Navicula, Amphora* and *Diploneis* (Sayadyan 2006). The ostracod assemblages in Lake Sevan during the Pleistocene were essentially similar to those of the Pliocene, but Bubikyan (1984) also recorded the first occurrence of *Prionocypris zenkeri* together with *Limnocythere inopinata* (including males which were recorded as *L. fontinalis*).

2.3 The Holocene of Lake Sevan

Water level of Lake Sevan varied considerably during the Holocene. However, Lake Sevan was an important resource for water and food throughout much of that time, attracting human occupation around its margins. Archaeological remains are found around the lake, including Bronze Age "cyclopean" castles (e.g. Lchashen and Hayravank), medieval buildings (e.g. Sevanavank and Hayravank), stone crosses, etc. Excavations at Lchashen and Norashen (Fig. 2.3b–c) have revealed Bronze Age wooden wagons (giving radiocarbon dates of 3500 and 3630 ± 100 years BP), and Iron Age boulders with cuneiform writing (Fig. 2.3d) dating to the time of the Urartian King Argishti (785–760 BC) and King Rusa II (730–714 BC), and ceramics dating to about 300 BC (Aslanian and Sayadian 1984). There is also evidence of Bronze and Iron

Fig. 2.3 The decline in the water level during the late twentieth century, resulted in the exposure of the former lake bed and the presence of many archaeological remains. **a** 'Sevan Island' is now connected to the mainland to form the Sevan Peninsula, the low-lying area in the middle distance once formed the lake bed. **b** Several hundred stone Bronze Age graves (or 'box houses') were revealed in the exposed land surface surrounding the lake, such as this one at Lchashen. **c** A mid-Bronze Age Dolman, near Lchashen, can also be seen on the shore of Lake Sevan, with the 1927 AD lake terrace behind. **d** A large stone at Lchashen bears cuneiform writing which dates to the time of the Urartian King, Argishti I (r. 786–764 BCE) and mentions the capture of 'Ishtikuni' (possibly the early name of Lchashen)

Age agricultural practices around the lake such as at Sotq, where wheat (*Triticum aestivum*), barley (*Hordeum vulgare)* and emmer (*Triticum dicoccum*) were cultivated (Hovsepyan 2013).

The final phase of lacustrine development began c. 2000 years BP when volcanic activity led to the damming of the River Hrazdan by lava flows and the lake adopted its modern morphology; molluscs overlying the Bronze Age archaeological horizon gave radiocarbon dates of 2020 ± 120 and 2090 ± 70 years BP (Aslanian and Sayadian 1984). The warm and wet climatic conditions at this time were replaced by more arid conditions throughout the eastern Mediterranean region, Turkey and Armenia by c. 1000 years BP (Eastwood et al. 2007).

The Holocene ecology of Lake Sevan is best demonstrated by reference to the sedimentary successions at Norashen, Dzknaget and boreholes in Great Sevan (Sayadyan

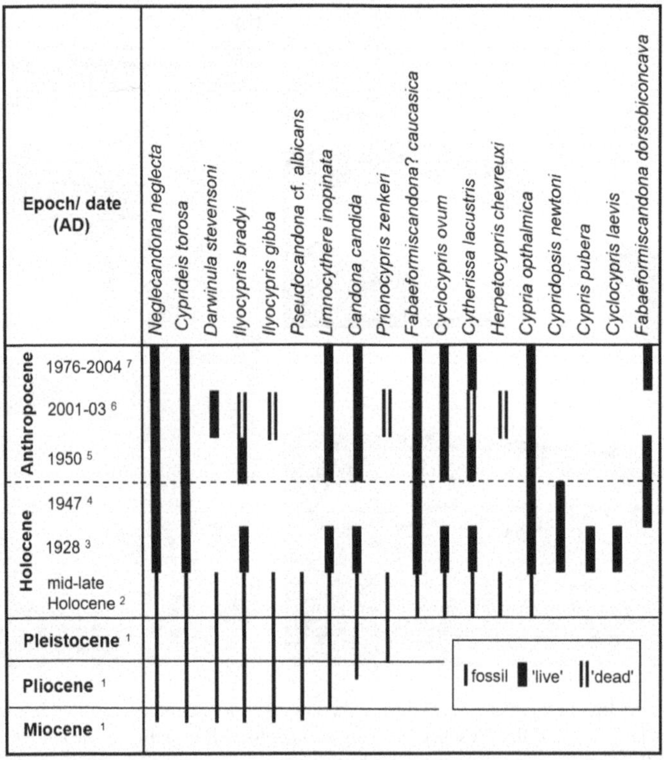

Fig. 2.4 The temporal distribution of freshwater ostracods from the Miocene to the Anthropocene of Lake Sevan. After [1]Bubikyan (1984); [2]Wilkinson et al. (2005); [3]Bronshtein (1928; the record of '*Limnocythere* sp.' is considered to be *L. inopinata*); [4]Bronshtein (1947; data given appears to be essentially a repetition of his 1928 paper, although he includes the first record of the deep water species *Fabaeformiscandona dorsobiconcava*); [5]Fridman (1950); [6]Wilkinson and Gulakyan (2010); [7]Jenderedjian et al. (2012)

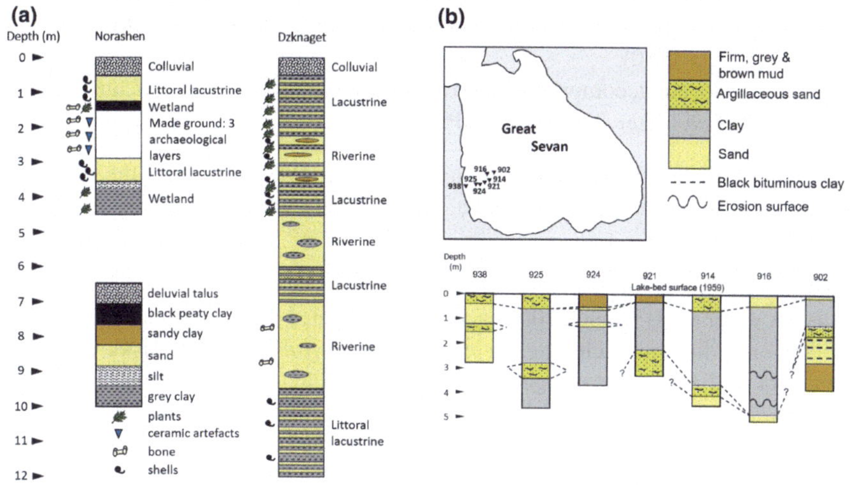

Fig. 2.5 Holocene stratigraphy and palaeoenvironments. **a** Exposures at Norashen and Dzknaget and **b** A transect of boreholes in the western part of Great Sevan (after Sayadyan et al. 1977; Aleshinskaya 1980; Aslanian and Sayadian 1984; Sayadyan 1991; Wilkinson et al. 2005)

et al. 1977; Aslanian and Sayadyan 1984; Wilkinson et al. 2005; Wilkinson and Gulakyan 2010; Fig. 2.5).

Late Pleistocene and post-glacial conditions resulted in the formation of deep water lacustrine conditions at Norashen, which was followed by four cycles of oscillating water levels:

> The deep water basal clay (c. 12,800 and 9300 years BP) passes up into peaty soil with plant remains and Cyperaceae pollen, culminating in medium-grained sand with a mollusc assemblage, dominated by *Lymnaea stagnalis*, characteristic of the littoral zone. An increase in water depth followed this phase of low lake level (c. 9300–8900 years BP).
> A second fall in water level led to littoral environments and the accumulation of sandy clay and medium-grained sand with molluscs dating to c. 8500 to 6200 years BP. The overlying unit contains early (3rd millennium BC) and late (2nd millennium BC) Bronze Age archaeological artefacts of a similar age to those at Lchashen (Sayadyan et al. 1977; Sayadyan 1978; Aleshinskaya 1980). The absence of the middle Bronze Age archaeology at both Norashen and Lchashen may suggest a short lived increase in lake level at that time. The overlying sand with abundant molluscs (dominated by *Planorbis planorbis* and *Lymnaea* (*Radix*) *auricularia*), pollen of Cyperaceae and *Sparganium* and abundant planktonic diatoms such as *Cyclotella ocellata* and *C. kuetzingiana radiosa* indicates an increase in water depth between 3400 and 2900 years BP.
> The third drop in lake level began c. 2950 years BP, and the peaty soil associated with low water level is estimated to date to c. 2500–2350 years BP before overlying transgressive sand accumulated, with molluscs and the abundant planktonic

diatoms indicating increased depth between about 2350 and 1700–900 years BP (Aleshinskaya 1980).

The uppermost unit, comprising lava blocks and rock fragments, indicates a fourth fall in lake level after c. 900 years BP.

Sayadyan et al. (1977) and Aslanian and Sayadyan (1984) discussed the Holocene succession of the Dzknaget River delta, on the northwestern margin of Lake Sevan (Fig. 2.5).

The lower unit, comprising clayey silt and sandy clay, was deposited at a time of deeper water (c. 2150 and 1700 years BP) as indicated by the presence of abundant planktonic diatoms including *Stephanodiscus astraea, Cyclotella kuetzingiana radiosa* and *C. ocellata*, together with sparse pollen of aquatic plants (*Typha, Sparganium* and species of Cyperaceae). This transgressive interval correlates with high lake-level phase indicated by the Norashen record.

The sand and sandy silt with pebble and gravel interbeds accumulated during a phase of shallowing between c. 1700 and 950 years BP. Aquatic plant pollen (Cyperaceae, *Typha* and *Sparganium*) occurs more commonly whereas the planktonic diatoms are much reduced in proportion and poorly preserved, but benthonic epiphytic taxa (*Meridion circulare, Cocconeis placentula, Amphora ovalis, Epithemia turgida* and *Rhopalodia gibba*) are abundant. Towards the top of the succession, unstable conditions seem to be indicated by the alternation of sand and organic-rich silts, together with variable proportions of aquatic plant pollen and of planktonic (*Cyclotella kuetzingiana, C. ocellata, Stephanodiscus atraea, Aulacoseira italica*) and benthonic (*Epithemia turgida, Meridion circulare, Fragilaria intermedia, Synedra ulna, Cocconeis placentula*) diatoms.

The surface gravelly sand with abundant soil and aerophilic diatoms (*Navicula mutica* and *Hantzschia amphioxys*) accumulated at the time of the Anthropogenic fall in water level since 1927 AD.

Shallow water Holocene sands, argillaceous and silty sands together with clays in Great Sevan, were penetrated by a series of boreholes (Fig. 2.5), which are believed, on the basis of sedimentation rates and geochemical evidence, to date from about 5000 years before present to the Holocene/Anthropocene transition (Aslanian and Sayadian 1984; Gulakyan and Wilkinson 2002; Wilkinson and Gulakyan 2010). Lake Sevan is crossed by several major geological faults along which seismic activity took place from time to time and there appears to have been an inverse relationship between organic carbon content of the lacustrine silts and seismic activity. It is postulated that this was caused by the agitation of the silts and release of the organic rich material into the water column. Gulakyan and Wilkinson (2002) attempted to use this to recognise the signatures of historical earthquakes preserved within the lake silts penetrated by boreholes. During the Holocene, sedimentation rate of silts was estimated at about 0.8–1 mm per year (Ljatti 1932; Afanasev 1933) and using sedimentation rate as an approximate time equivalence Gulakyan and Wilkinson (2002) postulated that the fall in the percentage of organic material observed within the sediment cores in the south-western part of Great Sevan were signatures of, respectively, the 1319 AD,

995 AD and 782 BC earthquakes. The earliest was the strongest earthquake when, according to Urartuian cuneiform inscriptions, the underground god, Teisheba, killed and burnt many people and destroyed cities and mountains (Satian and Chilingaryan 1994).

Wilkinson and Gulakyan (2010) indicated that ostracod diversity within Lake Sevan was low during the latest Atlantic-early Sub-Boreal, rarely exceeding three or four species, but it then increased to between seven and ten species probably reflecting ameliorating conditions and environmental heterogeneity during the Sub-Boreal climate. A total of 15 species were recorded from Holocene deposits, although a sixteenth species, *Fabaeformiscandona dorsobiconcava*, may have been present in profundal areas (Wilkinson et al. 2005; Wilkinson and Gulakyan 2010). However, of these, two, *Prionocypris zenkeri* and *Ilyocypris bradyi,* tend to favour spring waters and were probably introduced into the lake from surrounding fluviatile environments and a third, *Herpetocypris chevreuxi*, may have been derived from marshy areas around the lake margins. Species of Candoninae were common in the Holocene of Lake Sevan, particularly *Neglecandona neglecta* and *N. angulata,* but *Pseudocandona albicans* and *Fabaeformiscandona caucasica* were also present. The last named, is worthy of note as its distribution appears to be restricted to the Caucasus and Crimea (Bronshtein 1947). *Limnocythere inopinata* is one of the more common species having entered the Lake Sevan region during the Pliocene (Bubikyan 1984) and retained a sexual population throughout the Holocene (Wilkinson and Gulakyan 2010), unlike northern Europe where only females are known. *Cyprideis torosa*, *Cytherissa lacustris*, *Darwinula stevensoni*, *Ilyocypris gibba* and *Cyclocypris ovum* complete the ostracod community, although they were rare. *Limnocythere inopinata*, *Neglecandona neglecta* and *Ilyocypris bradyi,* three of the more common species, show a preference for shallow water less than about 20 m. However, at times water depths might have been much shallower (perhaps as shallow as c. 5–6 m) as indicated by the presence of *Darwinula stevensoni*. *Neglecandona neglecta* shows a preference for cool water.

2.4 The Inception of Ecological Degradation in Lake Sevan

2.4.1 Ecological Parameters

During the early twentieth century, when 'natural' oligotrophic conditions prevailed, the water level of Lake Sevan was more or less stable between c. 1915 and 1916 m asl (Fig. 2.6) and the maximum depths of Little and Great Sevan was 98.6 and 58.7 m respectively (Hovanesian and Bronozian 1994). As a result of the restricted lake basin area and the dry continental climate, renewal of the lake water is very slow at c. 50 years in its natural state. The main component of water loss was evaporation, and it was not until the 1950s that outflow via the River Hrazdan (1383 million m^3/year on average) exceeded evaporation (1041 million m^3/year on average; Babayan et al.

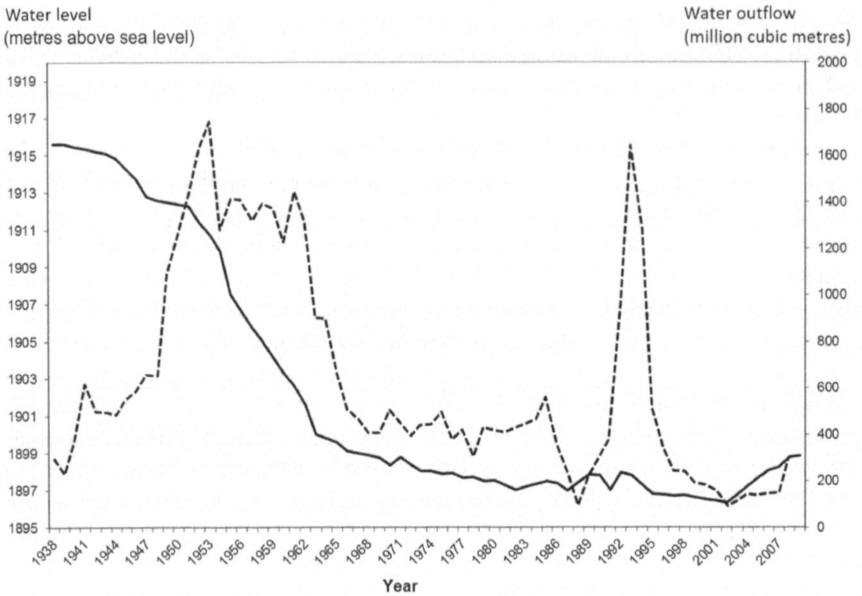

Fig. 2.6 Water levels (solid line) and rate of outflow (broken line) during Anthropocene (after Hovanesian and Bronozian 1994; Babayan et al. 2006)

2006). During the 1940s, with increased pressures from irrigation and industry, water level dropped by c. 0.2–0.3 m per year (Hovanisian and Bronozian 1994). Sedimentation in the littoral zone of Lake Sevan comprised silts (15–18%), sand (22–25%) and gravel (c. 60%), whereas calcareous, shelly muds and silts predominated further offshore and muds were characteristic of the profundal zone (Ljatti 1932). In a few areas in the deeper parts of the lake, calcium carbonate precipitated onto the lake floor.

Between 1927 and 1933 surface water temperatures averaged about 3–4 °C in Little Sevan and 4–5 °C in Great Sevan (Hovhannissian 1994), but there was seasonal variability. The seasonal increase in water temperatures during spring, when the young of many species appear, was slower than in the modern lake so that April-June surface temperatures were as much as 5 °C lower (Fig. 2.7). Summer surface waters reached their maximum temperature of about 16–17 °C during August, the hypolimnion being at 25–30 m depth (Gezalyan 1979). In Great Sevan the temperature of bottom waters during August and September was about 5 °C in 1938 (Barsukov et al. 1992). From September (until early January) water temperatures fell more slowly than in the modern lake. In Great Sevan, the hypolimnion began to break down in the early autumn and water started to mix in November, about a month later than in the modern lake and at a temperature 3–4 °C lower (Fig. 2.7). In Little Sevan, the low temperature (4–6 °C) hypolimnion which isolated the bottom of the lake from active circulation, was an important aspect in the ecosystem. During the winter, surface temperatures were such that the lake was

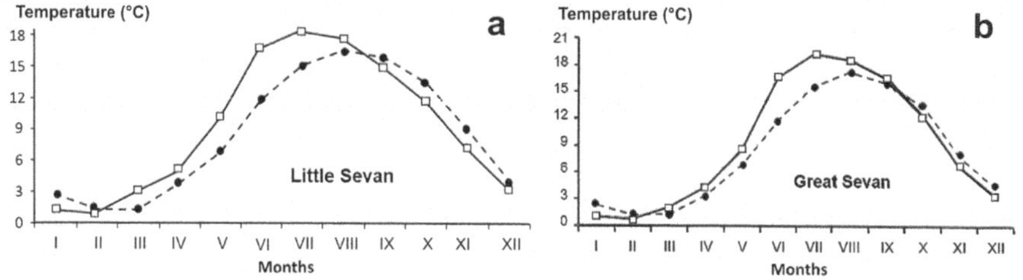

Fig. 2.7 A comparison of the average monthly surface water temperatures of Little **a** and Great Sevan **b** to demonstrate the effect of the fall in water level across the Holocene/Anthropocene transition. (1927–1933: dashed line and 1982–1985: solid line; after Hovhannissian 1994; Wilkinson and Gulakyan 2010)

Table 2.2 Concentration of major ions (mg/l) in the water of the Lake Sevan between 1893 and 2001

Epoch	Years	Ca^{2+}	Mg^{2+}	Na^+ + K^+	HCO_3^-	CO_3^{2-}	Cl^-	SO_4^{2-}
Anthropocene	2001[f]	32.5	63.2	113.0	402.6	–	76.6	19.4
	1976–1984[d]	24.6	57.3	76.7	425.5	40.3	66.2	16.4
	1976[e]	31.3	59.6	–	422.3	42.1	65.1	14.9
	1950–1970[d]	39.0	56.0	92.8	399.3	45.0	64.0	19.9
Holocene	1947–1948[c]	37.4	52.4	101.5	416.6	38.3	61.9	15.7
	1928–1929[b]	33.9	55.9	98.7	414.8	36.0	62.3	16.9
	1893–1894[a]	37.8	56.9	101.0	435.8	–	65.1	17.0

After [a]Stakhovski in Ljatti (1932); [b]Ljatti (1932); [c]Slobodchikov (1951); [d]Parparova (1979); [e]Hovhannissian (1994); [f]Wilkinson and Gulakyan (2010)

covered by ice only once every 15–20 years in the early twentieth century; heat capacity of the lake was 700 × 10^{12} kcal (Gezalyan 1979).

During the mid-twentieth century, Ljatti (1932) recorded the average dissolved oxygen content to between 5.6 and 11.8 mg/l, although there was seasonal variation, being highest during the spring, when the water column was mixed, and lowest in autumn. Table 2.2 illustrates some of the hydrochemical components of the water of Lake Sevan. Between 1893 and 1948 (the latest part of the Holocene as interpreted herein) there was only slight variation in water chemistry, although there is a trend for Mg to decline slightly during that time (Ljatti 1932; Slobodchikov 1951).

2.4.2 *Biotas*

Prior to the drop in water level, Lake Sevan was colonised by a diverse biotic community.

Macrophytes play an important role in the ecosystem of Lake Sevan, as they provide a food source, spawning habitat, substrate and refuge for many benthonic organisms. The biomass of macrophytes in the lake was 600,000 tons comprising more than 20 common species (Hovhannissian 1994). At this time there was a very high light pellucidity and macrophytes reached depths of 17 m. Stonewort (*Chara*) and moss dominated the lake bottom vegetation down to c. 14 m depth.

Phytoplankton and algae were represented by 26 species in Lake Sevan. Diatoms were particularly common, including *Asterionella formosa, Stephanodiscus astraea* and *S. hantzschii*, and green algae were also well represented, particularly *Gloeococcus schroeteri* and *Oocystis* spp. (Hovhannisyan et al. 2010). Hovhanissian and Gabrielyan (2000) reported that phytoplankton biomass was 0.3 g/m^3 and productivity was high, at approximately 100 gC/m^2 per year, due to the greater depth of the euphotic zone and the constant presence of algae.

Zooplankton comprised ten species of rotifers, cladocerans and copepods during the latest Holocene, all of which were common, although the total biomass was low. Copepods formed the largest part of the total biomass (about 60%) and the cladoceran *Daphnia* formed a large proportion (Savvaitova and Petr 1999).

Zoobenthos biomass was between 3.4 and 4.2 g/m^{-2} (wet weight) during the latest Holocene with a total of 170 species recorded in Lake Sevan (Arnoldi 1929; Domrachev 1940; Fridman 1950). Between 1928 and 1948 Oligochaeta made up about 50% of the total biomass, Amphipoda 14–18%, Chironomidae 11–17%, Hirudinea 8–12%, mollusks 5–13%, and the remaining zoobenthos 1–2%. These groups peaked in waters shallower than 15 m, although the last two groups were also abundant at, respectively, 25–30 m and 40–60 m. Jenderedjian et al. (2012) calculated that species that formed in excess of 80% of the total biomass of their taxonomic group were *Monhystera paludicola* (Nematoda), *Candona caucasica* and *Neglecandona neglecta* (Ostracoda), *Ilyocryptus sordidus* (Cladocera), *Potamothrix alatus paravanicus* (Oligochaeta), *Glossiphonia complanata* and *Herpobdella octoculata* (Hirudinea), *Lymnaea stagnalis* and *Valvata piscinalis* (Gastropoda), *Gammarus lacustris* (Amphipoda), *Agrypnetes crassicornis* (Trichoptera) and *Chironomus plumosus* and *Procladius* sp. (Chironomidae). The profundal zone was inhabited by 22 species of Cnidaria, Turbellaria, Ostracoda, Hirudinea, Gastropoda, Bivalvia, Amphipoda, and Chironomidae.

The latest Holocene ostracod community, which was first recorded by Bronshtein (1928), comprised (by original designation): *Cypris pubera, Cypridopsis newtoni, Cyclocypris laevis, Cyclocypris ovum, Cypria lacustris* [=*C. ophtalmica*], *Candona neglecta, Candona candida, Candona elpatiewskyi* v. *caucasica* [=*Fabaeformiscandona caucasica*], *Ilyocypris* sp. [?=*Ilyocypris bradyi* and possibly *I. gibba*], *Cytherissa lacustris, Cyprideis littoralis* (=*C. torosa*) and *Limnocythere* sp. [?*Limnocythere inopinata*]. The first three species have not been recognised by other authors and may be misidentified. *Candona dorsobiconcava* (now placed into *Fabaeformiscandona*) was added to this list by Bronshtein (1947). Fridman (1950) found only nine species living in the lake (Fig. 2.4) and Wilkinson and Gulakyan (2010) recorded that the latest Holocene community (now considered to be Anthropocene) to be dominated by *Neglecandona neglecta* and *Limnocythere inopinata* with subordinate *Candona candida and Fabaeformiscandona caucasica.*

Fish were abundant in Lake Sevan during the Holocene including four subspecies of the endemic Sevan Trout (*Salmo ischchan*):

> The "Winter" Sevan Trout (*Salmo ischchan ischchan*), which fed principally on *Gammmarus* formed two morphotypes; the shallow water morphotype spawned in the north-western part of the lake and the deeper water morphotype spawned in the south-eastern part of Great Sevan (Fortunatov et al. 1932; Dadikyan 1986).
> The "Summer" Sevan Trout (*Salmo ischchan aestivalis*), was present throughout the lake (Fortunatov 1927; Leshchinskaya 1950; Pavlov 1951).
> The Gegarkuni Sevan Trout (*Salmo ischchan gegarkuni*) spawned in the rivers Gavaraget, Tsakkar, Makenis and Masrik from September to January and in the upper reaches of the rivers Argichi and Gezal in autumn (Fortunatov 1927).
> The Bodzhak Sevan Trout (*Salmo ischchan danilewskii*) spawned in littoral waters of Little Sevan and western Great Sevan during the autumn (Fortunatov 1927; Pavlov 1951).

Of the other fish, the Sevan Khramulya (*Capoeta capoeta sevangi*) was common in some parts of the lake and larger rivers where they fed on aquatic macrophytes and detritus (Dadikyan 1986). The Sevan Barbel (*Barbus lacerta goktchaicus*) was not common, but inhabited the travertine and stony areas of the lake bed, spawning between June and August and fed on zooplankton, gammarids, insects and trout eggs (Dadikyan 1986). In 1924, the Ladoga whitefish (*Coregonus lavaretus ladoga*) and Lake Chud whitefish (*Coregonus lavaretus maraenoides*), were introduced into Lake Sevan for human consumption. They hybridised into the Sevan whitefish (*Coregonus lavaretus sevanicus*), which rapidly established itself and became common by the mid-twentieth century (Pavlov 1947; Mailyan 1957; Yuzhakova 1985). Armenian riffle minnow (*Alburnoides bipunctatus armeniensis*) and gudgeon (*Pseudorasbora parva*) also occupied the lake.

2.5 Ecological Degradation During the Anthropocene

2.5.1 Ecological Parameters and Anthropogenic Change

During the early twentieth century, water loss was predominantly by evaporation, but a plan to reduce this by lowering the water level by 50 m was proposed by Soukias Manasserian in 1910 (Mikirtitchian 1962; Meybeck et al. 1997). This plan would drain Great Sevan entirely and leave a reduced Little Sevan, but it was believed that the Sevan trout yield would increase by 8–10 times in the smaller lake. Whilst General Secretary of the Central Committee of the Communist Party of the Soviet Union, Stalin initiated this project with the dredging of the River Hrazdan in 1933 and the construction of a tunnel, which opened in 1949. As a result, and exacerbated by increased irrigation and population pressure, during the 1950s (i.e. early in the Anthropocene) the fall in water level accelerated to over 1 m per year (Fig. 2.6). Stalin's project was questioned in the mid-1950s and the Soviet government, under Khrushchev, ordered that the lake level should be maintained as high as possible, not for environmental reasons, but to increase energy generation. However, the rate of extraction reduced and, during the 1960s, water level stabilised at about 18 m below the original level. The average depth of Little Sevan had dropped from 50.9 to 39.5 m and that of Great Sevan had fallen from 37.8 to 22.8 m (Hovanesian and Bronozian 1994). Although slowed, and occasionally reversed, water level continued to drop. By the 1970s and 1980s the fall in water level and volume had the effect of changing the annual temperature regime and water chemistry, which had a knock-on influence on the food chain and trophic structure. By 2001 the lake surface had fallen to its lowest level, some 20 m lower than its 1927 level (Fig. 2.3a), the surface area was reduced by 12%, the average depth by 34.2% and the volume by 42.2%.

Sedimentation and water pellucidity during the Anthropocene are noticeably different compared to those of the Holocene. At the end of the twentieth century, following the fall in water level, sedimentation changed so that 85% of the lake bed area became blanketed by mud and silt (Resnikov 1984). Sands were confined to a very narrow zone along the shoreline and small embayments, whilst coarser gravels were restricted to the foot of the high mountain chain along the northern shore of Little Sevan. The widespread mud and silt has resulted in a decrease of pellucidity by 3.7 times for Little Sevan and 3.9 times for Great Sevan, with transparency falling from 14.3 m in 1930–1935 to 3.0 m in 1976–1978 and 4.5 m in 1985–1996 (Hovhannissian 1994). Pellucidity has been seen to deteriorate further when seismic activity and storms agitate the lake bed.

As a consequence of shallowing, the average water temperature increased to 4.2–5.0 °C in Little Sevan and 8–12 °C in Great Sevan by 1982–1985 (Hovhannissian 1994) and the heat capacity fell from 700×10^{12} kcal during the late Holocene (the early twentieth century) to 460×10^{12} kcal during the Anthropocene (in the late twentieth century). The seasonal temperature regime also changed (Fig. 2.7). Water temperatures start to increase in early March, although thermally the lake remains unstratified; a thermocline begins to develop at a depth of about 10 m in late May and reached about 15–20 m in June. The increase in water temperature during spring is much more rapid in the modern lake compared to that of 1927–1933; April-June surface temperatures are up to 5 °C higher (Fig. 2.7). By 1982–1985 maximum summer temperatures were reached a month earlier than in 1927–1933 with July surface temperatures being about 19 °C. In Great Sevan the temperature of bottom waters during August and September rose from 1938 levels (5 °C) so that after 1961 temperatures rarely fell to 7 °C and were generally 9 to 12 °C and up to 14 °C in some years (Barsukov et al. 1992; Fig. 2.7). Autumn water temperatures now fall more rapidly compared to 1927–1933, mixing commences about a month earlier and at temperatures 3–4 °C higher than previously. During the late twentieth century complete stratification breakdown took place in Great Sevan between October and December, with water temperatures averaging 11–12 °C (Gezalyan 1979). In Little

Sevan the drop of water level caused a 50% decrease of the hypolimnion volume. Winter-surface temperatures were such that the lake became ice covered every 3–4 years, but since 1971 it has had an ice cover most winters between late January and early April (when the ice was up to 30–40 cm thick). Winter bottom water temperatures drop to a low of 2 °C in Little Sevan and 1 °C in Great Sevan (Gezalyan 1979).

The pH of Lake Sevan was 9.25 in 1937, but Hovhannissian (1994) showed that it was generally about 8.7–8.8 between 1950 and 1991, although abrupt decreases took place in 1983 (pH 8.5) and 1988 (pH 8.2). These abrupt falls coincide with high magnitude earthquakes, 6.9 and 7.0 respectively, in Lake Sevan when agitation of low pH waters at the water-sediment interface (generally 8.0–8.5) mixed with higher pH waters in the rest of the water column. The drop in pH was accompanied by an increase in the proportion of compounds such as H_2S and CH_4 and the enrichment of, for example, V, Mn and Cu, perhaps as a result of the agitated sediment (Danielyan et al. 2011).

In 1928, although poorly circulating, Lake Sevan was oligotrophic with a high dissolved oxygen content throughout the year. The dissolved oxygen content showed seasonal variation with highest levels in spring and although it declined in autumn, the average oxygen minimum was between 5.6 and 11.8 mg/l (Ljatti 1932). Intense anthropogenic eutrophication resulted in a sharp decrease in dissolved oxygen concentrations (Fig. 2.8) heralded by the first bloom of cyanobacteria in 1964. However, by the 1970s the dissolved oxygen concentration in the bottom waters had become extremely low throughout the lake and between 1970 and 1973, anoxia was reached in the profundal zone during the late summer, when stratification was well established (Parparov 1990). Towards the end of the decade, dissolved oxygen concentrations were between 6.23 and 11.08 mg/l in the littoral zone of the lake and between 4.56 and 11.60 mg/l in the central part (Gezalyan and Khorlashko 1979).

Water pollution in Lake Sevan is a combination of anthropogenic factors such as agriculture, industry and population pressures, including sewage (Simonyan et al. 2016) and, to a lesser extent, natural pressures related to seismic activity (Gulakyan and Wilkinson 2002; Danielyan et al. 2011). Climate change is also felt in terms of reduced flow of the rivers and streams entering the lake and the increased air temperatures result in higher rates of evaporation. These may have deleterious effects on water quality in the future.

Anthropogenic pollution has also contributed to variations in water chemistry (Fig. 2.8). Although the quantity of NH_4^+, NO_2^- and NO_3^- varies throughout the year and from depth to depth, between 1929 and 1982–1989, mineral nitrogen flowing into the lake had increased from 229 to about 2290 tons. Between 1965 and 1994 the average annual inflow of nitrogen was 7000 tons and total mineral nitrogen concentration increased from 0.01 to 0.03 g/m^3 (Hovhannissian 1994; Hovanesian and Bronozian 1994; Simonyan et al. 2016). The concentration of mineral phosphorus, which washes into the lake from the catchment area decreased from 0.320 to 0.015 g/m^3, probably by its combination with calcium and its capture within the sedimentary succession (Hovanesian and Bronozian 1994). There are other concerns, for example, the Ministry of Nature Protection reported in 2000 that near the

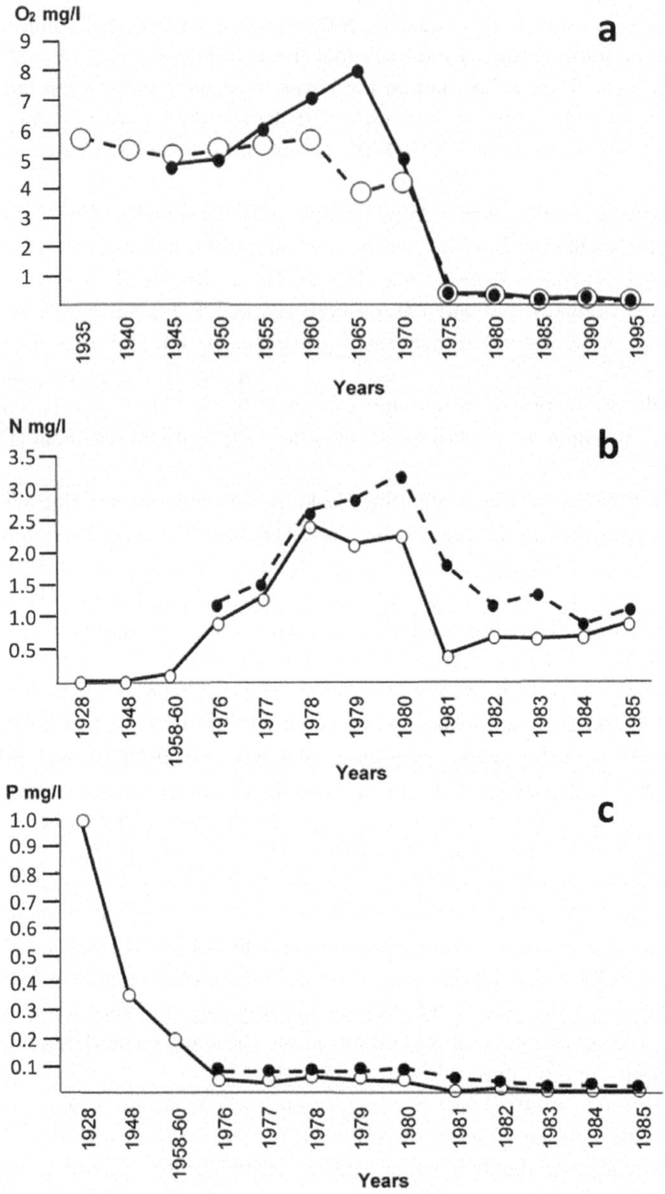

Fig. 2.8 Variations in water chemistry of Lake Sevan across the Holocene/Anthropocene transition to demonstrate the effects of anthropogenic change. **a** The major collapse in oxygen concentration in the deeper waters of Little (dashed line) and Great Sevan (solid line) during the late twentieth century. **b** The annual concentration of mineral nitrogen in surface (solid line) and bottom (dashed line) waters. **c** The annual concentration mineral phosphorus in surface (solid line) and bottom (dashed line) waters. Data from Hovhannissian (1994, 1996); Hovanesian and Bronozian (1994), Simonyan et al. (2016)

Sevan Peninsula, the admissible level of copper pollution was exceeded by 32 times, oil derivatives by 90 times; and there are concerns over other pollutants including microplastics and mercury in the river Masrik, a major tributary flowing into Lake Sevan.

Heavy metal accumulation within the lake is not only a risk to the environment, but also human health via the food chain and drinking water. The distribution and quantity of heavy metal pollution in the water and sediment of Lake Sevan varies according to, for example, water depth, season and the occurrence and magnitude of earthquakes. Since 1965, an average of 135 tons of heavy metals have entered Lake Sevan each year (Hovhannissian 1994; Simonyan et al. 2016). Fe formed the highest concentration in the lake, followed by Zn and (in the order of concentration) Mn, Cu, Co, Pb, Ni and Cd, together with small quantities of Cr (Hovhannissian 1994). A very high anomaly of Cl, SO_4, Mg and Cr occurred in the sediments of the central part of Great Sevan and very high Mo, Zn and Mn concentrations in the south-western part (Satian and Chilingarian 1994). The decreasing volume of lake water may account for the high concentration of heavy metals, which in 2005 comprised, in order of concentration, Fe, Mn, Zn, Ti, Cu, Cr, Co, Ni, Pb and Cd (Vardanyan and Ingole 2006).

Many rivers entering Lake Sevan flow through rural rather than industrial or densely inhabited areas so that they bear low concentrations of heavy metals. In contrast, the waters of the Sotq and Masrik rivers have high concentrations of V, Fe, Cu, Ni, As, Mo and Pb related to gold mining, in some places exceeding admissible levels (Gevorgyan et al. 2016). These are particularly concerning as they are known to cause deformities and malignant tumors in mammals.

Lake Sevan, and most rivers flowing into it, contain only low levels of persistent organic pollutants such as α-Hexachlorocyclohexane (α-HCH), Dichlorodiphenyldichloroethylene (DDE) and polychlorinated biphenyl (PCB). There has been a significant reduction in agricultural activities requiring the use of large quantities of pesticides, although an average of about 13 tons have been added annually since 1965 (Hovanesian and Bronozian 1994; Simonyan et al. 2016). The presence of α-HCH is a reflection of the agricultural use of the insecticide Lindane. During the Soviet period DDT was widely used in the Sevan catchment area, residues of which have degraded into DDE, which is washed into the lake during periods of raised precipitation and run-off. PCBs are present only in low concentrations because there is little large-scale industrial activity in the catchment area.

2.5.2 Anthropocene Biotas

The physical changes in the morphology of the Lake Sevan during the second half of the twentieth century resulted in considerable ecological degradation which is reflected in the trophic structure. The change was principally caused by anthropogenic causes, although natural phenomena also played a role. For example, seismic events resulted in changes in the biomass of zooplankton and encouraged "blooms" of blue-green cyanobacteria by increasing water circulation and enriching of surface waters with organic material (Legovich 1979; Simonyan 1991; Hovhannissian 1994; Gulakyan and Wilkinson 2002).

Anthropogenic change is demonstrated by the variability in the community structure of the zoobenthos. The anoxic phase of the 1970s caused changes in the vertical distribution, energy budgets and community structure. Detritivores formed the largest proportion of the total biomass of zoobenthos since records began, but their relative proportion increased from an average of about 80% during 1928–1971 to about 98% by 2004. The trophic groups that were particularly affected by anoxia, were the herbivores and predators which remained at low levels between 1970 and 2010 (Jenderedjian et al. 2012). Oligochaetes, particularly the dominant species *Potamothrix alatus paravanicus*, were more resistant to hypoxia compared to other zoobenthic groups and thrived in the lake (Jonasson and Thorhauge 1976; Jenderedjian 1990). Amongst the nematodes, those characteristic of the late Holocene of Lake Sevan (*Punctodora ratzeburgensis, Chromadorita leucarti* and species of Chromadoridae) were replaced by hypoxia- and pollution-tolerant *Tripyla glomerans* and *Monhystera paludicola* (Fridman 1950; Tsalolikhin 1976; Gagarin and Hakobyan 1991, 1992).

Macrophyte biomass collapsed during the late twentieth century from 900,000 tons per year between 1930 and 1935 to 8000 tons per year in 1976–1978 (Hovhanissian and Gavrielyan 2000). Diversity also collapsed from 20 common species to one common and 14 rare species as a result of habitat loss related to drainage of the littoral area and deteriorating water quality. By 1976 *Potamogeton natans, P. pusillus, P. densus* and *Zannichellia palustris* had reached a point of extirpation (Gambaryan 1979).

The reduction in the macrophyte community in the littoral zone is felt throughout the food chain and trophic structure causing a negative impact on the distribution of the biotas that used them for a food source, substrate, spawning ground and shelter. *Chara* and moss stands were particularly affected, and are confined mainly to sheltered bays (Babayan et al. 2006). This in turn caused a decline in the oligochaete *Rhyacodrilus coccineus*, which is particularly associated with organic matter derived from macrophytes, from 28% in 1938 (Fridman 1950) to 9% in 1984–1986 (Jenderedjian and Poddubnaya 1987). Hirudinea, gastropods and freshwater shrimps, which feed mainly on *Chara* and moss were also adversely affected (Stroykina 1957). Muds and silts now blanket areas that were once stone and sand substrates so that Trichoptera and Ephemeroptera larvae, which show a preference for these substrates, were also adversely affected. While some biotas were in decline, others, notably filamentous green alga such as species of Zygnemataceae spread during the mid-1980s, becoming abundant at depths of 7–12 m.

Phytoplankton are important indicators of eutrophication in Lake Sevan, their biomass being related to increased levels of nitrogen and decreased phosphorus in the lake water, which in turn encouraged the increase in phytoplankton from 0.2 to 0.5 g/m^3 between 1937 and 1962, to 2.0–3.0 g/m^3 throughout the remainder of twentieth century, and reaching 3.8 g/m^3 in 2005 (Kazaryan 1979; Legovich 1979; Mnatsakanyan 1984; Hovhannisian et al. 2010). However there is seasonal variation related to the Chlorine-*a* concentration and the nitrogen/phophorous ratio as shown by Hovhannisyan et al. (2010) who found that in 2005 there was a dominance of Chlorophyta in early summer, Cyanophyta in mid- and late summer and Bacillariophyta in autumn.

Diatoms were particularly common in the lake between 1947 and 1982. However, after 1983 green algae became the most abundant group, changes in the phytoplankton community being brought about by the anthropogenic alteration of the ecosystem. Species of eutrophication-tolerant *Melosira* entered the lake in the mid-1960s, formed a common element of the community by the 1970s and dominated the phytoplankton by 2005 (Kazaryan 1979; Hambaryan et al. 2007; Hovhannisyan et al. 2010). *Melosira granulata* is particularly abundant at a water depth of about 20 m during the autumn when its vegetative phase begins, and *Tabellaria* sp. has also been recorded (Hovhannisyan et al. 2010). Amongst the green algae, numerous members of the order Volvocales (e.g. *Ankistrodesmus, Hyaloraphidium, Lagerheimia, Golenkiniopsis, Treubaria* and *Tetrastrum*) appeared in the shallow waters and *Euglena* colonised the pelagic zone. Newly introduced unicellular green algae *Closterium* sp. and *Mougeotia* sp. became particularly abundant and the dinoflagellate *Peridinium* was also recorded (Kazaryan 1979). *Hyaloraphidium rectum*, which is considered a green algae by some and a lower fungus by others, is also abundant (Ustinova et al. 2000). Since 2002, water level has been rising and Hovsepyan et al. (2014) recorded that by 2010, 96 taxa of phytoplankton were recorded in Lake Sevan (42% diatoms, 33% green algae and 15% cyanobacteria).

Cyanobacteria have increased in number during the Anthropocene forming blooms particularly between 1964 and 1972. *Anabaena flosaquae* bloomed in 1964, but *Anabaena lemmermannii* became the main contributor to the blooms between 1966 and 1972 (Legovitch 1979). The raised bacterial concentration is in part a function of water temperature and partly enrichment of the surface waters with organic matter as a result of wind-induced water circulation in the shallower parts of the lake and occasional agitation by seismic activity (Gulakyan and Wilkinson 2002).

Zooplankton comprised ten common species during the early twentieth century, after which a gradual and then a more rapid change took place, so that between 1972 and 1980 biomass increased and rotifers became dominant (80% of the biomass). Further evolution of the ecosystem resulted in a decrease in biomass between 1981 and 1985 with rotifers forming about 60-65%, although that of *Daphnia* and *Cyclops* increased in proportion (Simonyan 1988). The reduction in phytoplankton and zooplankton communities caused perturbations throughout the food web and, in 1983, resulted in a major decline in numbers, size and reproduction of whitefish (Simonyan 1991).

By 2011, the national programme to raise the water level in Lake Sevan caused flooding of the coastal strip, rejuvenation of wetlands and the establishment of renewed habitats, resulting in increased diversity and biomass of the zooplankton, including the first record of 16 species of rotifers (including *Trichocerca stylata* and *Lecane* (*Monostyla*) *cornuta*) and eight cladoceran species (e.g. *Acanthodiaptomus denticornis, Arctodiaptomus bacilifer, Diaphanosoma brachyurum,* and *Keratella quadrata*; Krylov et al. 2013). The cladoceran *Daphnia (Ctenodaphnia) magna*, which had been recorded very rarely in the first half of the twentieth century, became abundant (Krylov et al. 2007, 2010, 2013; Gerasimov and Krylov 2012) and during 2012 common rotifers (e.g. *Conochilus unicornis* and *Synchaeta grandis*) and copepods (such as *Acanthodiaptomus denticornis*) were also recorded (Krylov et al. 2015).

There appears to be a close relationship between the biomass of fish and that of zooplankton. In 2004, the number of whitefish decreased sharply due to overfishing and infection of crucian carp by the parasite *Ligula intestinalis* and remained low through to 2011. The reduced number of fish were unable to consume the daphnia, the remains of which accumulated on the lake floor in huge numbers after 2011. This resulted in elevated levels of nitrogen and phosphorus encouraging the increase in cyanobacteria.

The zoobenthos of Lake Sevan has been studied many times so that a detailed data base has been built up (Jenderedjian et al. 2012 and references therein). The total biomass has varied considerably during the twentieth century; between 1948 and 1971 the biomass had increased to 15.1 g/m^2 (wet) and during the late 1970s it had reached 30.5–36.6 g/m^2 (wet), but declined rapidly to 9.9–10.9 g/m^2 (wet) in 1989–1991 (Jenderedjian et al. 2012). Diversity has always been observed to decrease from littoral to profundal depths, although this was particularly pronounced between 1984 and 2004 (Jenderedjian et al. 2012). Severe hypoxia had a huge impact on the zoobenthos, particularly between 1976 and 2004, when anoxia became an annual event. Between 1928 and 1948 there were at least 20 more species living at depths of 10–50 m compared to 1984–2004. The 22 species that inhabited the profundal zone between 1928 and 1948 comprised Cnidaria, Turbellaria, Ostracoda, Hirudinea, Gastropoda, Bivalvia, Amphipoda, and Chironomidae, whereas during the period of acute hypoxia this was reduced to six species (one species of Oligochaeta, two Nematoda and three Cyclopoida). The Anthropocene has been characterised by a number of extirpations within the biota of Lake Sevan. One such amongst the gastropod community is *Planorbis carinatus*, empty shells of which litter the beaches of Lake Sevan, yet no living specimens have been found since the drop in water level, although they have been found living in the Argichi River (Aghababyan et al. 2010). This species is particularly associated with aquatic macrophytes and its extirpation was probably caused by the loss of habitat.

The Oligochaete community of Lake Sevan forms excellent indices for dysaerobia as their vertical distribution is controlled by oxygenation of the water (Jenderedjian and Poddubnaya 1987; Jenderedjian 1996). Under the oligotrophic conditions of 1937 to 1939, *Potamothrix alatus paravanicus* was found in 85% of the samples examined by Fridman (1950), *Trichodrilus tubifex* in 1% and *Rhyacodrilus coccineus* and *Limnodrilus hoffmeisteri* in 2% of samples. During 1938–1948 oligochaetes represented 36.2% of the total biomass (Fridman 1950), which remained comparatively stable (2.0–4.3 g/m^2 wet) between 1955 and 1971, but increased to over 10 g/m^2 (wet) between 1978 and 1985, before decreasing again to 7.6 g/m^2 (wet) in 1989–1991. The acute hypoxia of the lake during the 1970s resulted in a major decline such that by 1984–1991 an almost monospecific assemblage comprising *Potamothrix alatus paravanicus* occupied the lake; the combined proportion of the other four members of the community at that time (*L. claparedeianus, T tubifex, R. coccineus* and *L. hoffmeisteri*) amounted to about 2%. Jenderedjian et al. (2012) recorded that during 1984–2004 the zoobenthos biomass peaked at 20–25 m water depth, and that Oligochaeta made up 70–75% of the total. Of the remainder Chironomidae comprised 22–27%, mollusks 2%, and the remaining groups less than 1%.

Ostracods are considered in a little more detail as the preservation of their biomineralised carapace can be used to compare the community throughout the evolution of Lake Sevan and the impact of the Anthropocene changes of water depth and associated environmental parameters including water temperature, oxygenation, changes of substrate, decline in macrophytes, and anthropogenic eutrophication on the Ostracoda. Jenderedjian et al. (2012) listed nine species in Lake Sevan during 1976–2004: *Neglecandona neglecta, C. candida, Fabaeformiscandona caucasica, F. dorsobiconcava, Cyprideis torosa* (as *C. littoralis*), *Cytherissa lacustris, Limnocythere inopinata, Cyclocypris ovum* and *Cypria ophtalmica* (as *C. lacustris*). Wilkinson and Gulakyan (2010) recorded eight species in a similar ostracod community in 1999–2000, although *Cytherissa lacustris* was not found living at that time, only dead specimens being present, and *Darwinula stevensoni* was found living amongst the vegetation in the littoral zone, although it was very rare.

Two species, *Cytherissa lacustris* and *Ilyocypris gibba,* may have undergone extirpation as a consequence of the eutrophication of the lake. *Cytherissa lacustris* is widespread throughout Europe, Asia and North America, where it favours oligo-mesotrophic lakes, but oxygenation appears to be significant in its distribution as it is badly affected by eutrophication (Danielopol et al. 1985, 1990). That appears to be the case in Lake Sevan because living specimens were recorded during the 1920s and 1940s (Bronshtein 1928; Fridman 1950) and it was listed as present 1976–2004 (Jenderedjian et al. 2012). However, Wilkinson and Gulakyan (2010) found no evidence of it living in the lake in 1999–2000, although well preserved, but empty carapaces were found, sometimes in moderately large numbers. If the specimens recorded by Jenderedjian et al. (2012) were living, then this may be a sign of ecological recovery.

Wilkinson and Gulakyan (2010) recorded dead carapaces of *Ilyocypris gibba* from Lake Sevan, but found no evidence that the species was living in the lake in 1999–2000. In addition, several species that populated the lake during the late Holocene were not found living, although dead carapaces occur, very rarely, in modern sediments: *Ilyocypris bradyi, Pseudocandona* cf. *albicans, Prionocypris zenkeri* and *Herpetocypris chevreuxi*. It is concluded that these dead specimens were either derived from older sediments, possibly reworked during seismic events disturbing the lake-bed sediments, or introduced into the lake from neighbouring fluviatile environments. *Ilyocypris bradyi,* for example, shows a preference for water bodies fed by springs, and Meisch (2000) suggested that when present in lakes, it is probably derived from nearby sources of spring water. Wilkinson and Gulakyan (2010) postulated that *Fabaeformiscandona dorsobiconcava*, which inhabited the deeper parts of the lake during the Holocene/Anthropocene transition, may have gone into extinction as a result of the acute hypoxia established during the 1970s, although Jenderedjian et al. (2012) has since listed it amongst their assemblages.

2.6 Ecological Recovery

Recent attempts to improve and at least partially reverse the ecological degradation of Lake Sevan began in 2002 with the introduction of legislation and management policies (Lind and Taslakyan 2005). Schemes to increase water levels by transferring water into the lake via the Arpa-Sevan Tunnel had begun in the 1980s with little success, but with the Arpa-Voratan extension in 2003, water levels began to increase. It was calculated that the lake would recover from its eutrophic condition if water levels were raised by c. 6 m at which level the hypolimnion should be re-established (Garibyan 2007). By 2014 water level had risen by 3.84 m from 1896.32 m to 1900.16 m asl. Between 2002 and 2015, water transparency improved to 15.0 m in Great Sevan and 13.5 m in Little Sevan during the summer and, respectively, 12.5 and 11.0 m during the autumn, although there is seasonal variability caused by, for example, spring homothermy and mixing of water layers (Hakobyan and Jenderedjian 2016). The increased water level broadened the littoral zone, increased the biospace available for macrophytes which grow down to a depth of c. 10 m and reintroduced marginal wetland habitats. Improvements in the environment of Great Sevan are seen in the wider distribution and increased biomass of bivalves by 2015. The biomass of chironomids increased considerably exceeding oligochaetes (in terms of the average biomass of zoobenthos) in the shallower waters of Great Sevan during the spring and autumn of 2015. Amphipods and mayflies are more common and new taxa have also been introduced including the damselflies *Ischnura pumilio* and *Coenagrion pulchellum* and the gastropod *Costatella (Physa) acuta*. In the deeper waters of Little Sevan, oligochaetes (as a proportion of the total biomass of zoobenthos) increased from 52.5% in 2012 to 75.9% in October 2013 and 93.8% in 2014, although in November 2015 there was a decrease to 69.4%. There are encouraging signs that the phase of recovery is progressing, although the political will is required for this to continue. However, this is only part of the problem because aspects such as sewage, chemical and heavy metal concentration must be controlled and problems associated with the pollution and rotting vegetation in the newly flooded areas have also to be overcome.

2.7 Conclusions

The Palaeogene Paratethys Sea shallowed during the Oligocene and Miocene so that fluviatile and lacustrine conditions were established in the Sevan Basin during the Pontian and throughout the Pliocene. However, the modern morphology of Lake Sevan did not evolve until the Pleistocene and Holocene when periods of volcanic activity dammed the basin.

The diverse biota which had been established in the pristine, oligotrophic, Holocene lake thrived through to at least 1928. Anthropogenic ecological degradation, the inception of which began in the 1930s, accelerated rapidly during the second half of the twentieth century when water extraction for hydroelectric power generation, irrigation and consumption exceeded replenishment, leading to changes in water level, water temperatures, dissolved oxygen concentrations, sedimentation, etc. The profound and accelerating environmental degradation, which began in approximately 1950, marks the start of the Anthropocene in the Sevan Basin. The degradation of the physical environment was felt throughout the trophic web resulting in a decline in richness and diversity, and an increase in extinction and extirpation of macrophytes, zoobenthos and zooplankton. Higher in the food chain, fish stocks declined and introduced 'exotic' species out-competed elements of the endemic population. A phase of recovery commenced in 2002, when the Government of Armenia adopted the EU Water Framework Directive's approaches in water resource management, monitoring and planning.

Acknowledgements The author wishes to thank Dr S. Z. Gulakyan for his collaboration during the original project on Lake Sevan, and his assistance with the translation of the Armenian and Russian data sets. The author publishes with the permission of the Executive Director of the British Geological Survey (U.K.R.I.)

References

Afanasev GD (1933) Sediment of Lake Sevan. The basin of Lake Sevan (Geokcha), vol 3, part 2, pp 53–154, Leningrad (in Russian)

Aghababyan K, Ananian V, Avetisyan A, Badalyan J, Danchenko A, Hakobyan N, Hambardzumyan A, Harutyunova L, Hovhannesyan V, Kalashian M, Karagyan G, Khachatryan MM, Mirumyan L, Aghasyan A, Aghasyan L, Amiryan S, Danielyan F, Eghiazaryan E, Gabrielyan B, Yavruyan E (2010) The red book of animals of the Republic of Armenia. Ministry of Nature Protection of the Republic of Armenia, Yerevan

Aleshinskaya ZV (1980) Results of the radiocarbon dating of the Sevan Lake Holocene deposits. Isotopic and geochemical methods in biology, geology and archaeology, Abstracts, Tartu, pp 5–7

Aleshinskaya Z, Pirumova L (1982) Morphological studies of diatoms from Pliocene lacustrine deposits of Armenia. In: Menner B, Druschiza B (eds) The nature of the historical development of fossil organisms. University of Moscow, Moscow, pp 97–109 (in Russian)

Arnoldi LV (1929) Bottom productivity in Lake Sevan. Trudy Sevanskoi Gidrobiologicheskoi Stantsii 1:1–96 (in Russian)

Aslanian AT, Sayadian YV (1984) Excursion 010: Neotectonics of Armenia. In: Aslanian AT (ed) Excursions to the Armenian Soviet Socialist Republic. Guide Book for the 27th International Geological Congress, pp 53–86

Aslanian AT, Vehuni AT (1984) General geological features of the Armenian SSR: stratigraphy. In: Aslanian AT (ed) Excursions to the Armenian Soviet Socialist Republic. Guide Book for the 27th International Geological Congress, pp 8–24

Babayan A, Hakobyan S, Jenderedjian K, Muradyan S, Voskanov M (2006) Lake Sevan: experience and lessons learned brief. Lake Basin management initiative regional workshop for Europe, Central Asia and the Americas, Saint Michael's College in Vermont, USA, 18–21 June 2003, pp 347–362

Barsukov VL, Beliaev AA, Bakalin VA, Igumnov VA, Ibragimova TL, Serebrennikov VS, Sultankhojaev AN (1992) Geochemical methods for earthquake prediction. Moscow, Nauka, 213pp (in Russian)

Bronshtein ZS (1928) Contribution to the identification of the ostracod fauna of the Caucuses and Persia. Works North-Caucuses Hydrobiol Station 2:67–119 (in Russian with a German summary)
Bronshtein ZS (1947) Faune de l'URSS. Crustacés, volume 2, numéro 1: Ostracodes des eaux douces. Zoologicheskiy Institut Akademii Nauk SSSR, n.s. 31, 1–339 (English translation, Freshwater Ostracoda, 1988, Oxonian Press, New Delhi)
Bubikyan SA (1984) Recent Ostracoda from the bottom beds of the south-western part of Lake Sevan. Izvestiya Akademii Nauk Armyanskoi SSR, Nauki o Zemle 37:10–24 (in Russian)
Cotton LJ, Zakrevskaya EY, van der Boon A, Asatryan G, Hayrapetyan F, Israyelyan A, Krijgsman W, Less G, Monechi S, Papazzoni CA, Pearson PN, Razumovskiy A, Renema W, Shcherbinina E, Wade BS (2017) Integrated stratigraphy of the Priabonian (upper Eocene) Urtsadzor section, Armenia. Newsl Stratigr 50(3):269–295
Dadikyan MG (1986) Fishes of Armenia. AN Arm. SSR, Erevan 245pp (in Russian)
Danielopol DL, Geiger W, Tölderer-Farmer M, Orellana CP, Terrat MN (1985) The Ostracoda of Mondsee: spatial and temporal changes during the last 50 years. In: Danielopol DL, Schmidt R, Schultz E (eds) Contributions to the paleolimnology of the Trumer Lakes (Salzburg) and the Lakes Mondsee, Attersee and Traunsee (Upper Austria). Limnologisches Institut der Österreichischen Akademie der Wissenschaften, Mondsee, pp 99–121
Danielopol DL, Olteanu R, Löffler H, Carbonel P (1990) Present and past geographical ecological distribution of *Cytherissa lacustris* (Ostracoda, Cytherideidae). In: Danielopol DL, Carbonel P, Colin JP (eds) *Cytherissa* (Ostracoda) - The *Drosophila* of Paleolimnology. Bulletin de l'Institut de Géologie du Bassin d'Aquitaine 47:97–118
Danielyan K, Gabrielyan B, Minasyan S, Chilingaryan L, Melkonyan G, Karakhanyan A, Tozalakyan P, Vanyan A, Pirumyan G, Ghukasyan E, Sarksyan L (2011) Integral evaluation of environmental status of Lake Sevan, Yerevan. "SHD"/UNEPCom, pp 1–42 (in Russian)
Domrachev PF (1940) The influence of the drop of water level of Lake Sevan (Goktcha) on its biological productivity. Proc All-Union Geogr Soc 42:395–399 (in Russian)
Eastwood WJ, Leng MJ, Roberts N, Davis B (2007) Holocene climate change in the eastern Mediterranean region: a comparison of stable isotope and pollen data from a lake record in southwest Turkey. J Quat Sci 22:327–341
Filatov NN (1983) Dynamics of lakes. Gidrometeoizdat, Leningrad 166pp (in Russian)
Fortunatov MA (1927) Lake Sevan trout co-species *Salmo ischchan* Kessler. Trudy Sevanskoi Gidrobiologicheskoi Stantsii 1:1–131 (in Russian)
Fortunatov MA, Fortunatova KR, Kulikova EV (1932) Evaluation of natural resources in Lake Sevan. Trudy Sevanskoi Gidrobiologicheskoi Stantsii 3:1–182 (in Russian)
Fridman GM (1950) Benthonic fauna of Lake Sevan. Trudy Sevanskoi Gidrobiologicheskoi Stantsii 10:7–92 (in Russian)
Gagarin VG, Hakobyan SH (1991) On the Nematoda fauna of Lake Sevan. Biol Bull Internal Waters Leningrad 90:25–28 (in Russian)
Gagarin VG, Hakobyan SH (1992) Ecological review of free living Nematoda of Lake Sevan. Biol Bull Internal Waters Leningrad 92:36–38 (in Russian)
Gambaryan MG (1979) Temperature regime of Lake Sevan. Trudy Sevanskoi Gidrobiologicheskoi Stantsii 17:123–129 (in Russian)
Garibyan M (2007) In: Agyemang TK (ed) Lake Sevan water levels. Hydrometeorological Agency, Ministry of Nature Protection, Sevan, Armenia
Gerasimov YV, Krylov AV (2012) Effect of fish on the zooplankton of Lake Sevan (Armenia). Kaliningrad, pp 90–91 (in Russian)
Gevorgyan GA, Mamyan AS, Hambaryan LR, Khudaverdyan SK, Vaseashta A (2016) Environmental risk assessment of heavy metal pollution in Armenian river ecosystems: case study of Lake Sevan and Debed River catchment basins. Polish J Environ Stud 25:2387–2399
Gezalyan MG (1979) Temperature regime of Lake Sevan. Trudy Sevanskoi Gidrobiologicheskoi Stantsii 17:5–23 (in Russian)
Gezalyan MG, Khorlashko LI (1979) Oxygen regime of Lake Sevan. Trudy Sevanskoi Gidrobiologicheskoi Stantsii 17:24–37 (in Russian)

Golovenkina N (1977) Microscopical investigation of diatomites from the Sisian region, Armenian SSR. Leningrad University News 3:39–47 (in Russian)
Gulakyan SZ, Wilkinson IP (2002) The influence of earthquakes on large lacustrine ecosystems, with particular emphasis on Lake Sevan, Armenia. Hydrobiologia 472:123–130
Hakobyan SH, Jenderedjian KG (2016) Current state of the zoobenthos community of Lake Sevan. Armenia Nat. Acad. Sci., Nat Sci 2:18–22
Hakobyan MA, Djomin YL (1982) Numerical modeling currents of the Lake Sevan. Metrol Hydrol 8:68–74 (in Russian)
Hambaryan LH, Hovhannisyan RH, Hovsepyan AA (2007) Dynamics of Lake Sevan phytoplankton abundance in 2005. Agroscience 7:355–360 (in Russian)
Hovanesian R, Bronozian H (1994) Restoration and management of Lake Sevan in Armenia: problems and prospects. Lake Reserv Manage 9:178–182
Hovhannissian RH (1996) Evolution of eutrophication processes of Lake Sevan and how to control them. Proceedings of International Conference, Lake Sevan: problems and strategies of action. Sevan, Armenia, 13–16 October 1996, Yerevan, pp 81–85
Hovsepyan R (2013) First archaeobotanical data from the basin of Lake Sevan. Veröffentlichungen des Landesamtes für Denkmalpflege und Archäologie Sachsen-Anhalt 67:93–105
Hovhanissian R, Gabrielyan B (2000) Ecological problems associated with biological resource use of Lake Sevan, Armenia. Ecol Eng 16:175–180
Hovhannissian RO (1994) Lake Sevan Yesterday, Today. Armenia Nat. Acad. Sci., Yerevan, 478pp (in Russian with extended summaries in English and Armenian)
Hovhannisyan RH, Hovsepyan AA, Hambaryan LR (2010) Phytoplankton distribution and abundance in relation to physico-chemical factors in Lake Sevan, Armenia. Armenia Nat. Acad. Sci., Nat Sci 2:28–37
Hovsepyan AA, Khachikyan TG, Hambaryan LR (2014) Influence of Lake Sevan catchment basin phytoplankton community structure on the same of the lake. In: Shin H-S, Nalbandyan M (eds) Sustainable management of water resources and conservation of mountain lake ecosystems of Asian countries. Proceedings of AASSA regional workshop, 25–29 June 2014. Institute of Geological Sciences, Armenia Nat. Acad. Sci., Yerevan, pp 102–112
Jenderedjian KG (1990) Seasonal dynamic of *Potamothrix hammoniensis* in model semi-transect of Lake Sevan. Biology, ecology and productivity of water invertebrates. Navuka i Technika, Minsk, pp 214–217 (in Russian)
Jenderedjian K (1996) Energy budget of Oligochaeta and its relationship with the primary production of Lake Sevan, Armenia. Hydrobiologia 334:133–140
Jenderedjian KG, Poddubnaya TL (1987) Species composition and distribution of Oligochaetes in Lake Sevan. Biol J Armenia 40:36–42 (in Russian)
Jenderedjian K, Hakobyan S, Stapanian MA (2012) Trends in benthic macroinvertebrate community biomass and energy budgets in Lake Sevan, 1928–2004. Environ Monit Assess 184:6647–6671
Jonasson PM, Thorhauge F (1976) Population dynamics of *Potamothrix hammoniensis* in the profundal of Lake Esrom with special reference to environmental and competitive factors. Oikos 27:193–203
Kazaryan AG (1979) Phytoplankton of Lake Sevan. Trudy Sevanskoi Gidrobiologicheskoi Stantsii 17:75–87 (in Russian)
Krasheninnikov VA, Grigorian SM, Martirosian YA, Ptuchian AE, Zaporozhets NI (1986) Section Landzhar (USSR, Armenia). In: Pomerol C, Premoli-Silva I (eds) Terminal Eocene events. Developments in Palaeontology and Stratigraphy vol 9, pp 133–136
Krylov AV, Akopyan SA, Nikogosyan AA (2007) Modern species composition of the zooplankton of Lake Sevan in Autumn. Inland Water Biol 4:48–54
Krylov AV, Akopyan SA, Nikogosyan AA, Hayrapetyan AO (2010) Zooplankton of Lake Sevan and its tributaries. In: Ecology of Lake Sevan in the period of high water level. The Results of Research of the Russian–Armenian Biological Expedition on the Hydroecological Survey of Lake Sevan (Armenia) (2005–2009)). Nauka, Makhachkala, pp 168–200

Krylov AV, Gerasimov YV, Gabrielyan BK et al (2013) Zooplankton in Lake Sevan during the period of high water level and low fish density. Inland Water Biol 6:203–210
Krylov AV, Romanenko AV, Gerasimov YV, Borisenko ES, Hayrapetyan AO, Ovsepyan AA, Gabrielyan BK (2015) Distribution of plankton and fish in Lake Sevan (Armenia) during the process of mass growth of cladocerans. Inland Water Biol 8:54–64
Ljatti SJ (1932) The lake bed of Sevan. Materials on the investigation of Lake Sevan and its basin. Tiflis, 42pp
Legovich NA (1979) "Blooms" in the water of Lake Sevan (Observations 1964–1972). Trudy Sevanskoi Gidrobiologicheskoi Stantsii 17:51–74 (in Russian)
Leshchinskaya AS (1950) Reproductive biology of trout in Lake Sevan. Trudy Sevanskoi Gidrobiologicheskoi Stantsii 10:93–175 (in Russian)
Lind D, Taslakyan L (2005) Restoring the fallen Blue Sky: management issues and environmental legislation for Lake Sevan, Armenia. Environs 29:31–103
Mailyan RA (1957) Whitefish of Lake Sevan. Trudy Sevanskoi Gidrobiologicheskoi Stantsii 15:136–196 (in Russian)
Meisch C (2000) Freshwater Ostracoda of Western and Central Europe. Süßwasserfauna von Mitteleuropa, 8/3 Crustacea: Ostracoda. Spektrum Akademischer Verlag, Heidelberg, 521pp
Meybeck M, Akopian M, Andréassian V (1997) What happened to Lake Sevan? SILNews 23
Mikirtitchian L (1962) Lake Sevan development projects in Soviet Armenia. Studies on the Soviet Union, New Series, Munich II:92–106
Milanovskii EE (1968) Recent Tectonics of the Caucasus. Nedra, Moscow, 484pp (in Russian)
Mnatsakanyan AT (1984) Limnology of mountainous lakes. Acad. Sci. Armenian SSR, Yerevan, pp 172–173 (in Russian)
Neill I, Meliksetian K, Allen MB, Navasardyan G, Kuiper K (2015) Petrogenesis of mafic collision zone magmatism: the Armenian sector of the Turkish–Iranian Plateau. Chem Geol 403:24–41
Nevesskaya LA, Goncharova IA, Iĺina LB, Paramonova NP, Khondkarian SO (2003) The Neogene stratigraphic scale of the Eastern Paratethys. Stratigr Geol Correl 11:105–127
Parparov AS (1990) Some characteristics of the community of autotrophs of Lake Sevan in connection with its eutrophication. Hydrobiologia 191:15–21
Parparova RM (1979) Hydrochemical conditions in Lake Sevan in 1976. Trudy Sevanskoi Gidrobiol Stantsii 17:38–50 (in Russian)
Pavlov PI (1947) Spawning grounds and the impact of the fall in water level on trout population in Lake Sevan. Trudy Sevanskoi Gidrobiologicheskoi Stantsii 12:93–140 (in Russian)
Pavlov PI (1951) Contribution to the Sevan trout biology. Trudy Sevanskoi Gidrobiologicheskoi Stantsii 12:93–140 (in Russian)
Poretzky W (1953) Fossil diatoms from Nurnus and Arzni, Erevan region, Armenian SSR. Diatom Symposium, Leningrad, pp 55–107 (in Russian)
Resnikov SA (1984) Biogenic elements of sediments in Lake Sevan. Exp Field Res Hydrobionts Lake Sevan 19:1–17
Satian MA, Chilingaryan GV (eds) (1994) Geology of Sevan. Armenia Nat. Acad. Sci., Yerevan, 181pp (in Russian)
Savvaitova KA, Petr T (1999) Fish and fisheries in Lake Sevan, Armenia, and in some other high altitude lakes of Caucasus. In: Petr T (ed) Fish and fisheries at higher altitudes: Asia. FAO Fisheries Technical Paper 385, Rome, pp 279–304
Sayadyan YV (1978) Postglacial times in Armenia and adjacent regions. Studia Geomorphologica Carpatho-Balcanica 12:77–93
Sayadyan YV, Aleshinskaya ZV, Khanzadyan EV (1977) Late glacial deposits and archaeology of the Sevan Lake. In: Sayadyan YV (ed) Geology of the Quaternary period (Pleistocene). Acad. Sci. Armenian SSR, Yerevan, pp 91–109
Sayadyan YV (1991) Archaeological relic and Sevan Lake history. The history of lakes Sevan, Issyk-Kul, Balkhash, Zaisan, and Aral, Leningrad, Nauka, pp 31–37
Sayadyan, YV (2006) Upper Miocene, Pliocene, and Quaternary stratigraphic reference sections of large intermontane depressions in Armenia. Dokl Earth Sci 407:217–219

Shcherbinina EA, Iakovleva AI, Zakrevskaya, EY (2017) Middle Eocene to Early Oligocene nannofossils and palynomorphs from the Landzhar Section, Southern Armenia: zonal stratigraphy and palaeoecology. Stratigr Geol Correl 25:557–580

Simonyan AA (1988) Zooplankton in changing conditions of a water body (case study: Lake Sevan). Ph.D. thesis, Leningrad, pp 43 (in Russian)

Simonyan (1991) Zooplankton of Lake Sevan. Yerevan, 298pp (in Russian)

Simonyan A, Gabrielyan B, Minasyan S, Hovhannisyan G, Aroutiounian R (2016) Genotoxicity of water contaminants from the basin of Lake Sevan, Armenia Evaluated by the Comet Assay in Gibel Carp (*Carassius auratus gibelio*) and Tradescantia Bioassays. Bull Environ Contam Toxicol 96:309–313

Slobodchikov BY (1951) Hydrochemical regime of Sevan Lake according to the data of 1947–1948. Trudy Sevan. Gidrobiologicheskoi Stantsii 12:5–28

Stroykina VG (1957) Nourishment of *Gammarus* in Lake Sevan. Trans Sevan Hydrobiol Station 15:89–107 (in Russian)

Tsalolikhin SY (1976) Free living Nematodes as indicators of pollution in freshwaters. Methods Biol Anal Freshwaters 50:84–86 (in Russian)

Ustinova I, Krienitz L, Huss VAR (2000) *Hyaloraphidium curvatum* is not a green alga, but a lower fungus; *Amoebidium parasiticum* is not a fungus, but a member of the DRIPs. Protist 151:253–262

Vardanyan LG, Ingole B (2006) Studies on heavy metal accumulations in aquatic macrophytes from Sevan (Armenia) and Carambolim (India) lake systems. Environ Int 32:208–218

Waters CN, Zalasiewicz JA, Williams M, Ellis MA, Snelling AM (eds) (2014) A stratigraphical basis for the Anthropocene? Geological Society, London, Special Publications, vol 395, 309pp

Waters CN, Zalasiewicz J, Summerhayes C, Barnosky AD, Poirier C, Gałuszka A, Cearreta A, Edgeworth M, Ellis EC, Ellis M, Jeandel C, Leinfelder R, McNeill JR, Richter D de B, Steffen W, Syvitski J, Vidas D, Wagreich M, Williams M, Zhisheng A, Grinevald J, Odada E, Oreskes N, Wolfe AP (2016) The Anthropocene is functionally and stratigraphically distinct from the Holocene. Science 351(6269), https://doi.org/10.1126/science.aad2622

Wilkinson IP, Gulakyan SZ (2010) Holocene to recent Ostracoda of Lake Sevan, Armenia: biodiversity and ecological controls. Stratigraphy 7:301–315

Wilkinson IP, Poirier C, Head MJ, Sayer CD, Tibby J (2014) Microbiotic signatures of the Anthropocene in marginal marine and freshwater palaeoenvironments. In: Waters CN, Zalasiewicz JA, Williams M, Ellis MA, Snelling AM (eds) A stratigraphical basis for the Anthropocene? Geological Society, London, Special Publications, vol 395, pp 185–219

Wilkinson IP, Bubikyan SA, Gulakyan SZ (2005) The impact of late Holocene environmental change on lacustrine Ostracoda in Armenia. Palaeoecol Palaeogeogr Palaeoclimatol 225:187–202

Yuzhakova GG (1985) Growth and sexual maturation of Sevan whitefish. Trudy Sevanskoi Gidrobiologicheskoi Stantsii 20:188–198 (in Russian)

Zalasiewicz J, Waters C, Williams M, Aldridge DC, Wilkinson IP (2018) The stratigraphical signature of the Anthropocene of England and its wider context. Proc Geol Assoc 120:482–491

Zalasiewicz J, Williams M, Barnosky T, Wilkinson IP (2019) Late Quaternary extinctions. In: Zalasiewicz J, Waters C, Summerhayes C, Williams M (eds) The Anthropocene as a geologic time unit. Cambridge University Press, Cambridge, pp 115–119

Chapter 3
Past and Current Changes in the Largest Lake of the World: The Caspian Sea

Suzanne A. G. Leroy, Hamid A. K. Lahijani, Jean-Francois Crétaux, Nikolai V. Aladin and Igor S. Plotnikov

Abstract The Caspian Sea (CS), located between Europe and Asia, is the largest lake in the world; however, its physical environment and its floor have oceanic characteristics. The CS is composed of a very shallow north sub-basin with a very low salinity mostly below 5 psu. The middle and southern sub-basins are deep and have a salinity of c. 13 psu. To the east, the Kara-Bogaz-Gol, a hypersaline lagoon, is connected to the middle sub-basin. The CS is endorheic and therefore very sensitive to changes in hydrography and climate. Because of its long history of isolation following the disconnection of the Caspian Sea from the Paratethys c. 6 million years ago, this ancient lake has many endemic species. The harsh environment of its brackish waters and the repeated salinity changes over the millennia, however, do not allow for a high biodiversity. The benthos is more varied than the plankton. The history of water-level changes remains poorly known even for the last centuries. Nevertheless, the amplitude was of >150 m in the Quaternary, several tens of meters in the Holocene and several meters in the last century. Many factors affect its natural state, such as petroleum pollution (an industry dating back to Antiquity), nutrient increase

S. A. G. Leroy (✉)
Aix Marseille Univ, CNRS, Minist Culture, LAMPEA, UMR 7269, 5 rue du Château de l'Horloge, 13094 Aix-en-Provence, France
e-mail: suzleroy@hotmail.com; leroy@mmsh.univ-aix.fr

H. A. K. Lahijani
Iranian National Institute for Oceanography and Atmospheric Science, #3, Etamadzadeh St., Fatemi Ave, 1411813389 Tehran, Iran
e-mail: lahijani@inio.ac.ir

J.-F. Crétaux
Laboratoire d'Etudes en Géophysique et Océanographie Spatiale, UMR5566, Université de Toulouse, LEGOS/CNES, 14 Ave. Edouard Belin, 31400 Toulouse, France
e-mail: jean-francois.cretaux@legos.obs-mip.fr

N. V. Aladin · I. S. Plotnikov
Laboratory of Brackish Water Hydrobiology, Zoological Institute, Russian Academy of Sciences, Universitetskaya nab. 1, 199034 St. Petersburg, Russia
e-mail: Nikolai.Aladin@zin.ru

I. S. Plotnikov
e-mail: Igor.Plotnikov@zin.ru

S. Mischke (ed.), *Large Asian Lakes in a Changing World*, Springer Water,
https://doi.org/10.1007/978-3-030-42254-7_3

(alongside >14 million inhabitants along the coast), invasive species (e.g. the comb jelly *Mnemiopsis leidyi*), overfishing (including sturgeon) and modifications of its coastline (e.g. sand extraction). In comparison to other ancient lakes, the CS surface temperature has suffered from the fastest increase on record. Owing to the complex natural state of the CS, it is not easy to identify the Holocene-Anthropocene transition, although it may be suggested that is was approximately AD1950 when intense human activity started to modify the lake.

Keywords Hydrography · Brackish water · Water-level change · Endemism · Anthropogenic impact

3.1 Introduction

The Caspian Sea (CS) is currently the largest lake on Earth with an area slightly greater than Germany. It lies at the geographical border between Europe and Asia (Fig. 3.1). Originally it was created tectonically when the Paratethys Sea started splitting in smaller water bodies some 5.6 million years ago (Hinds et al. 2004). It may therefore be counted as an ancient lake. It can be considered to have been an endorheic water body for the last c. 2.6 million years, with only occasional outflow to the Black Sea and inflow from the Aral Sea. This situation has led to the development of endemism in many groups of animals and plants, living in the water or depending on it (e.g. the seals; Dumont 1998). The close nature of its water body also favours wide changes in water levels in reaction to changes in climate and in the main feeding rivers, such as the Volga (from the North) and the Amu-Darya (from the South-East, with decreasing frequency until perhaps the fifteenth century). The CS water level (CSL) has often changed dramatically: during its geological lifetime by more than 150 m, possibly several hundreds of metres (Kroonenberg et al. 1997; Mayev 2010; Forte and Cowgill 2013), during the Holocene by several tens of meters (Kakroodi et al. 2012), during the last millennium by >10 m (Naderi Beni et al. 2013) and during the last century by >3 m (Arpe and Leroy 2007; Chen et al. 2017; Fig. 3.2). However, CSL changes remain poorly identified, and their reasons not completely understood, even for the last decades.

Humans have lived along the shores of the CS since the Late Pleistocene at least, e.g. no less than four sites of Neanderthal occupation with Mousterian tools are known (Dolukhanov et al. 2010). Recently increasing human pressure is felt around the CS, as economic development is taking place. Nowadays the impact on the coastal environment and economical activities around and in the CS cause significant socio-economic problems. For many years, the main activity in the area was related to the fishery industry, as the CS holds the most appreciated sturgeon species in the world, with the export of caviar. In the last decades, the petroleum (oil and gas) industry has boomed, especially with large offshore infrastructures, e.g. offshore Baku (Azerbaijan's capital) and in the Kashagan offshore oil field (N CS).

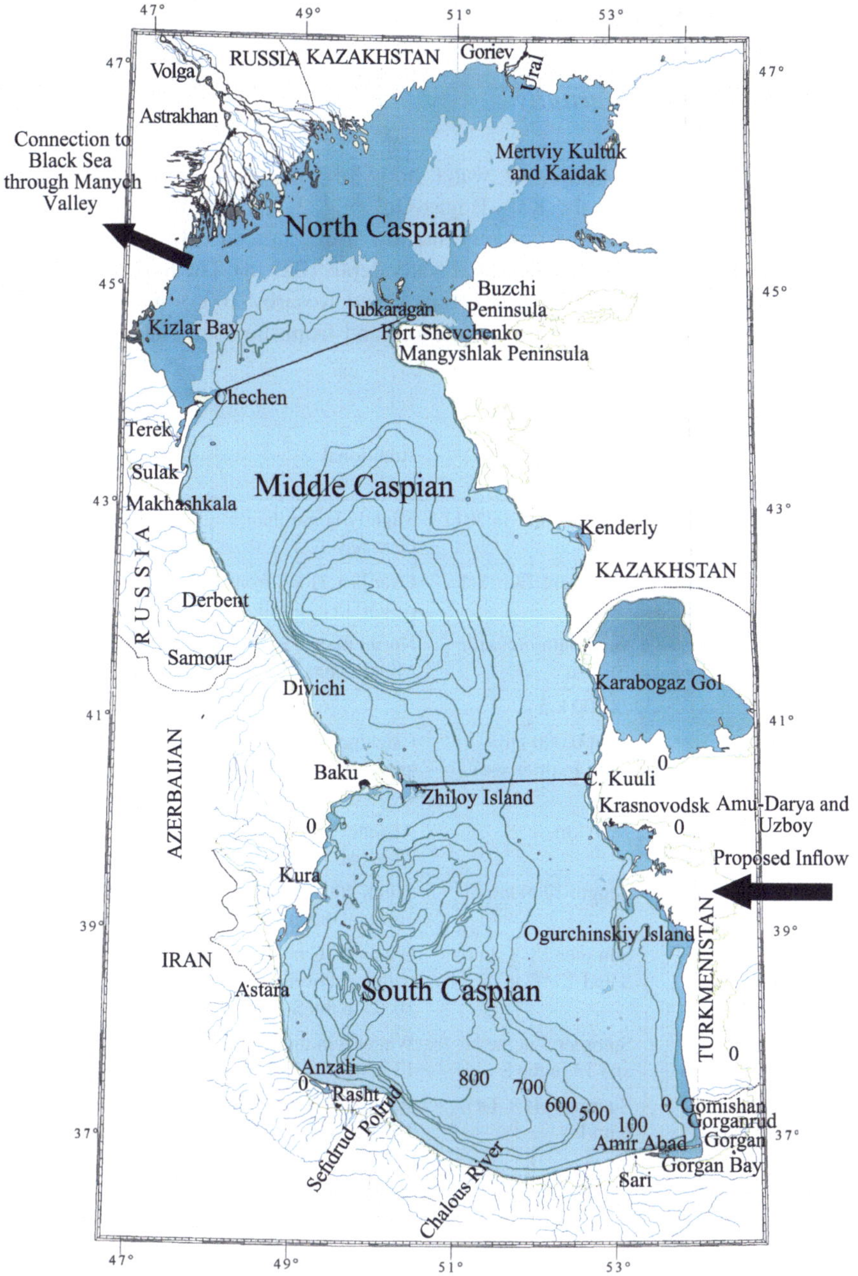

Fig. 3.1 Location map with bathymetry. The black lines through the Caspian Sea show the limit between its three sub-basins. The green contour is the zero in the Baltic datum

3.2 Modern Setting

3.2.1 Physical Geography

The CS is the largest endorheic water body of the world. It has an area of ~386,400 km^2, excluding the Kara-Bogaz-Gol (Fig. 3.1; Table 3.1). It is made of three sub-basins, deepening from very shallow in the north (5–10 m deep) to the deepest sub-basin in the south (maximal water depth 1025 m). The middle sub-basin has a maximum depth of 788 m (Kostianoy and Kosarev 2005) and is separated from the north by the Mangyshlak Threshold and from the south sub-basin by the

Table 3.1 The Caspian Sea in numbers

Caspian Sea	Information	Additional explanation	Source
Size	386,400 km^2 in 2017	Slightly larger than Germany	Arpe et al. (2018)
Level	At present 28 m bsl	Over last 21 ka, from c. +50 to c. −113 m	Several sources, see text
Depth	Maximum in South: 1025 m	North: max. 25 m Middle: max 788 m	Kostianoy and Kosarev (2005)
Volume	78,200 km^3		Dumont (1998)
Catchment	~3,500,000 km^2 (with E. drainage)	Catchment/surface area ratio: from ~7:1 to 10:1	Kostianoy and Kosarev (2005), Chen et al. (2017)
Length of coastline	7500 km	At water level of 27 m bsl	Kostianoy and Kosarev (2005)
N-S length and W-E width	Length 1200 km	Width 200–450 km, average 350 km	Kostianoy and Kosarev (2005)
Surface water temperature	Summer 28 °C in the S and 23 °C in the N	Winter 0 °C in the N with ice cover and 10 °C in the S	Kostianoy and Kosarev (2005)
Salinity	Summer 3 in the N and 13 in the S	Winter 5 in the N and 13 in the S	Kostianoy and Kosarev (2005)
Surface pH and dissolved oxygen	Summer pH 8, DO 6 ml/l	Winter pH 8, DO 10 ml/l in N to 7 ml/l in S	Kostianoy and Kosarev (2005)
pH and dissolved oxygen at 100 m	Summer pH 8, DO 6 ml/l in N, 4 ml/l in S	Winter, pH 8, DO 7 ml/l in N, 4 ml/l in S	Kostianoy and Kosarev (2005)
River inflow	290 km^3/year	84% Volga	Lahijani et al. (2008; this paper)
Number of inhabitants in the catchment	>80 millions	In five countries	Lahijani et al. (2008)

Apsheron Sill (150 m; Kuprin 2002). An ecogeographical classification of the CS based on its physical parameters shows the justification of the division into three sub-basins, as well as the role of the distance to the shore (Fendereski et al. 2014). The volume of the CS is 78,200 km^3 and this makes it three times larger than Lake Baikal (Messager et al. 2016). The north sub-basin is very shallow, and any small vertical change is translated in vast inundation or emersion along the north coast (the Caspian Depression) reaching a horizontal amplitude of c. 1000 km during the last glacial-interglacial cycle.

The CS is located in a depression bordered by the Caucasian mountains in the West, the central Asian plateaus and desert in the East, the Russian and Kazakh plains in the north, and by the Alborz Mountains in the south (Figs. 3.1 and 3.3). The CS watershed is approximately 3.5 million km^2 and it covers nine countries: Armenia, Azerbaijan, Georgia, Iran, Kazakhstan, Russia, Turkey, Turkmenistan and Uzbekistan.

The CSL was at 28 m below world sea level (bsl), geodetical station Kronstadt, Baltic, Russia, in 2018.[1] Recent CSL changes are not only rapid, e.g. a hundred times faster than global ocean level rise in the twentieth century, but also large (Kroonenberg et al. 1997). The instrumental record, starting in 1837, shows a 3-m amplitude with a minimum in 1977 and maxima in 1929 and 1995 (Voropayev 1997; Leroy et al. 2006), thus ranging between 26 and 29 m below sea level (Fig. 3.2a). Changes of the CSL can, moreover, be very abrupt as observed after 1930 when the level dropped by 2 m in approximately ten years and also when the water level increased by 2 m between 1979 and 1996. Since then, the CSL has dropped again by 1 m (June 2018). Between 1996 and 2016, the water level fell by 7–10 cm per year (Arpe et al. 2014; Chen et al. 2017). Since 2016, the CSL seems to have stabilized to equilibrium at ~−28 m (Fig. 3.2b). As it is an endorheic lake, the CSL variations are mostly controlled by water discharge from the rivers, the direct precipitation and evaporation over the CS and the discharge to the Kara-Bogaz-Gol (Kosarev et al. 2009).

Around 130 rivers flow into the CS from the north, west and southern coasts. They supply annually 250 km^3 of water and 70 million tons of sediments. The Volga discharge into the CS is highly variable and in total, it represents between 80 and 90% of the surface water inflow (Fig. 3.2c).

Because of its great meridional extension, the CS straddles several climatic zones (Fig. 3.3). The northern part of the drainage basin lies in a zone of temperate continental climate with the Volga catchment well into the humid mid-latitudes. The emerged Volga delta (the Caspian Depression) forms a low-lying area and is the most arid region of European part of the Russian Federation (UNEP 2004). The western coast features a moderately warm and dry climate, while the southwestern and the southern regions fall into a subtropical humid climatic zone. The eastern coast is desert.

[1]Since 1961, the water level reference is the Kronstadt gauge in the Baltic Sea. It is a historical reference for all former Soviet Union regions.

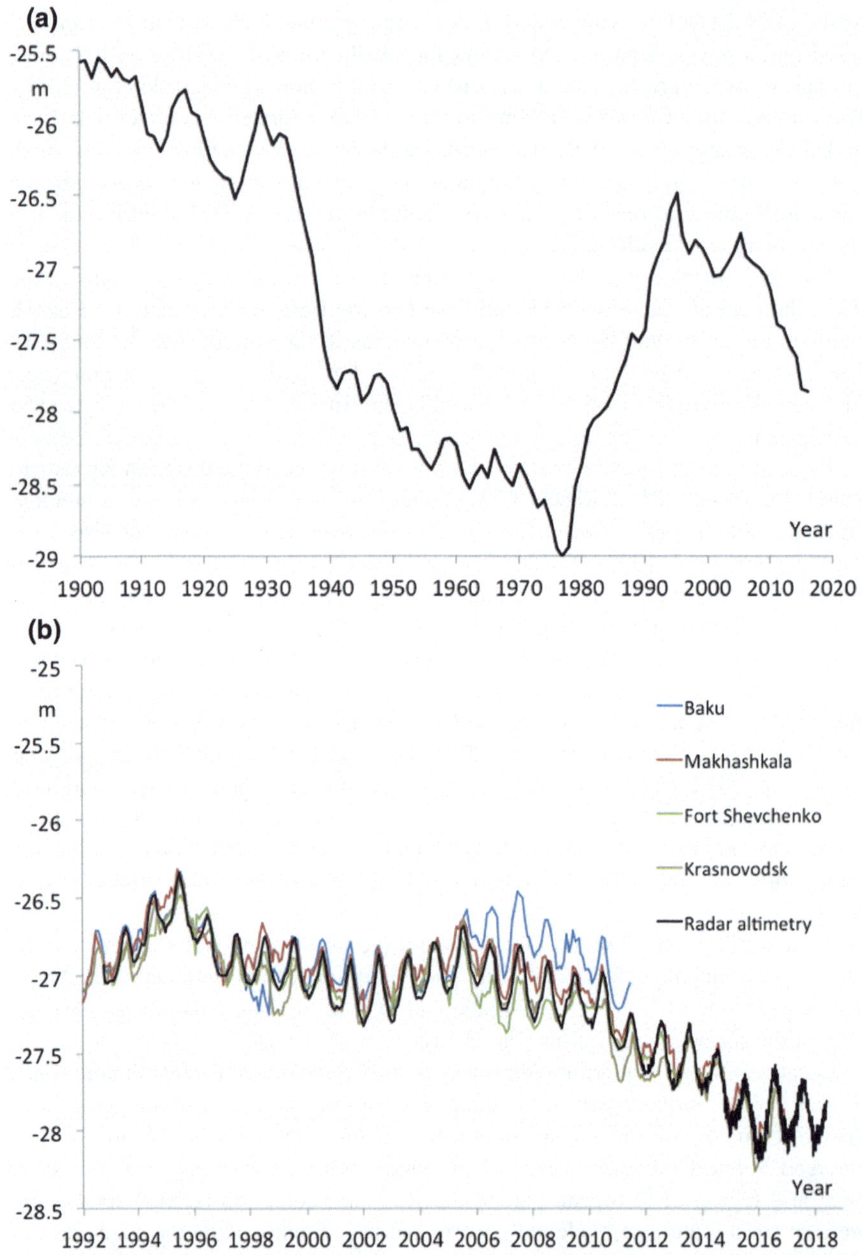

Fig. 3.2 **a** Instrumental data of Caspian Sea levels since AD1900 from in-situ average gauge data. **b** Monthly Caspian Sea level variations from 1992 to 2016 from in-situ gauge data collected in Baku, Makhashkala, Port Shevchenko and Krasnovodsk, and from radar altimetry using a constellation of Topex/Poseidon, Jason-1, Jason-2, Jason-3 and Sentinel-3A satellites. **c** Volga River discharge in percent of the total surface inflow. **d** Discharge in km^3 for each river excluding the Volga

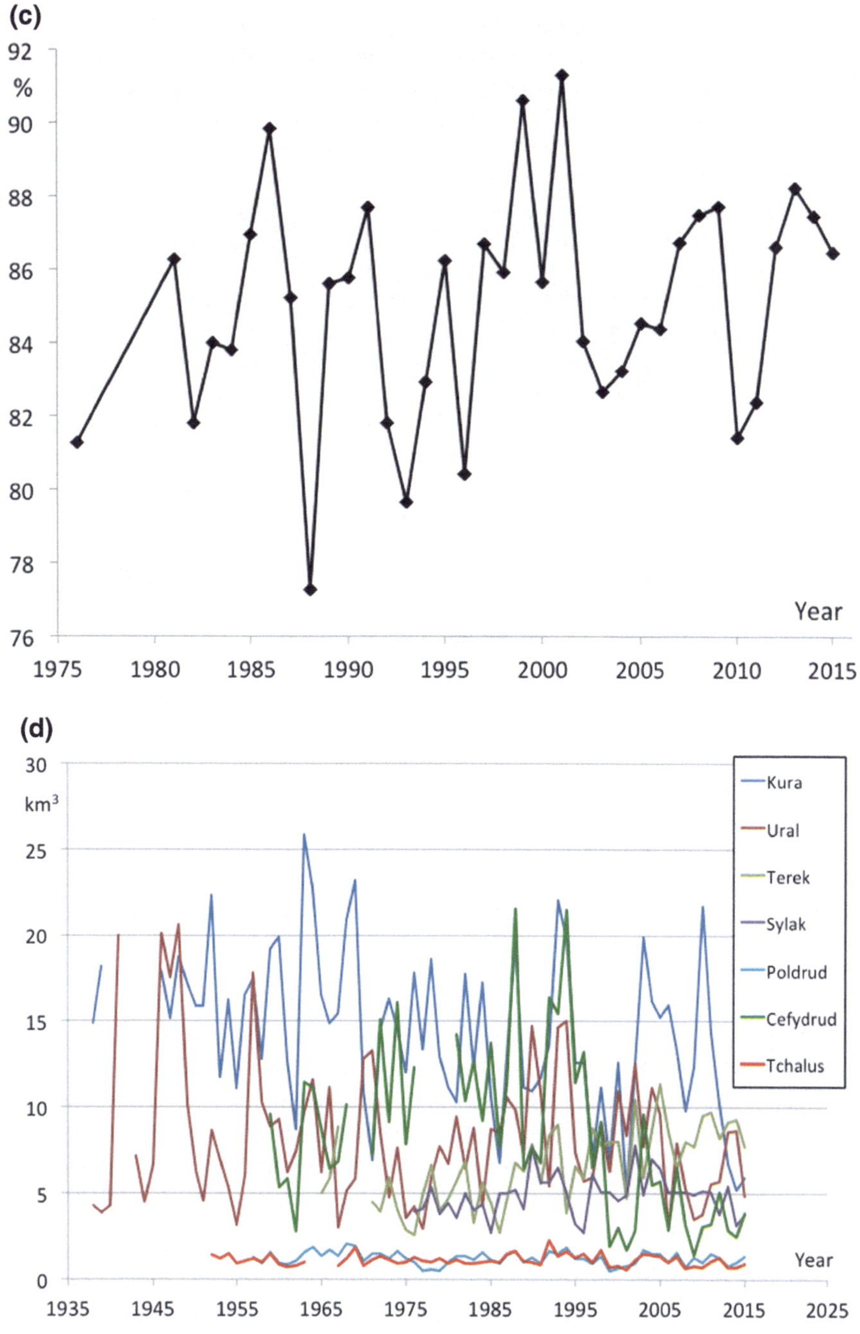

Fig. 3.2 (continued)

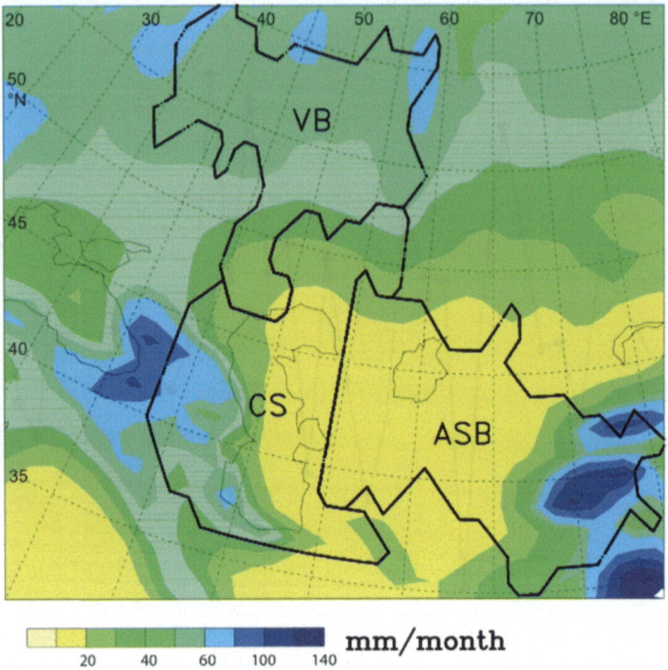

Fig. 3.3 Precipitation in mm per month (colour scale) and drainage basins (bold black line) over the Volga (VB) and the Amu-Darya and Syr-Darya (ASB) basins (provided graciously by K. Arpe)

At present, precipitation received over the drainage basin of the Volga River drives changes in the CSL, especially summer precipitation (Arpe and Leroy 2007; Arpe et al. 2012; Fig. 3.3). Nowadays, it is possible to forecast to some extent the CSL based mainly on the precipitation across the catchment area, considering that it takes a few months for the water to come down the Volga and feed the CS (Arpe et al. 2014).

Explanation of long-term CSL change has puzzled the scientists of many countries: Shiklomanov et al. (1995), Arpe and Leroy (2007), Arpe et al. (2000, 2012, 2018), Lebedev and Kostianoy (2008), Ozyavas and Khan (2008), Ozyavas et al. (2010) and Chen et al. (2017). Their objective was often to forecast future evolution of the CSL. The link between CSL and climate can be seen in two ways. Firstly the analysis of the causes of the observed CSL is made in terms of the inter-annual variability of the water-balance components (Ozyavas et al. 2010; Chen et al. 2017). Secondly the relationship between global large-scale phenomena like El-Niño southern Oscillation (ENSO) and atmospheric circulation in general (Arpe et al. 2000; Arpe and Leroy 2007), or extreme phenomena like drought events (Arpe et al. 2012), including feedback mechanisms from the CSL (Arpe et al. 2018), is analysed. It has, for example, been demonstrated in Arpe et al. (2000) that CSL long-term changes have been attributed partly to ENSO enhancing the teleconnection between regional water-level variability with a global index such as the Southern Oscillation Index calculated from sea surface temperature (SST) over the Pacific Ocean.

3.2.2 *Physical Limnology*

The modern salinity of the surface waters shows a gradient from freshwater in the large Volga delta to 13 psu in the middle and southern sub-basins (Kostianoy and Kosarev 2005). The Kara-Bogaz-Gol, a small basin connected to the east of the CS, serves as an overflow and is hypersaline, up to 20 times that of the CS (Giralt et al. 2003; Leroy et al. 2006). The shallow gulfs of Mertviy Kultuk and Kaidak in the east of the northern sub-basin also have a high salinity in the range of 30 psu (Fig. 3.1).

The chemical composition of the CS water is largely similar to that of oceanic water although with some differences (Kosarev and Yablonskaya 1994). The CS water is poorer in sodium and chlorine ions and is richer in calcium and sulphate ions. This difference in the ratio of salts has arisen due to the isolation of the CS from the world ocean and transformation under the influence of river runoff.

The surface water temperature gradients are from 23 to 30 °C in summer and from 0 to 10 °C in winter, with c. three months of ice cover in the north sub-basin (Kostianoy and Kosarev 2005). Beyond the complex surface and bottom currents, the overall water transfer is from the north to the south, as a result of the inflow of the Volga in the north and evaporation in the south. Satellite altimetry has indeed observed an amplitude of annual change of the CS water topography which varies by approximately 10 cm from north to south, and a CSL rise which is higher in the north than in the south by 3 cm/year (Cazenave et al. 1997; Lebedev and Kostianoy 2008; Kouraev et al. 2011).

The CS, as a large lake, has the biodiversity and bioproductivity of a lake. However, the physical water environment behaves as that of an open water due to its size and depth. The CS currents and water-mass circulations are strong and wave patterns are complex. The rate of the CS deep-water circulation has changed significantly during water-level fluctuations (Kostianoy and Kosarev 2005; Sapozhnikov et al. 2010). The physiography of the CS, riverine input, geographical distribution of salinity and temperature determine the formation of water masses and circulation processes (Terziev et al. 1992). The differences in water masses, mainly created by climate, lead to relatively distinctive physical and chemical specifications. To a lesser extent, geological forces may cause a mixing of water masses, e.g. caused by the dispense of materials and energy through the water column. Four water masses encompass the CS water column: (1) north CS water mass, (2) surface-water mass of the middle and south Caspian sub-basins, (3) deep-water mass of the middle Caspian sub-basin and (4) deep-water mass of south Caspian sub-basin (Fig. 3.4b; Lahijani et al. 2018a). The shallow water of north Caspian sub-basin and the huge freshwater influx of the Volga River form a relatively small water mass that is nevertheless crucial for circulation. This water mass has considerable annual variability in terms of salinity and temperature (Terziev et al. 1992). The upper layer of the middle and south Caspian

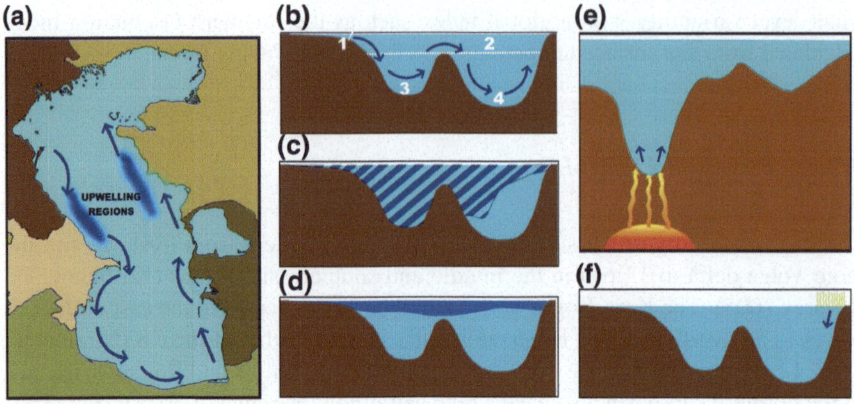

Fig. 3.4 Caspian circulation. **a** General circulation and main upwelling areas. **b** Circulation of water masses due to winter freezing of the N Caspian Sea; numbers denote to: 1 – North Caspian water mass, 2 – surface-water mass of the middle and south Caspian sub-basins, 3 – deep-water mass of middle Caspian sub-basin, and 4 – deep-water mass of south Caspian sub-basin. **c** Depth penetration of winter convection of 1969. **d** Depth penetration of winter convection of 1975. **e** Negligible geothermal impact of oceanic crust in the south Caspian sub-basin. **f** Local extensive evaporation and dense saline water formation in the southeastern part of the south Caspian sub-basin

sub-basins creates a separate water mass that is governed by the overall climate. Wind, wave, current and temperature cause mixing in a layer of around 100–150 m thickness in the middle CS and of around 50–100 m thickness in the south CS (Terziev et al. 1992). The Apsheron Sill separates the two deep-water masses of middle and south sub-basins and prevents free mixing between them. The middle and south Caspian deep-water masses differ in terms of their physical and chemical characteristics. The middle Caspian deep-water mass has lower temperature and salinity and higher dissolved oxygen content compared to the southern one. Various mechanisms trigger water exchange between these two deep-water masses. The first mechanism starts from the north Caspian sub-basin, as freshwater of the Volga and Ural rivers enter into the CS. The highest river discharge happens during spring and early summer when snowmelt increases runoff and moves southward via the bottom of the western coasts mainly due to Coriolis forcing and other river influxes (Ibrayev et al. 2010). It moves like a long counter-clockwise wave that can be detected by tide gauges as elevated water-level up to 45 cm in July (Terziev et al. 1992). Its salinity increases in the arid climate of the east coast and reaches the north CS waters. The mentioned mechanism and direct contact of the north Caspian water mass and middle and south Caspian surface-water mass cause water exchange between them. Vertical exchange in the middle CS is partially caused by wind forcing that forms a cyclonic gyre in the centre and upwelling along the east and west coasts of the middle Caspian sub-basin leading to contrasting temperatures (Fig. 3.4a). Moreover, the intense evaporation at the eastern coast may cause local warm but saline and dense water to penetrate into deeper waters (Fig. 3.4f). The higher geothermal impact of the ocean crust beneath the south Caspian sub-basin could increase water temperature causing upwelling, although this is negligible in comparison to other mixing mechanisms (Kosarev 1975; Fig. 3.4e).

Extensive water exchange among the water masses occurs during the cold season through two mechanisms: winter convection and north Caspian freezing. The winter convection mixes the water column vertically, the extent of which depends on the severity, distribution, frequency and duration of the driving cold air masses. During mild winters, the winter convection penetrates down to around 200 m in the middle and 100 m in the south Caspian sub-basins, which is not strong enough to mix all water masses (Fig. 3.4d). During severe winters, it reaches down to the middle Caspian bottom waters and to the depth of around 400 m in the south Caspian sub-basin, which enables vertical mixing of oxygen and other biochemical elements (Terziev et al. 1992; Ghaffari et al. 2010; Fig. 3.4c). The vertical mixing is forced by wind, evaporation and winter convection; this however, is not affected by the current water-level fluctuations. The Arctic-type mixing occurs annually in the CS and it affects the whole Caspian water mass (Kostianoy and Kosarev 2005). The strength of this type of mixing is closely related to the severity of winter and the Caspian water-level status (Fig. 3.4b).

During water-level lowstand, mixing would be stronger and deeper; however, in the highstand, it could hardly reach deep-water masses. The Caspian water-levels during highstands and lowstands are correlated to the increase and decrease of Volga River inflow, respectively. The salinity of north Caspian water mass increases during water-level fall (decrease of Volga discharge) and decreases during water-level rise (increase of Volga discharge).

In winter, when the surface water of the north CS is frozen, salt is released to the deeper water mass and a saline water plume flows southwards to mix with the middle Caspian water mass (Kosarev 1990). Cold (c. 0 °C) and saline water becomes denser close to $\sigma_T = 11$ to $\sigma_T = 11.5$ and sinks downward into the middle Caspian water mass. The plume pushes the middle Caspian deep-water mass into the south Caspian deep-water mass where water flowing from the south, via the eastern coast, further increases salinity. The depth penetration is strongly controlled by water level. In the course of water-level rise, dense water formation is weaker inhibiting downwelling. During water-level fall, the north Caspian water mass is saltier. It can release more salt during freezing which makes the adjacent water denser, which thus penetrates deeper. High nutrient and oxygen-bearing north Caspian water mass triggers a circulation pattern that engages the whole CS water (Kosarev and Tuzhilkin 1995). This mechanism enhances the biochemical condition for marine productivity and favours off-shore fishery after a few years of stabilized lowstand.

3.3 The Past of the CS and CS Level Changes

3.3.1 Geological Background

The geological structure of the CS basement is heterogeneous, including the south Russian platform and the Scythian-Turanian Plate to the north and the Alpine folded zone to the south. Three-layer structure of the crust under the CS was disconnected by rifting since the Late Triassic when thick sedimentary sequences overlaid the basaltic rocks (Ulmishek 2001; Brunet et al. 2007). The crust of the south Caspian sub-basin shows more oceanic characteristics. Rifting and extensive subsidence provided accommodation space for sediments with an estimated thickness of c. 25 km (Brunet et al. 2003). The geological setting of the CS and the past water-level changes are reflected in the morphology of CS floor and coastal areas. The CS was a part of the Paratethys Sea until the Pliocene and experienced a common history with the Black, Azov, Mediterranean and Aral seas. The Alpine Orogeny during the Cenozoic led to uprising of the Caucasian territory and consequently separated the CS and Black Sea at the Manych Valley (between north Caucasian Mountains and southern Don River, Fig. 3.1) around five million years ago. The CS began its independent geological, hydrological, and biological history as an enclosed basin during the Pliocene (Forte and Cowgill 2013). The CSL has changed drastically since the isolation from the world ocean. Old river valleys and deltas are preserved on the CS bottom morphology, which represents extreme lowstand conditions. Throughout the Quaternary, the CSL changed widely, possibly in relation with Milankovich cycles, but with a clear overprint of large changes in river palaeogeography. During the Akchagylian transgression (very late Pliocene and very Early Pleistocene), a brief connection to the Arctic has been highlighted (Agalarova et al. 1940; Richards et al. 2018; Hoyle et al. 2020).

Mud volcanoes of the CS are mainly concentrated in the southern part and are very active due to a high sedimentation rate associated with compressional forces and seismicity. Their eruptions bring old sediments, gas and water into the CS and form dome-shaped features up to 400 m in height on the CS lake bed and in the coastal area (Huseynov and Guliyev 2004). Annually around 2 million tons of sediments associated with 0.001–0.01 km^3 of fossil water and 0.36 km^3 of gas is brought to the lake floor by mud volcanoes and gas is released to the atmosphere (Glazovsky et al. 1976). If the onshore-mud volcanoes are included, the total annual gas emission by the mud volcanoes reaches 1 km^3 (Huseynov and Guliyev 2004). Offshore mud-volcano eruptions have environmental consequences including high amounts of hydrocarbons and heavy metal concentration around the cone and mass death of biotas (Glazovsky et al. 1976; Ranjbaran and Sotohian 2015).

Bottom sediments of the CS have different origins including terrigenous, biological and chemical. The terrigenous sediments are supplied mainly through riverine discharge as well as by aeolian transport, coastal erosion and mud volcanoes. Biological components comprise calcareous and siliceous shells and organic materials. Chemical sediments are mainly formed by precipitation of calcium carbonate and sulphur minerals (in deep anoxic environments during early diagenesis). Rates of sedimentation are closely related to the proximity of the sink and source areas of sediments and the hydrodynamics of the environment. The highest sedimentation rate is measured around the river mouths of the Kura (23 mm per year) and Sefidrud (14 mm per year) in the south Caspian sub-basin (Lahijani et al. 2008). In contrast, the deep middle and south Caspian sub-basins have the lowest sedimentation rate (0.1 mm per year). The areas with strong wave and current actions experience continuous erosion that exposes bedrocks. The sedimentation rate in the geological past was strongly dependent on the synsedimentary tectonics such as subsidence, rifting and closure from and connection to the adjacent basins. The past geological history of the CS is a key issue to determine the rich hydrocarbon resources that are distributed in the three sub-basins. The north CS hydrocarbon resources originated from the Palaeozoic, while the south CS resources are attributed to the Late Cenozoic (Ulmishek 2001).

3.3.2 *Changes Since the Last Glacial Maximum*

Transgressions are usually accompanied by freshening of water masses and cold climate; while regressions primarily correspond to increased salinities and warm climate (Chalié et al. 1997), although many exceptions exist. For example, during glacial periods, this simple relationship is complicated by a large influx of meltwater (including via rivers that are now dry) originating from the North and East of the CS catchment area (Grosswald 1993) and climatic feedbacks (Arpe et al. 2018).

No agreed terminology and stratigraphy exist for the CSL, but common points can be summarised broadly as follows. The Khvalynian is a highstand in the Late Pleistocene (Fig. 3.5). A large part of the Holocene equates roughly to the intermediate levels of the Neocaspian (or Novocaspian). However, no consensus exists about the exact date for the transition between the Khvalynian and the Neocaspian. In addition, the Khvalynian and the Neocaspian (separated by a deep lowstand) were both interrupted by several lowstands, some very deep (>100 m), with an inconsistent terminology.

3.3.2.1 The Last Glacial Maximum

Available CSL change reconstructions from the Last Glacial Maximum (LGM) to the present have been compiled and compared (Fig. 3.5). Most of the dates are with uncalibrated radiocarbon ^{14}C dates, as many dates are from before the calibration era (Table 3.2). The CSL in the LGM is poorly known and reconstructions differ amongst investigations.

A lowstand, named the Atelian (or Enotayevian), which reached 50 to 113 m bsl, is reconstructed by several authors around the LGM; and this is when the Early Khvalynian is now mostly recognised to have started rather than much earlier (Varushchenko et al. 1987; Chepalyga 2007; Svitoch 2009; Yanina et al. 2018). In addition, a level lower than the modern CSL has been reconstructed by climate modelling for the LGM (Arpe et al. 2011, 2018). However, higher than present-day water levels are proposed by other authors: 15 m above sea level (asl) by Klige (1990) and 25 m asl by Toropov and Morozova (2010). If we accept very low levels for the Atelian lowstand, the transition to the highstand of the Early Khvalynian had to have been extremely rapid with a rise of 2.5 cm per year (Svitoch 2009), resulting in dramatic

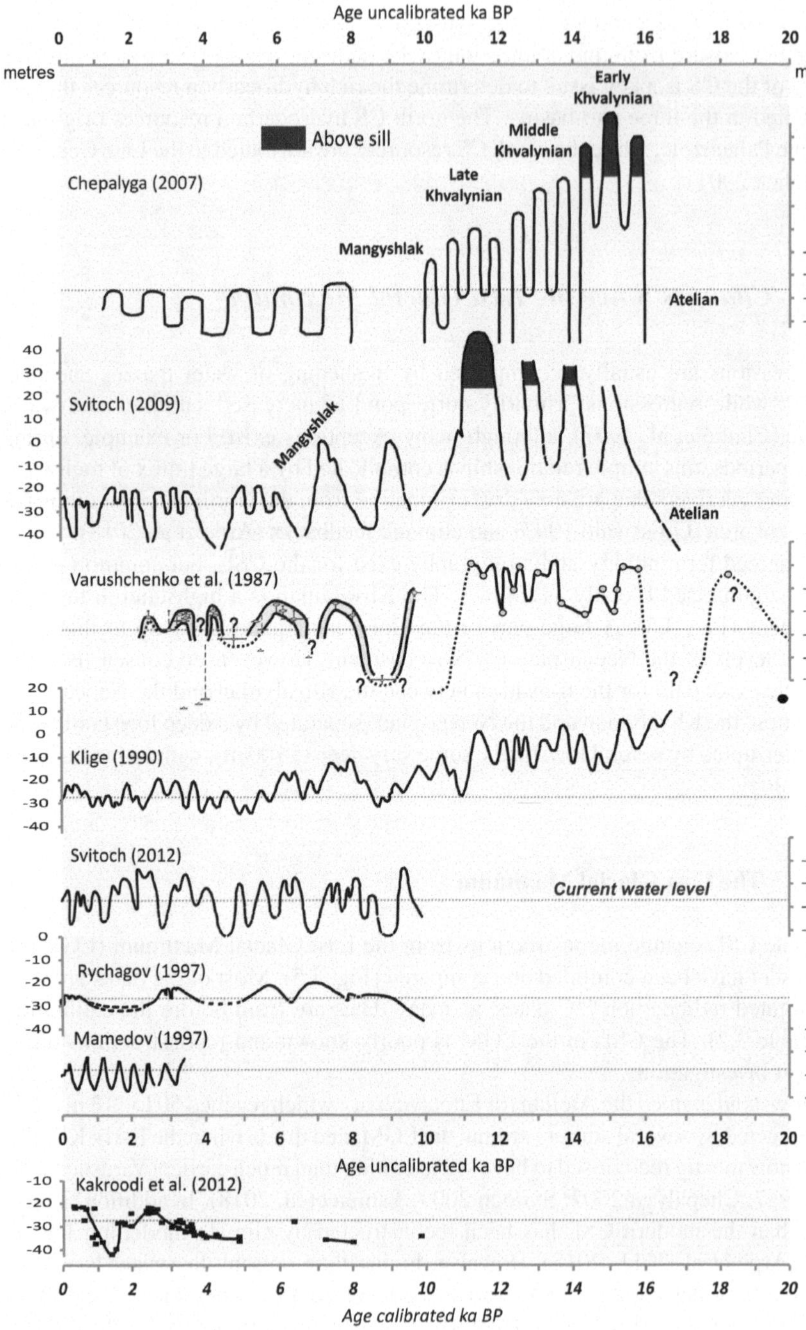

Fig. 3.5 Different Caspian Sea level reconstructions since the Last Glacial Maximum. In black, water levels above the sill causing flow to the Black Sea (upper two panels)

Table 3.2 Published Caspian Sea level curves (shown in Fig. 3.5), the reconstruction methods and the availability of metadata

Authors	Proxy for water level
Chepalyga (2007)	Geomorphology and dating of mollusc shells, scheme largely theoretical for the pre-Holocene part
Svitoch (2009, 2012)	Geomorphology and dating of mollusc shells, mostly collected in the 1960s, but some recently. Onshore and offshore areas
Varushchenko et al. (1987)	Compilation and integration of geology, geomorphology and archaeology. >35 points dated using shells with metadata (in their Tables 6 and 7)
Klige (1990)	Not documented
Rychagov (1997)	Original data. Levelling of terraces for highstands and base of alluvium in river mouths for lowstands. Dating of mollusc shells
Mamedov (1997)	Critical review of dating methods. Archaeological evidence and radiometric dating for the last four millennia. Theoretical scheme based on 450–500 year long cycles
Kroonenberg et al. (2005), Hoogendoorn et al. (2005, 2010), Kakroodi et al. (2012), Richards et al. (2014)	Geomorphology of coastal outcrops, seismic profiles and offshore cores. With metadata

flooding in the north of the north sub-basin with an horizontal advance of as much as 5–10 km per year (Chepalyga 2007).

3.3.2.2 The Deglaciation

After the LGM, the deglaciation brought a large amount of Eurasian ice-sheet meltwater to the CS through the Volga drainage basin (Tudryn et al. 2016), although part of the meltwater may also have flown through the Aral Sea and then to the CS (Chepalyga 2007). Large ice-dammed freshwater lakes formed along the southern edge of the Eurasian ice sheet (Mangerud et al. 2004). The timing of this palaeolake development is poorly known. Their drainage into the CS caused the Late Pleistocene Khvalynian highstand, especially high in the Early Khvalynian (Chepalyga 2007; Kroonenberg et al. 1997; Rychagov 1997) when the CS filled up to c. 50 m asl and the Precaspian plain was flooded (Fig. 3.5).

This means that the CSL was above the Manych sill (between the CS and the Black Sea; Fig. 3.1), which is at c. 26 m asl nowadays, but this sill is considered to have been higher in the Late Pleistocene (Mamedov 1997; Svitoch 2008, 2009). Water that spilled over the sill flowed into the Black Sea, where the inflow from the

CS is marked by red clay layers (16.3–15 calibrated kiloyears before present (cal. ka BP)), although a debate still exists regarding their origin (Bahr et al. 2008; Tudryn et al. 2016). By 13.8 cal. ka BP the Eurasian ice-sheet had become so reduced that its meltwater failed to reach the Volga drainage basin (Tudryn et al. 2016). This is when the meltwater channels were finally abandoned and the meltwater moved north-west, i.e. through the English Channel as a river (Tudryn et al. 2016).

3.3.2.3 The Mangyshlak Lowstand

After the Younger Dryas climatic deterioration, an important regression, usually called the Mangyshlak lowstand, has been reconstructed in many records. This lowstand was perhaps as far down as 113 m bsl, although its water level reconstruction varies significantly among authors. It caused the Volga River to carve deeply incised valleys into the emerged shelf, and formed a lowstand delta in the middle CS. The Mangyshlak lowstand may have lasted between c. 3000 years (on the shelf of the shallow north sub-basin at c. 12.5–9.5 cal. ka BP; Bezrodnykh and Sorokhin 2016) and c. 1000 years (recorded in deep-water cores at 11.5-10.5 cal. ka BP; Leroy et al. 2013c, 2014).

The causes of the rapid regression and following rapid transgression are unknown, but factors other than climate alone must have been involved, such as changes in hydrography (Leroy et al. 2013c). Mayev (2010) suggests a rise of 20 cm/year at the end of this period, i.e. twice the rate of the end of the twentieth century. It is during this lowstand that the Amu-Darya would have switched its flow from the south CS sub-basin to the Aral Sea (Boomer et al. 2000).

3.3.2.4 The Neocaspian Phase

Around 10.5 cal. ka BP, the CS filled up again to the present intermediate level or even higher (Leroy et al. 2013c, 2014). Some authors consider that the Neocaspian period started at this time (Varushchenko et al. 1987); others consider that the high levels can only be part of a later final phase of the Khvalynian highstand lasting until c. 7 ^{14}C ka BP and the subsequent lowstand is a "late" Mangyshlak dating of c. 6.5 ^{14}C ka BP (Svitoch 2009). Svitoch (2009, 2012), after reviewing radiocarbon dates of coastal deposits, suggests that the Neocaspian phase (characterized by the mollusc *Cerastoderma glaucum*) starts after 3.9 ^{14}C ka BP, with an earlier high level, the Gousan at 6.4–5.4 ^{14}C ka BP. Thus, the preceding high phase (8.6–7.3 ^{14}C ka BP) is still attributed to the Late Khvalynian on the basis of its mollusc assemblages. Dinoflagellate-cyst analyses tend to confirm this. The latter have highlighted a major assemblage change at 4 ka (Leroy et al. 2007, 2013a, c, 2014). Most authors agree that the Neocaspian period began no later than 6.5 ^{14}C ka BP. During this period, the water levels are generally moderately high with some fluctuations, although their number and their dates are equivocal (Varushchenko et al. 1987; Mamedov 1997; Rychagov 1997).

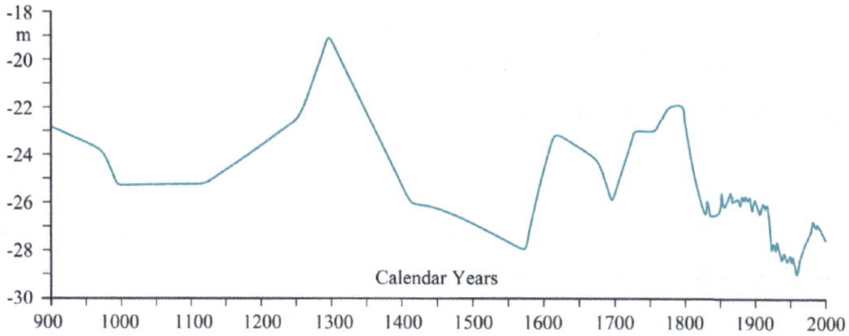

Fig. 3.6 Caspian Sea level curve based on historical data and instrumental measurements modified from Naderi Beni et al. (2013)

A more consensual reconstruction exists around the highstand of 2.6–2.3 cal. ka BP (Kroonenberg et al. 2007; Kakroodi et al. 2012). The last important regression is the Derbent regression that was relatively pronounced, down to 34 m bsl at least which corresponds roughly to the Late Antiquity-early Middle Ages (Sauer et al. 2013; Fig. 3.6). During the Little Ice Age, the levels were generally, but not continuously, high (Naderi Beni et al. 2013; Fig. 3.6). Since a maximum in the early nineteenth century at −22 m, the CSL decreased with a minimum at −29 m in 1977.

The last major connexion with the Aral Sea, via the Uzboy (now dry) and the Amu-Darya, took place c. 5 ka ago when *C. glaucum* invaded the Aral Sea (Boomer et al. 2000). Much more recently, the flow of the Amu-Darya has been briefly diverted to the CS due to human activities: in the thirteenth century by Gengiskan troups destroying the dam built in the tenth century, again in the fourteenth and finally in the fifteenth century by others (Herzfeld 1947; Naderi Beni et al. 2013; Krivonogov et al. 2014).

3.3.2.5 Limitations to Reconstruct CS Level

It is important to be aware of the following: (1) that most of the published CSL curves do not match with each other (Fig. 3.5), (2) that usually no metadata on levels and dates are available to check the validity of the points on the curves with the noticeable exceptions of Varushchenko et al. (1987) and Kakroodi et al. (2012; Table 3.2), (3) that many schemes are based on theoretical assumptions rather than based on real data (existence of a cyclicity, e.g. Mamedov 1997; Chepalyga 2007), and (4) that most of the records have been obtained from the coast, hence it is impossible to estimate the depth and duration of the lowstands precisely and from the south sub-basin with the confounding influence of basin subsidence and mountain uplift.

Only a transect of long cores from the coast to depths below the lowest stand would provide clear information, but this is far from reach at the moment. In brief, a reliable record of water-level changes in the CS since the LGM is not yet available. This situation is changing now with a modest increase in the number of offshore cores providing continuous records (Ferronsky et al. 1999; Kuprin et al. 2003; Leroy et al.

2013c, 2014; Bezrodnykh and Sorokhin 2016). Moreover, no stratotypes[2] are defined for the main highstands and lowstands. So, confusion is easily introduced with names having different meanings, as no rules govern the use of terms such as the Mangyshlak lowstand. Moreover, chronozones and biozones are often mixed up. This is partially due to the continued difficulty of using and calibrating radiocarbon dates in the CS. The correct identification of the reservoir effect with a range currently between 290 and 747 years is plagued with problems due to, for example, the changing sources of water in space and time, and the hard-to-constrain influences from methane seepage in this zone of petroleum production (Leroy et al. 2018).

3.4 The Biota of the Caspian Sea

3.4.1 Biodiversity and Endemism

The CS has a high diversity of biotopes, biotic and abiotic conditions (Zenkevich 1963); the main cause is the range of salinity that varies from fresh in deltas to hypersaline in the Kara-Bogaz-Gol. Due to this large salinity range, freshwater, brackish, euryhaline and hyperhaline hydrobionts can inhabit the water. Owing to the similarity of chemical compositions between marine and Caspian waters, many marine organisms thrive. As a result of the complexity of its geological origin and connexion with other water bodies, the modern fauna and flora of the CS consists of five main components: (1) Caspian origin (species that survived the closure of the Paratethys Sea), (2) freshwater reservoirs of the pre-glacial and the glacial periods, (3) Mediterranean origin, (4) Arctic origin (overlap between the edge of the Eurasian ice-sheet and the drainage basin of the Volga; Tudryn et al. 2016), and (5) a few recent Atlantic and freshwater invaders (Fig. 3.7a; Zenkevich 1963; Kosarev and Yablonskaya 1994). The total biodiversity of the CS is nevertheless lower than that of the Black Sea and of the Barents Sea (Zenkevich 1963).

The main reason for the relatively low biodiversity is probably the salinity change over time: for freshwater species the salinity is often too high, for marine species it is often too low. Only brackish-water species from both marine and continental water bodies are favoured (Mordukhai-Boltovskoy 1979). Most aquatic biota have a great plasticity to salinity: marine species that can tolerate low salinities (down to 13 psu) and freshwater and very low salinity species that can tolerate high salinities (up to 13 psu). In comparison to freshwater lakes, the diverse salinity conditions in the CS serve to increase biodiversity rather than reduce it (Aladin and Plotnikov 2004; Plotnikov et al. 2006). The fauna and flora belong to the Pontocaspian biota: i.e. found in parts of the current Caspian, Black and Aral seas, satellite lakes and river mouths. They are mostly remnants of the larger Paratethys Sea and are characterised by unusual and fluctuating salinities. Fish and Crustacea (especially benthos) have

[2]The type section of a layered stratigraphic unit that serves as the standard of reference for the definition and characterization of the unit (www.stratigraphy.org).

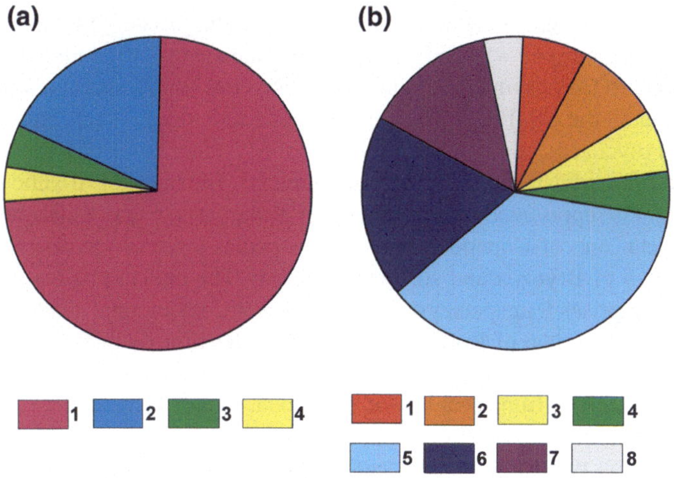

Fig. 3.7 Faunal composition of free-living Metazoa of the Caspian Sea in percentages (redrawn from Kosarev and Yablonskaya 1994). **a** by origin: 1—Autochthonous, 2—Freshwater, 3—Mediterranean, 4—Arctic. **b** by systematic groups: 1—Turbellaria, 2—Nematodes, 3—Rotatoria, 4—Annelida, 5—Crustacea, 6—Mollusca, 7—Pisces and Cyclostomata, 8—others

the largest species numbers with 63% of all modern species (Fig. 3.7b). This is because of their osmoregulation capacities that allow them to live in a wide range of salinities, from fresh to hypersaline waters (Zenkevich 1963). Owing to its isolation since the beginning of the Quaternary, the CS shows a high endemism level (Dumont 1998; Grigorovich et al. 2003; Marret et al. 2004). This is due to a complicated and long history of formation of the Caspian fauna and flora with successive episodes of isolation and connections. Nowadays, descendants of many ancient organisms, whose ancestors had penetrated into the region some millions of years ago, inhabit the CS.

The phytoplankton of the northern CS includes more than 400 species (Chrysophycea – 1, Euglenophycea – 7, Dinoflagellata – 58, Cyanobacteria – 90, Chlorophyta – 138, Bacillariophycea – 149; Aladin et al. 2001). However, only a few species predominate. Diatoms are widespread all over the CS. They take a leading place by number of species. Marine diatom *Pseudosolenia calcaravis* represents the basic part of the phytoplankton. In the CS, this alga is an invader that appeared before 1935 (Karpevich 1975; Karpinsky 2010). Thirteen Caspian endemics were previously recognised in the phytoplankton (Proshkina-Lavrenko and Makarova 1968), but in recent years only *Thalassiosira inserta*, *Th. caspica* and *Chaetoceros subtilis* were found. More recent investigations on dinoflagellates have highlighted a range of forms, species and even genera, typical of the CS (Marret et al. 2004; Leroy et al. 2018; Mudie et al. 2017). The species composition of periphyton amounts to some 200 species of algae (Phaeophycea – 8, Rhodophyta – 11, Chlorophyta – 12, Cyanobacteria – 33, Bacillariophycea – 126). Of the 87 species of algal macrophytes recorded (Zaberzhinskaya 1968), green algae (Cladophoraceae – 11, Ulvaceae – 10, including *Enteromorpha* – 9) and charophytes (Characeae – 10) dominate.

According to Derzhavin (1951) and Zenkevich (1963), 476 species of free-living Metazoa inhabit the CS (Fig. 3.7b). The fauna of Arctic origin is represented by one species of Polychaeta, one of Copepoda, four of Mysidacea, one of Isopoda, four of Amphipoda, two of fish and one species of mammal. The fauna of Atlantic–Mediterranean origin is represented by one species of Turbellaria, one of Coelenterata, two of Polychaeta, one of Copepoda, two of Cirripedia, three of Decapoda, three of Mollusca, two of Bryozoa and six species of fish. The endemic fauna is the most diverse. It comprises four species of Porifera, two of Coelenterata, 29 of Turbellaria, three of Nematoda, two of Rotifera, two of Oligochaeta, four of Polychaeta, 19 of Cladocera, three of Ostracoda, 23 of Copepoda, 20 of Mysidacea, one of Isopoda, 68 of Amphipoda, 19 of Cumacea, one of Decapoda, two of Hydracarina, 53 of Mollusca, 54 of fish and one mammal species. Many species are of freshwater origin, particularly from the Rotifera, Cladocera, Copepoda and Insecta (Derzhavin 1951; Mordukhai-Boltovskoy 1960; Zenkevich 1963). According to the more recent data of Chesunov (1978), the number of species of free-living Metazoa in the CS is larger, up to 542.

In the zooplankton, 315 species and subspecies are registered, 135 species of them are ciliates (Agamaliev 1983; Kasimov 1987, 1994). The main part of the zooplankton population comprises species of Caspian origin. Crustaceans *Eurytemora grimmi, E. minor, Limnocalanus grimaldii, Acartia clausi, Heterocope caspia, Calanipeda aquaedulcis, Evadne anonyx, Podonevadne camptonyx, P. angusta, P. trigona, Polyphemus exiguus, Apagis* spp., *Cercopagis* spp., and *Pleopis (Podon) polyphemoides* are common in the zooplankton. Freshwater complex – rotifers *Brachionus* and cladocerans *Moina*, *Diaphanosoma*, and *Bosmina* occupy low salinity areas. Larvae of benthic organisms – of barnacles *Balanus* spp. and molluscs – are abundant especially in spring and summer plankton of coastal zone (Bagirov 1989). According to Derzhavin (1951) and Zenkevich (1963), the Crustacea include 114 autochthonous species. The Mysidae include one species of *Diamysis*, one of *Limnomysis*, four of *Mysis*, one of *Hemimysis*, one of *Katamysis*, one of *Caspiomysis*, one of *Schistomysis*, and 12 of *Paramysis* (Stepanjants et al. 2015). Of them, 13 species are endemics of the CS. In the Pseudocumatidae are 19 species. Isopods have three species, including *Mesidothea entomon* of Arctic origin. In the amphipods, the Gammaridae have 60 species and the Corophiidae have eight species. The Decapoda have five species (Vinogradov 1968; Karpevich 1975; Aladin et al. 2002), including ancient native crayfishes *Astacus leptodactylus* and *A. pachypus*, prawns *Palaemon elegans* and *P. adspersus* introduced incidentally by humans in the 1930s during acclimatization of mullets, and one species of small crab *Rhithropanopeus harrisii* that was inadvertantly introduced by vessels in the 1950s.

The diversity of Mollusca species is also large: according to Bogutskaya et al. (2013), the unusual number of 124 species and subspecies, including 101 endemics, are recorded from the CS. Four families of gastropods are present: Neritidae with two species of *Theodoxus*, Hydrobiidae with 85 species, including species of genera *Andrusovia, Pseudamnicola, Caspia, Caspiohydrobia*, *Turricaspia* with 17 endemics, and 38 endemic species of *Pyrgula,* Bythiniidae with one endemic species and Planorbidae with three endemics from genus *Anisus*. The bivalves include several species and subspecies from two families. The Dreissenidae have five species: *Dreissena polymorpha* (subspecies *D. p. polymorpha* and endemic *D. p. andrusovi*), endemics *D. rostriformis* (subspecies *D. r. compressa, D. r. distincta, D. r. grimmi, D. r. pontocaspica*), *D. caspia* and *D. elata* (the last two have become extinct in the last century) and the invader *D. bugensis* that was introduced relatively recently with ballast waters. The Cardiidae have up to 21 species: *Adacna* with nine species (invader *A. colorata*, endemics *A. acuticosta, A. albida, A. caspia, A. laeviuscula, A. minima, A. polymorpha, A. semipellucida, A. vitrea), Monodacna* with one species (endemic *M. caspia*), *Didacna* with eight species (endemics *D. baeri, D. barbot-demarnii, D. longipes, D. parallela, D. profundicola, D. protracta, D. pyramidata, D. trigonoides*), *Hypanis* with one species (*H. plicata*) and *Cerastoderma* with two species (*C. glaucum, C. rhomboides*; Bogutskaya et al. 2013; Leroy et al. 2018).

Four species of Porifera are recorded: two species of *Metschnikovia*, *Protoschmidtia flava* and *Amorphina caspia.* The coelenterate *Moerisia pallasi* is an endemic species. Three species of Polychaeta belonging to the Ampharetidae are recorded.

Derzhavin (1951) and Zenkevich (1963) stated that the ichthyofauna consisted of 78 species in the middle of the twentieth century. The family Petromyzonidae had one species, Acipenseridae five, Clupeidae nine, Salmonidae two, Esocidae one, Cyprinidae 15, Cobitidae two, Siluridae one, Gadida one, Gasterosteidae one, Syngnathidae one, Atherinidae one, Percidae four, Gobiidae 30, Mugilidae two, Pleuronectidae one, and Poeciliidae one species. According to more recent data of Kazancheev (1981), the total ichthyofauna amounts to 76 species and 47 subspecies, referring to 17 families and 53 genera. In comparison with other southern seas (Azov, Black, Mediterranean), the ichthyofauna of the CS is extremely poor and consists of representatives of autochthonous (63 species and subspecies), Mediterranean (5), Arctic (2) and freshwater (56) faunistic complexes. Some species have populated the Caspian because of human activities. The distinctive feature of the Caspian ichthyofauna is its high endemism, observed from the category of genus down to the subspecies level. Early separation of the CS from the world ocean has ensured a high level of endemism of its ichthyofauna. According to Kazancheev (1981), the number of endemics at the genus level account for 8.2%, at species level for 43.6%, and at subspecies level for 100%. The greatest number of endemic forms belongs to Clupeidae and Gobiidae. In general, the CS is inhabited by four endemic genera, 31 endemic species and 45 endemic subspecies.

Sturgeons – *Acipenser gueldenstaedtii, A. nudiventris, A. persicus, A. stellatus* and *Huso huso* – are the most valuable commercial fish species in the CS. In the 1950s, the main dams blocking migratory routes of anadromous fish species were built on many rivers of the Caspian basin. Main estuaries and deltas are also now heavily polluted. The total area of the spawning grounds of all sturgeons shrank by around 90%. In order to compensate for losses of natural spawning and to protect sturgeon stocks, special sturgeon-rearing stations were built. From the early 1960s, sturgeon catches steadily increased for almost 20 years. After 1980, a sharp decline in catches took place. This was caused both by the increased catch of previous years and by a decrease in the number of mature fish as a result of a decrease in natural reproduction due to loss of spawning grounds. By the early 1990s, the catches were almost halved and they continue to decline. Poaching significantly increased (Bogutskaya et al. 2013).

The Caspian seal, *Pusa caspica,* is the only aquatic mammal. It is a small species (Goodman and Dmitrieva 2016) that is almost exclusively fish-eating. In winter, seals concentrate in the northern CS along the margins of the ice cover. Almost all of whelping, mating and molting take place on ice. In summer, seals migrate to feed in the middle and southern CS, with some part of the population remaining in the northern CS (Aladin et al. 2001). Phylogenetic studies indicate that the seals originated from the Arctic during past connections 3–2 million years ago (Palo and Väïnolä 2006). At the end of the nineteenth century, the seal population exceeded one million. During the twentieth century, hunting pressure increased and the population declined, being halved by the 1950s. Since that time, sealing quotas have been reduced, but population decline has continued due to many factors. At the end of the twentieth century, about 25% of Caspian seal population had died out due to various diseases. By 2005, the population was estimated at 34,000 (Goodman and Dmitrieva 2016).

Thus the CS has a high level of endemism, but is not especially biodiverse. The main characteristic of many species is their great plasticity to salinity.

3.4.2 Recent Changes in the Caspian Sea Biota

Although there is evidence for the anthropogenic introduction of species during the Middle Holocene, greater change took place in the twentieth century. For example, widespread invasion from the Black Sea to the CS by the mollusc *Cerastoderma*, occurred in the Middle Holocene (Grigorovich et al. 2003), although earlier dates were suggested (8–6 ^{14}C ka BP; Tarasov and Kazantseva 1994). Some researchers (Starobogatov 1994) deny the natural colonization of the CS by this species, since a strong current in the Manych Gulf was only directed away from the CS. Thus, they assume that the penetration was associated with the activities of ancient humans.

In the twentieth century, many invertebrate and fish species were introduced to the CS due to anthropogenic activity (Aladin et al. 2002). In the 1920s, algae *Pseudosolenia calcar-avis* and bivalve *Mytilaster lineatus* were accidentally introduced. This mollusc was introduced (Bogutskaya et al. 2013) by unusual manner – with biofouling of military vessels transported to the CS by railway. Bivalve *Mytilaster lineatus* is unwanted in the CS. It forces out native bivalves and is not much consumed by fish. Later, people deliberately introduced several species into the CS including mullets *Mugil auratus, M. saliens* and polychaete worm *Hediste diversicolor* (in the 1930s), flounder *Pleuronectes flesus luscus* and bivalve *Abra segmentum* (in the 1940s). *Hediste diversicolor* and *Abra segmentum* were introduced deliberately for consumption by fish because for them these invertebrates are valuable food. Together with mullets, two species of shrimps – *Palaemon elegans* and *P. adspersus* were introduced accidentally. All these species have adapted to the conditions of the CS.

In the middle of the twentieth century, after the Volga-Don channel had been built, a new group of species was introduced into the CS. Some of them were transported in ballast water of vessels; others were attached to their bottom. Barnacles *Balanus impovisus, B. eburneus* and bryozoan *Membranipora crustulenta*, were introduced with biofouling. Planktonic crustacean *Pleopis polyphemoides*, jelly fish *Blackfordia virginica*, algae – *Ceramium diaphanum, C. tenuissimum, Ectocarpus confervoides* f. *fluviatilis*, *Polysiphonia variegata* and crab *Rhithropanopeus harrisi* were introduced with ballast waters. The introduction of invasive species into the CS through the Volga-Don channel continues. At the end of the twentieth century planktonic copepods *Calanipeda aquaedulcis, Acartia clausi, A. tonsa* and ctenophore *Mnemiopsis leidyi* were introduced with ballast waters. As for the ctenophore, this species is a clear example of a negative impact on the biodiversity of the CS (Ivanov et al. 2000; Nasrollahzadeh et al. 2014). It consumes zooplankton and hence outcompetes zooplankton-feeding fish. The decrease of zooplankton thus encourages the bloom of phytoplankton. This is probably the most dangerous alien species for the CS ecosystem.

An increase in the dinocyst *Lingulodinium machaerophorum* has been observed since the late 1960s and it has been attributed to the increase in the temperature of surface waters (Leroy et al. 2013b). Since 2005, harmful algal blooms, including potentially toxic dinoflagellates, have been observed owing to a combination of higher temperatures and increased nutrient availability.

3.5 The Changing Physical World

The Caspian deep-water environment has benefited from the water-level fall until 2016 and the consequent accelerated circulation that causes deep-water ventilation and nutrient exchange. However, most changes are felt at the surface.

3.5.1 Recent Changes to the Lake Surface

3.5.1.1 CSL Changes from in-situ to Satellite-Altimetry Observations

The first measurement of water level was made in Baku in 1837, followed by many other sites around the margins of the CS during the nineteenth–twentieth century. Annual water-level variations of four base stations, Baku, Makhashkala, Port Shevchenko and Krasnovodsk, were then averaged to provide the "official" CSL (Figs. 3.1 and 3.2a, b; Terziev et al. 1992).

Figure 3.2a shows the CSL from 1900 to 2016. It was obtained using level gauges installed around the CS. Initially the measurements were made in different datum systems (Black Sea, Baltic Sea, local relative system). However, due to improved accuracy in levelling, particularly in the 1970s, allowing vertical crustal movements (~2.5 mm/year) to be determined, new zero marks have been defined, with different height systems for the east and west coasts (Pobedonostsev et al. 2005). This led to some differences in recorded water-level changes between gauges that caused ambiguities in the water-level-time series between one gauge and another. Even the recent data over the four historical gauges are not coherent all the time (Fig. 3.2b). For this reason, satellite altimetry offers an interesting complementary source of information.[3]

Over the first period between 1900 and 1929, the CSL slightly decreased with an average level around −26 m (Fig. 3.2a). In the 1930s, several large reservoirs along the Volga were filled. It led to a sharp decrease of 1.8 m by 1941, then more slowly until 1977 when the CS reached its lowest level over the last hundred years at −29 m. However, not all of this decline can be assigned to the building of dams. A good deal of the CSL variation can be simulated from the mean precipitation over the CS catchment area (Arpe and Leroy 2007a). Between 1977 and 1995 the CSL rose rapidly by 2.5 m, after which it again fell; so that by 2016 it had stabilized at a level of approximately −28 m. In such a large inland sea, the water-level changes observed over the last decades represent a very large amount of water-storage change. The two and half meter of increase between 1977 and 1995 is equivalent to additional water storage of ~900 to 1000 km^3. Such variability needs to be analyzed and understood since it has serious consequences. Understanding the mechanism behind these changes would allow forecasting the rise or fall of the level in the future, which is a topic of serious public concern for riparian countries (Arpe et al. 2014).

3.5.1.2 Water Balance of CS

[3]Since the middle of the 1990s, a noticeable progress occurred in the use of radar and laser altimetry for continental hydrology. It is important to note that this technique was initially designed for oceanography; but it very quickly became clear that satellite altimetry is an attractive technique for monitoring the water levels of lakes (Crétaux et al. 2016). Essentially, this technique has benefited from a continuous service since the launch of the satellite Topex/Poseidon in 1992, and this will continue in the years to come with various new missions.

The combined global altimetry data set has more than 2-year-long history and is intended to be continuously updated in the coming decade. A given lake can be flown over by several satellites, with potentially several passes, depending on its surface area. Thus, combining altimetry data from several in-orbit altimetry missions increases the temporal resolution and the accuracy of the water-level estimation which depends on several factors: range, orbit and correction errors (Crétaux et al. 2009). Comparisons of average water level from satellite altimetry and in-situ data for a set of 24 lakes of various locations and sizes have been established by Crétaux et al. (2016). Accuracy has been estimated by the calculation of the Root Mean Square (RMS) of the differences between both types of data: it ranges from 3 cm for large lakes to few tens of cm for small lakes, currently achievable using nadir altimeters.

Analysing the CSL changes is generally interpreted as balance between water budget components of the CS and the Kara-Bogaz-Gol resulting from dynamical response to the variations in inflow (precipitation P, river and underground discharge R and G) and output (evaporation E and discharge to the Kara-Bogaz-Gol, K). It can be expressed by the following equation (Eq. 1):

$$\frac{dV}{dt} = R + (P - E) + (G - K) \tag{1}$$

Many past studies were performed to analyse the water balance of the CS. It is generally based on in-situ data for some components (P, R, K) and/or model based on global climatological data (P, E, R, G) and/or remote sensing data (P, dV/dt). G is not measured or modelled in the most recent studies and is assumed to be a constant of 4 km^3/year (Shiklomanov et al. 1995). Ozyavas et al. (2010) however demonstrated that groundwater inflow could cause significant errors in the water balance and is presumably not constant year over year.

Discharge of the rivers is principally driven by the P-E ratio over the CS watershed. It can also be obtained from direct measurement. The calculation of evaporation over the CS can be inferred from a model based on the generalized Penman equation (Van Bavel 1966); but needs a large number of additional variables not always easy to measure (air and water temperature, air humidity, wind speed, wave height, net radiation, latent heat of vaporization). Ozyavas et al. (2010) used this model and found an average value of approximately 1100 mm/year of direct evaporation, while in Kouraev et al. (2011), average evaporation over the twentieth century is estimated to 970 mm/year. For many studies, the size of the CS, used to calculate the evaporation, is kept constant. However, as shown by Arpe et al. (2018) due to the high level variation of the CS and the high variability of its extent, in particular in the shallow northern part, the total evaporation over time can also significantly vary. Moreover, feedback of the change of CS size in the local precipitation should also be taken into account as shown by Arpe et al. (2018). Indeed, E-P is controlled by the presence of the CS itself since precipitation over the CS region is partially generated by water evaporated from the CS. In the context of CSL drop (as observed since 1995), a positive feedback (reduction of humid air in the CS watershed due to lower total evaporation) is combining with a negative feedback (lower size implies lower direct evaporation) and could reach an equilibrium state where gains are equal to losses (Mason et al. 1994). The volume change of the CS can be measured using water level changes from satellite altimetry and average lake extent.

In the paper of Chen et al. (2017), similar to Arpe et al. (2014), a model of water balance was setup using the climate model Climate Forecast System based on the National Centers for Environmental Prediction datasets. Considering only $R + (E - P)$ components of the right member of Eq. 1, using the Climate Forecast System model and in-situ discharge data of the Volga River, Chen et al. (2017) reconstructed the volume change of the CS. Then they compared it to the volume variations inferred from in-situ data and satellite altimetry. Their conclusion was that over the 37 years of analysis, the increased evaporation rate is the main cause of

CSL changes; while precipitation changes and river discharge play a secondary role, although not negligible. In that study however, the inter-annual variability of small river discharge, of Kara-Bogaz-Gol outflow and of underground water inflow have not been considered. This conclusion in Chen et al. (2017) seems to contradict the study made by Arpe and Leroy (2007a), which showed that CSL changes are at first order driven by the discharge over the Volga River basin, which is directly linked and correlated to total rainfall over the watershed of this river. However, it is also shown by Arpe and Leroy (2007a) that the post-1995 drop in CSL cannot be attributed to a drop in the Volga River discharge. Arpe and Leroy (2007a), Chen et al. (2017) and Arpe et al. (2018) consider the relationship between evaporation and CSL to be significant.

The link between CSL change and Kara-Bogaz-Gol is an important contribution to the water balance and is not always considered in literature as a specific water body with its own variability. For example, Chen et al. (2017) did not take the *K* term into account. Ozyavas et al. (2010) calculated the water balance of the CS from 1998 to 2005. They considered that the quasi-constant value of *K*, corresponding to a water balance of ~5 cm/year, is insignificant. This was partly due to the resolution of the model-generated data, being too low to resolve the Kara-Bogaz-Gol. In Arpe and Leroy (2007a), discharge to the Kara-Bogaz-Gol is considered as a constant outflow ~5% of the total surface runoff to the CS.

In most studies, the Volga River discharge to the CS is considered to be 80% of the total from all rivers. A recent dataset (www.caspcom.com) gives a discharge-time series of the eight main rivers inflowing in the CS (Volga, Ural, Terek, Kura, Sulak, Sefidrud, Polrud and Chalous; Fig. 3.1). It is obvious that discharge is not stable (Fig. 3.2d). Indeed, surface inflow from the Volga over the last 40 years has varied between 80 and 90%, and the total inflow from other rivers is very irregular from year to year (Fig. 3.2c, d). From wet to dry years, inter-annual variability of rivers excluding the Volga ranges from ~10 to ~20 cm of equivalent CSL. So, it is clear that the accurate calculation and understanding of the water balance of the CS can only be done with exhaustive data on small rivers (Arpe et al. 2014).

The Kara-Bogaz-Gol depression is shallow and at the beginning of the twentieth century when the CSL was high, the difference in water level between the CS and the Kara-Bogaz-Gol was low (c. <1 m). The long-term fall of the CS between 1930 and 1977 resulted in a decline of the discharge to the Kara-Bogaz-Gol. From 1930 to 1980, it decreased from ~30 km^3/year to 0. It therefore accounted for ~7 cm of decrease of the CS in 1930 to zero in 1980. In March 1980, in order to stop the filling of the Kara-Bogaz-Gol from the CS, the narrow strait between them was blocked by a solid and sandy dam. It immediately resulted in a full evaporation of the bay in only four years. At the end of 1984, the dam was broken and the water came back very rapidly to its full extent. From 1993 to 1995, the annual water flow reached 37–52 km^3/year (Kosarev et al. 2009) and the level rose by 5 m in three years to reach its maximum value around −27 m. Then the Kara-Bogaz-Gol level variations slightly decreased down to −28.5 m in 2017 with small annual oscillations linked to seasonal climate change.

By some assumption on the E-P term over the Kara-Bogaz-Gol (~800 mm/year, Kosarev et al. 2009), and using size and volume change of Kara-Bogaz-Gol extracted from satellite altimetry and imagery available in Hydroweb (n. d.), the inflow from the CS to the Kara-Bogaz-Gol can be recalculated. Therefore, one may calculate the corresponding water-equivalent changes of the CS, which is the opposite of the term K in Eq. 1. From 1993 to 1996, the total discharge K is 9 cm/year, and after 1996 it is only 4 cm/year. If we calculate the total loss of water from the CS to the Kara-Bogaz-Gol, it represents the significant number of >1 m from 1993 to 2017.

3.5.1.3 Caspian Sea Ice and Surface-Water Temperature

Ice formation[4] over large areas on the northern CS occurs every winter and ice stays for several months. Ice over the middle CS can also form to a smaller extent. The presence of ice has several impacts on navigation, fisheries and oil industries especially in the Russian and Kazakh coastal areas. Using active and passive microwave remote sensing data has allowed measuring time series of ice formation, break-up and duration of ice period, in the eastern and western parts of the northern CS, as well as estimating the mean and maximum ice extent over this area. Kouraev et al. (2011) reported results with remote sensing data, which exhibited a warming trend since the mid-1980s followed by a short cooling period in 1993–1994, and then again a warming trend characterised by diminution of ice duration and of maximum ice coverage for the western and eastern parts of the northern CS.

The same trend was observed in the CS SST (Ginzburg et al. 2005; Leroy et al. 2013b). Seasonal and inter-annual variability of SST were measured over four distinct regions (North, Middle, South and Kara-Bogaz-Gol) in that study. Trends for the period 1982–2000 were calculated for each hydrological season (February, April, August, and October) and compared to historical period from 1940 to 1980. Their results exhibited a significant global-warming signal with much higher rate of change in the recent period of time. In the middle and southern CS, warming trends of 0.05

[4]The study of ice cover in this region started in the nineteenth century using observations from coastal stations. It then became a monitoring system of the Soviet Union and measurements were collected on a regular basis using aerial surveys. After 1970s, ice-cover observations from airplane have drastically decreased for financial reason. It has then been compensated by using satellite imageries in the visible and infrared parts of the spectrum (Buharizin et al. 1992). However, these surveys from satellite imagery in the visible spectrum are strongly affected by cloud formation and cover, particularly in winter, and these types of information were consequently not dense enough to perform a full survey of ice presence over the CS. Since the mid-1980s, microwave-satellite observations, providing reliable, regular data on ice without masking from cloud, have been used. Since 1992, another source of data was from a synergy between active (radar altimetry) and passive (radiometer used to correct altimetry measurements) instruments. A discrimination method, developed and tested over the CS and the Aral Sea in Kouraev et al. (2003, 2004), was applied on the data of Topex/Poseidon, Jason-2, Jason-3 and Envisat missions. Combining results from this technique after 1992 (when the Topex/Poseidon satellite was launched) with microwave observation on SMMR and SSM/I missions allowed better spatial and temporal resolution than using satellite altimetry alone. This method has been successfully validated using independent in-situ measurements (Kouraev et al. 2003).

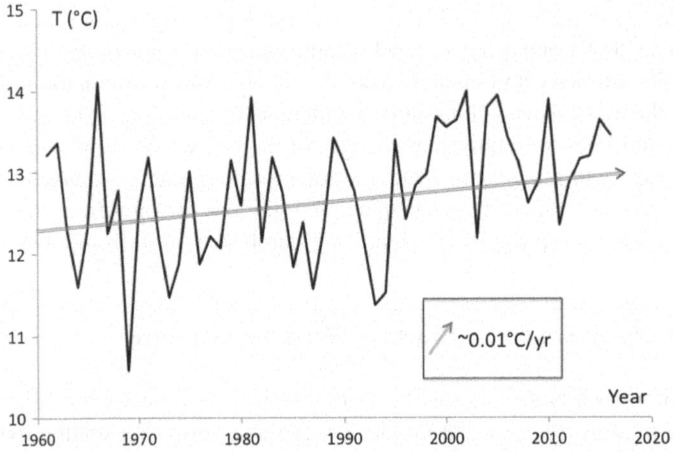

Fig. 3.8 Sea surface temperature at the station of Makhashkala in Russia on the western coast of the CS. Data from in-situ gauge over the period from 1960 to 2016. The arrow shows the 0.01 °C increase per year. *Source* http://www.caspcom.com

and 0.1 °C/year were observed respectively. SST was also compared with the ENSO and NAO phases. A strong correlation between SST in the central Pacific and the CSL has been observed, which is the sign of a linkage between ENSO and the CSL. No influence from NAO has been found (Arpe et al. 2000).

In a more recent paper (Khoshakhlagh et al. 2016), using 30 years of Advanced Very High Resolution Radiometre measurements, the SST over the four regions of the CS was analysed and trends investigated using a Mann-Kendall test. Based on this study, it can be unambiguously stated that increasing SST trend has been seen in all the four regions, mostly during March to November months. In the middle and the southern CS however, it has also been shown that no trends at all are observed in wintertime. Nazemosadat and Ghasemi (2005) have moreover shown that SST fluctuations have direct impact on winter and spring rainfalls, and that warm SST causes increasing rainfall in spring and, in contrast, cold SST causes low rainfall in spring, especially in the western areas of the CS. As an example, increasing rate of temperature in the middle part of the CS from data at the station of Makhashkala from 1961 to 2018 exhibits a trend of ~+0.012° C/year (Fig. 3.8).

3.5.2 *Pollution of the Caspian Sea*

The CS and its bottom sediments are the final destination for a wide variety of contaminants originating from the coast and the water itself through natural processes and anthropogenic activities. Sediments from the catchment basin mainly by riverine influx (particularly the Volga), water-bottom-mud volcanoes, submarine groundwater discharge, wind-blown sediments, and biological and biochemical activities are the main natural sources. However, offshore-and onshore-oil industries, agricultural activities and coastal urban areas are the main anthropogenic sources of the contaminants. Many contaminants could have both a natural and an anthropogenic source, while others, i.e. persistent organic compounds, only originate from human activities. Their concentration and distribution in the water column and bottom sediment are related to the proximity to source area and hydroclimatic conditions of the CS.

3.5.2.1 Heavy Metals

Heavy metal investigations demonstrate that fine-grained sediments are enriched compared to coarse sediments (de Mora et al. 2004a; Kholodov and Lisitsina 1989; Lahijani et al. 2018b). Their relationship to the sediment grain-size could be grouped as follows: the first group (Cr, Zr) displays association with silty sediments. They are mainly accompanied with Piconite and Zircon. The second group (Cu, Co, Ni, Ga, Mo, Mn, Fe) and, in some regions, Pb demonstrate a relationship with clay sediments. The third group (Ti, V, Ge) have mutual dependence, either with fine-grained or coarser-grained sediments. They are present in a wide variety of minerals that originate from the catchment basin (Klenova et al. 1962; Lahijani et al. 2018b).

Elevated amounts of some heavy metals are attributed to the natural source of mineral-rich bottom sediments (Kholodov and Lisitsina 1989). Higher concentration of Ba in the middle Caspian sub-basin is likely due to oil industry. Higher amount of Ag, Cd and Pb in west, south and north Caspian coastal sediment might be the signature of mining in the catchment basin. Baku Bay has the highest pollution level of Hg in bottom sediments (de Mora et al. 2004a).

3.5.2.2 Oil Pollution

Oil pollution is a major concern in the modern CS. Although exploitation dates back to Antiquity, natural leakage of oil and gas into the CS were higher than those due to exploitation until mid-nineteenth century. Deep wells started to be dug only in the nineteenth century, with a boom around AD1872 following a change in the law and improved technology. Since that time, both extensive oil exploitation and transportation through Volga River caused drastic change in the CS pollution (Tarasov 1996). During the Soviet period, the oil and gas industries were mainly limited in the west and north Caspian coasts, while after the dissolution of the Soviet Union, all riparian countries engaged in this industry.

Contaminants from petroleum industry are preserved in the bottom sediments as polycyclic aromatic hydrocarbons that have similar characters as those from biological activities. They can be classified in three groups based on their origins: pyrolytic (from combustion), fossil (as raw petroleum) and natural (diagenesis from biological activities). However, the limits between the three groups are not sharp. Pyrolytic compounds have elevated amounts in the Baku Bay and the north-west CS, whereas fossil compounds are very high in Baku Bay and diagenetic compounds in the southern Caspian lake bed (Tolosa et al. 2004).

3.5.2.3 Nuclear Pollution

Some pollution from nuclear experiments on the Mangyshlak Plateau (Kazakhstan) in 1969 and 1970 is felt in the CS (Nordyke 2000). For example, radiocarbon dating of surface sediment shows a distinct modification in ^{14}C activity (Leroy et al. 2018).

3.5.2.4 Organic Compounds

The production of many organic compounds including plastics, surfactants, drugs, fertilizers and organochlorinated compounds such as PCB, HCB, HCH and DDT, started at an industrial scale during the twentieth century. The PCB and DDT are among the persistent organic compounds that have severe and adverse impacts on the environment. Some components of DDT and HCH have elevated amounts in the north, west and south Caspian coastal sediments; the total amounts of PCB and HCB, however, are much lower than the world standards (de Mora et al. 2004b). The CS is the final receptacle for the agricultural, industrial, rural and urban wastewater that causes elevated nutrient levels in the CS water (Ivanov 1986).

Events of blue-green cyanobateria blooms in the last 20 years have been observed in the CS and can be attributed to the extra nutrient flux by rivers (Karpinsky et al. 2005; Leroy et al. 2013b; Naghdi et al. 2018). The status of the CS is oligotrophic, but in some coastal and shallow areas, owing to the input of nutrients, it has become meso-eutrophic (Nasrollahzadeh et al. 2014).

3.5.2.5 Extreme Impact on Environment and Humans

Finally, environmental degradation has created dead zones such as in the Baku Bay (Zonn 2005; Zonn et al. 2010). Contaminants transfer to the biota through dietary and non-dietary mechanisms. Accumulation of contaminants higher in the food chain of the Caspian biota directly impacts upon human health. Indeed, contaminants have been detected in the CS fish, some of which are used for human consumption, e.g. kutum (Nejatkhah-Manavi and Mazumder 2018).

3.5.3 Changes to the Coast Line

Rapid CSL changes impose various impacts on coastal and offshore environments. Shoreline migration, river mouth avulsion, disappearance of old wetlands and emergence of new ones (Leontiev et al. 1977; Kaplin and Selivanov 1995; Kroonenberg et al. 2000; Haghani et al. 2016; Haghani and Leroy 2016, 2020), and modification in the rate of deep-water exchange are some examples of impacts (Lahijani et al. 2018a). The setting and the origin of lagoons determine their final fate. Accordingly, the lagoons that formed under highstand conditions (such as the maximum in 1995) desiccated due to the current sea-level fall (post-1995). They are shallow-water lagoons that form behind beach ridges along sandy beaches with gentle to moderate slopes, where a low-lying area is fed by freshwater, groundwater and/or seawater inlets. They are strongly governed by water level. The shallow depth (1–2 m) is sensitive to water-level changes. Shortage of freshwater from agricultural fields and higher evaporation due to increasing temperature are superimposed on the water-level falling trend that cause drying up of almost all shore-parallel lagoons after the last rise. A good example is the Gomishan Lagoon, a wetland mentioned in the Ramsar convention,[5] situated in a low-lying area on the southeastern flank of the CS, which is almost dry since 2015 (Fig. 3.1).

The bar and spit-lagoon complexes that have weak water connection to the CS, or a connection in only one direction, are intensely influenced by freshwater inputs. The inlets provide limited exchange of offshore water or are incapable of transferring offshore water into lagoons. Extensive freshwater exploitation for agricultural uses along with dam construction on rivers lead to changes in hydrological regime around these lagoons and consequently accelerate lagoon shrinking and desiccation. Rapid sea-level fall facilitated freshwater outflow by not supporting surface-water level and groundwater table stabilization. They are undergoing major shrinkage, which requires water-resources management.

Other anthropogenic changes to the coastline include sand mining in spits, coastal dunes and islands, modifications of the coast for touristic purposes such as the canals in the AZAWA resort in Turkmenistan, the large-scale infrastructures recently erected or planned on Baku's waterfront and the damming of the Kara-Bogaz-Gol inlet in 1980–1983 (Leroy et al. 2006).

Thus, the impacts of natural water-level changes may be mitigated or exacerbated by human intervention and current global warming.

3.6 Discussion

In light of these considerable environmental and water-level changes, a deep investigation of their causes and impacts on coastal zones and water masses is still needed.

[5] International treaty for the protection and sustainable use of wetlands.

Only limited observational, instrumental and geological data are available, however, which reduces the capacity of forecasting models.

3.6.1 Critical Appraisal

3.6.1.1 Long-Term Data

Most CSL research has been made onshore, not only because of the easier access, but also because offshore sediment often enters the domain of petroleum exploration, which is under a certain degree of confidentiality. In addition, offshore coring requires the collaboration of large research teams, large budgets and planning in countries too often affected by political volatility. Academic research and its need for open-access information to the scientific community are incompatible with this situation common along large parts of the CS coast.

As an example, seismic profiles are rarely made available (Hoogendoorn et al. 2010; Putans et al. 2010), although at times they were obtained before coring, e.g. the European INCO-Copernicus project "Understanding the Caspian Sea erratic fluctuations" of the 1990s, but could not be published (Leroy et al. 2014). Also more research should be undertaken into dating of sediment, not only on the reservoir effect on radiocarbon dates, but also in tephrochronology that offers an excellent regional potential (Leroy et al. 2018).

3.6.1.2 Observational Data

The network of operating stations along the coast has sharply declined after a peak in the mid-1950s (Osmakov 2009). Furthermore in the last decades, the situation is not improving. No stations are maintained to international standards and no transnational uniform geodetic levelling network exists along the coasts, with the consequence that the measurements of water level at different sites are not comparable (Fig. 3.2b).

The satellite-altimetry technique demonstrates a great potential in this type of application to the CS, since the water-level variability can be measured over the whole basin beyond political barriers (Kostianoy et al. 2019). Moreover, satellite altimetry provides continuous measurements over open-water regions that have never been covered by direct water-level observations.

3.6.1.3 Models and Forecasts

In 1977 and the following years, the increase in CSL is linked to an increase in the number of ENSO events. Also, recently the recognition of the role of the small rivers and especially of rivers from the South, such as Kura and Sefidrud, is taking place, but not implemented yet in the models. In brief, one can say that the fluvial influx to the CS, hydrometeorological conditions over the water and catchment and to a lesser extent bottom-morphology changes are the main factors that control the water characteristics (Arpe et al. 2014).

With this in mind, the short-term water-level forecast seems relatively successful (Arpe et al. 2014), although, the longer-term forecast remains difficult (Arpe and Leroy 2007a). This is a real societal challenge for the riparian countries.

3.6.2 Threats to the Caspian Sea

Threats to the CS are numerous and some changes occur very rapidly, making it difficult to adapt. The following five are the most important.

Firstly, the regulation of rivers has an anthropogenic impact on the biodiversity by the construction of reservoirs, damming for hydroelectricity generation and irrigation. Every year ~3% of the annual flow of the Volga is lost due to evaporation from the surface of reservoirs (Zonn 2001). Used for irrigation, water from the rivers is also a loss to the CS if precipitation is not falling over its catchment. Shallower deltas and obstacles such as dams hinder fish migration, a particular problem in spring when spawning begins. Thus, many anadromous and semi-anadromous fish species lose their natural spawning habitats. This affects sturgeons and salmonids most severely. Populations of salmonids have disappeared almost completely. Fish farms maintain populations of sturgeons and salmons in Iran and Russia (http://sturgeon.areeo.ac.ir; Vassilieva 2004). Natural spawning grounds are available only in the Ural River and the Iranian rivers where dams are absent (Aladin et al. 2001).

Secondly, overfishing *and illegal fishing* in the CS strongly affect the populations of most commercial fish species. Sturgeons are endangered most of all. After the USSR collapse, fish such as Caspian lamprey, Volga shad, Caspian trout and whitefish were included in the Red Books; whereas in the 1920–1940s, they were still important commercial species. For species with a short life cycle, e.g. sprats, overfishing is less dangerous because their reduced abundance can recover within a few years (Aladin et al. 2001). Anthropogenic water pollution negatively influences sturgeon fish causing toxicoses (Aladin et al. 2001).

Artificial canals have connected the CS via the Volga and then, via other rivers, with the Sea of Azov, Baltic and White seas. They are a route for the third main threat: *uncontrolled penetration of many alien species* by shipping, either in ballast water or by biofouling. The influence of alien species on the CS biota and ecosystems is often negative. For example, the invasive bivalve mollusc *Mytilaster lineatus* is a competitor for the autochthonous *Dreissena* and cannot be used as food for fish because of its thick shell. The most dangerous alien species in the CS is clearly ctenophore *Mnemiopsis leidyi*.

Fourthly, the *CSL changes* are often rapid, e.g. water-level changes of 10 cm/year in the last century. For comparison, the rate was perhaps up to 20 cm/year in the Holocene. They have a clear influence on biodiversity (Dumont 1995). Following the high levels of the late Little Ice Age, the CSL decrease of 1920s–1970s (Fig. 3.6) had a negative impact on aquatic life. Shallow waters of the north CS and deltas of rivers suffered the most. Shallow bays, such as Kaidak and Mertviy Kultuk, dried and the populations died. The increase of 1978–1996 had a negative impact both on the biodiversity and coastal facilities such as those for industry and agriculture. Many pollutants were discharged into the CS. First of all, flooding of oil production and transportation facilities had a damaging effect on the biodiversity. Nevertheless, the positive impacts of long-term CSL rise are the improvement of spawning-ground conditions, increased spawning-ground areas, reinforced water exchange between different sections of the CS, extension of freshwater buffer zone and increase of potential productivity in the north CS.

Lastly the impact of *climate changes* on the biological diversity of the CS is not well studied. Studies of past climatic-change impacts have been only based on biotic proxies. Climatic changes impacted on the biodiversity of the ancient CS significantly, although indirectly, through impact on water level and salinity (Rodionov 1994). These changes nevertheless significantly altered the biodiversity. During Quaternary transgressions, freshwater species dominated, while the abundance of marine species reduced. Marine species survived only in the most saline parts of the CS. During regressions, the situation was the opposite. Marine species dominated, while freshwater species survived only in deltas and areas adjacent to rivers. However in some periods of the Quaternary history, changes of climate caused such extreme or fast changes of the CSL, salinity and temperature that many ancient species were lost. At present, climate change also has an indirect impact on the CS biodiversity. This impact is very weak and less obvious compared to those of earlier geological periods (Aladin et al. 2001), although the human factor may considerably accelerate natural changes.

3.6.3 An Ideal Baseline?

If one wishes to restore the CS to its pristine state before human influence, it is important to choose a realistic baseline. The start in the rise of the water temperature would be a good place to put this baseline. Indeed, the rate of 0.1 °C/year places the CS amongst the lakes with the strongest warming (Hampton et al. 2018). However, we do not know when it started as the oldest instrumental information we have is only from 1982. An increase in the number of phytoplankton cysts of the potentially toxic dinoflagellate *Lingulodinium machaerophorum* in the late 1960s has been suggested to be due to warming (Leroy et al. 2013b).

A point in water level during the twentieth century cannot be chosen as an ideal situation to return to, as fluctuations have been wider just before the start of instrumental measurement, i.e. during the natural state of the lake (Fig. 3.6). Thus, the CSL seems to be in permanent flux, never in a stable state. Other factors, developed hereafter, have significantly altered the CS and should be considered to define the start of the Anthropocene in the CS (Zalasiewicz et al. 2017).

The Caspian catchment has experienced human occupation since Late Pleistocene. However the signature of human impact on the CS is negligible until nineteenth century. A milestone is the start of the pollution by hydrocarbons in c. 1872 in Azerbaijan when improved technology led to higher well production and when legislation changed. Intensive petroleum exploitation and transport through the Volga-Baltic as well as forest removal for shipping fuel are key points in the Caspian environment. They caused hydrocarbon pollution, changed sediment discharge and in 1920s, introductions of alien species. Human activity intensified with industrialization and water regulation in Russia-Soviet Union mainly in the first half of the twentieth century. Despite the century-long human impact on the CS, the environment damage is recoverable, and its biological resources are well preserved.

The main turning point in the CS started after World War II when the catchment basin and its north, south and west coasts were subject of water regulation, agricultural development and urbanization. They led to water shortage in river mouths that prevented fish reproduction and released more nutrients and pollutants into the sea. Discharge of nitrogen and phosphorus by catchment population became significant (Hampton et al. 2018). Physical changes in the rivers, e.g. dams, barriers beneath bridges, sand excavation and overfishing, affected natural fish reproduction. Intensive occupation of the coastal environment for agriculture and urbanization drastically restricted the environment for the Caspian endemic species and wildlife refuges. It thus seems that the physical changes in the Caspian catchment and coast since the mid-twentieth century are the main concern about the Caspian environment.

Finally, as it is difficult to influence CS water levels and water temperature; it is more realistic to aim at environmental remediation by controlling pollution, biological invasions and the impact of dam construction. If a base level has to be suggested, the middle of the twentieth century would be the best candidate. It moreover fits with the definition of the Anthropocene by Zalasiewicz et al. (2017).

3.7 Conclusions

Many changes have been experienced by the CS in the past. However, the current environmental changes are faster and more profound as a result of the multiple aspects of human intervention. The environment receives anthropogenic stresses that are superimposed to natural ones.

The CS is a fascinating topic of study owing to its complexity and the many surprises it reveals. It is however foremost the source of living for more than 14 million people (Rekacewicz 2007) and its catchment has 43 urban centres larger than 300,000 inhabitants (Hampton et al. 2018). It is important to preserve its state for a sustainable future. The CS Anthropocene could be suggested to have started in the middle of the twentieth century as in the rest of the world.

Acknowledgements We would like to thank S. Kroonenberg for advice on Russian literature on CSL and K. Arpe for information on climate and providing Fig. 3.3. We are grateful to L. López-Merino for the preparation of Fig. 3.5. The work on modern biota by NVA and ISP was supported by the program of the Presidium of the Russian Academy of Sciences "No. 41. Biodiversity of natural systems and biological resources of Russia". We are grateful to the three reviewers who have contributed to improve the manuscript.

References

Agalarova DA, Djafarov DI, Khalizov DH (1940) Atlas of microfauna from Tertiary deposits of the Apsheron Peninsula. Azgostoptekhizdat Baku (in Russian)

Agamaliev FG (1983) Infusoria of the Caspian: taxonomy, ecology and zoogeography. Nauka, Leningrad (in Russian)

Aladin NV, Plotnikov IS (2004) Hydrobiology of the Caspian Sea. In: Nihoul JCJ, Zavialov PO, Micklin PP (eds) Dying and Dead Seas. Climatic versus anthropic causes. NATO Science Series IV: Earth and Environmental Sciences, vol 36. Springer, Dordrecht, pp 185–225

Aladin N, Plotnikov I, Bolshov A, Pichugin A (2001) Biodiversity of the Caspian Sea. https://www.zin.ru/projects/caspdiv/biodiversity_report.html. Last accessed 7 July 2018

Aladin NV, Plotnikov IS, Filippov AA (2002) Invaders in the Caspian Sea. In: Leppakoski E, Gollasch S, Olenin S (eds) Invasive aquatic species of Europe. Distribution, impacts and management. Kluwer, Dordrecht, pp 351–359

Arpe K, Leroy SAG (2007) The Caspian Sea level forced by the atmospheric circulation, as observed and modelled. Quat Int 173–174:144–152

Arpe K, Bengtsson L, Golitsyn GS, Mokhov II, Semenov VA, Sporyshev PV (2000) Connection between Caspian sea level variability and ENSO. Geoph Res Lett 27:2693–2696

Arpe K, Leroy SAG, Mikolajewicz U (2011) A comparison of climate simulations for the last glacial maximum with three different versions of the ECHAM model and implications for summer-green tree refugia. Clim Past 7:91–114

Arpe K, Leroy SAG, Lahijani H, Khan V (2012) Impact of the European Russia drought in 2010 on the Caspian Sea level. Hydrol Earth System Sci 16:19–27

Arpe K, Leroy SAG, Wetterhall F, Khan V, Hagemann S, Lahijani H (2014) Prediction of the Caspian Sea level using ECMWF seasonal forecast and reanalysis. Theor Appl Climat 117:41–60

Arpe K, Tsuang B-J, Tseng Y-H, Liu X-Y, Leroy SAG (2018) Quantification of climatic feedbacks on the Caspian Sea Level variability and impacts from the Caspian Sea on the large scale atmospheric circulation. Theor Appl Climat 136(1–2):475–488

Bagirov RM (1989) The Azov and Black Sea species introduced to the Caspian benthos and biofouling. Abstract of Ph.D. thesis, Baku: Institute of Zoology (in Russian)

Bahr A, Lamy F, Arz HW, Major C, Kwiecien O, Wefer G (2008) Abrupt changes of temperature and water chemistry in the late Pleistocene and early Holocene Black Sea. Geochem Geophys Geosyst 9:Q01004

Bezrodnykh YP, Sorokhin VM (2016) On the age of the Mangyshlakian deposits of the northern Caspian Sea. Quat Res 85:245–254

Bogutskaya NG, Kijashko PV, Naseka AM, Orlova MI (2013) Identification keys for fish and invertebrates, vol 1. KMK, Moscow (in Russian)

Boomer I, Aladin N, Plotnikov I, Whatley R (2000) The palaeo-limnology of the Aral Sea: a review. Quat Sci Rev 19:1259–1278

Brunet MF, Korotaev MV, Ershov AV, Nikishin AM (2003) The South Caspian Basin: a review of its evolution from subsidence modelling. Sed Geol 156:119–148

Brunet MF, Shahidi A, Barrier E, Muller C, Saidi A (2007) Geodynamics of the South Caspian Basin southern margin now inverted in Alborz and Kopet Dagh (Northern Iran). Geophys Res Abstracts 9:08080 (European Geosciences Union)

Buharizin PI, Vasyanin MF, Kalinichenko LA (1992) A method for short-term forecasting of the pack ice boundary in the northern Caspian. Meteorol Hydrol 4:74–81. Moscow (in Russian)

CASPCOM, no date. http://www.caspcom.com/. Last accessed 18 June 2018

Cazenave A, Bonnefond P, Dominh K, Schaeffer P (1997) Caspian sea level from TOPEX/POSEIDON altimetry: level now falling. Geophys Res Lett 24:881–884

Chalié F, Escudié A-S, Badaut-Trauth D, Blanc G, Blanc-Valleron M-M, Brigault S, Desprairies A, Ferronsky VI, Giannesini P-J, Gibert E, Guichard F, Jelinowska A, Massault M, Mélières F, Tribovillard N, Tucholka P, Gasse F (1997) The glacial-postglacial transition in the southern Caspian Sea. CRAS Paris 324(IIa):309–316

Chen JL, Pekker T, Wilson CR, Tapley BD, Kostianoy AG, Crétaux J-F, Safarov ES (2017) Long-term Caspian Sea level change. Geophys Res Lett 44:6993–7001

Chepalyga AL (2007) The late glacial great flood in the Ponto-Caspian basin. In: Yanko-Hombach V, Gilbert AS, Panin N, Dolukhanov PM (eds) The Black Sea flood question. Springer, pp 119–148

Chesunov AV (1978) New species of free-living nematodes from the Caspian Sea. Zool Zhurnal 57(4):505–511 (in Russian)

Crétaux JF, Calmant S, Romanovski V, Shabunin A, Lyard F, Berge-Nguyen M, Cazenave A, Hernandez F, Perosanz F (2009) An absolute calibration site for radar altimeters in the continental domain: Lake Issykkul in Central Asia. J Geodesy 83(8):723–735

Crétaux J-F, Abarca Del Rio R, Berge-Nguyen M, Arsen A, Drolon V, Clos G, Maisongrande P (2016) Lake volume monitoring from space. Surv Geophys 37:269–305

De Mora S, Sheikholeslami MR, Wyse E, Azemard S, Cassi R (2004a) An assessment of metal contamination in coastal sediments of the Caspian Sea. Mar Poll Bull 48:61–77

de Mora S, Villeneuve JP, Sheikholeslami MR, Cattini C, Tolosa I (2004b) Organochlorinated compounds in Caspian Sea sediments. Mar Poll Bull 48:30–43

Derzhavin AN (1951) Animal world of Azerbaijan. Izdatelstvo AN Azerbaydzhanskoy SSR, Baku (in Russian)

Dolukhanov PM, Chepalyga AL, Lavrentiev NV (2010) The Khvalynian transgressions and early human settlement in the Caspian basin. Quat Int 225:152–159

Dumont H (1995) Ecocide in the Caspian Sea. Nature 337:673–674

Dumont HJ (1998) The Caspian lake: history, biota, structure, and function. Limn Ocean 43:44–52

Fendereski F, Vogt M, Payne MR, Lachkar Z, Gruber N, Salmanmahiny A, Hosseini SA (2014) Biogeographic classification of the Caspian Sea. Biogeosciences 11:6451–6470

Ferronsky VI, Polyakov VA, Kuprin PN, Lobov AL (1999) The nature of the fluctuation of Caspian Sea level (based on results of the study of bottom sediments). Water Res 26(6):652–666

Forte AM, Cowgill E (2013) Late Cenozoic base-level variations of the Caspian Sea: a review of its history and proposed driving mechanisms. Palaeogeogr Palaeoclimatol Palaeoecol 386(15):392–407

Ghaffari P, Lahijani HAK, Azizpour J (2010) Snapshot observation of the physical structure and stratification in deep-water of the South Caspian Sea (western part). Ocean Sci 6:877–885

Ginzburg AI, Kostianoy AG, Sheremet NA (2005) Sea surface temperature. In: Kostianoy AG, Kosarev AN (eds) The Caspian Sea environment. Springer, Berlin, Heidelberg, New York, pp 59–81

Giralt S, Julià R, Leroy S, Gasse F (2003) Cyclic water level oscillations of the KaraBogazGol-Caspian Sea system. Earth Plan Sci Lett 212(1–2):225–239

Glazovsky NF, Batoyan VV, Brusilovsky SA (1976) Mud volcanism as a source of supply of matter to the Caspian Sea. In: Complex study of the Caspian Sea 5. Moscow State University, Moscow, pp 189–200 (in Russian)

Goodman S, Dmitrieva L (2016) *Pusa capsica*. The IUCN Red List of Threatened Species 2016: e.T41669A45230700. http://dx.doi.org/10.2305/IUCN.UK.2016-1.RLTS.T41669A45230700.en. Last accessed 7 July 2018

Grigorovich IA, Therriault TW, MacIsaac HJ (2003) History of aquatic invertebrate invasions in the Caspian Sea. Biol Inv 5:103–115

Grosswald M (1993) Extent and melting history of the late Weichselian ice sheet, the Barents-Kara continental margin. In: Peltier WR (ed) Ice in the Climate System. NATO ASI Subseries I, vol 12. Springer, Berlin, 1–20

Haghani S, Leroy SAG (2016) Differential impact of long-shore currents on coastal geomorphology development in the context of rapid sea level changes: the case of the Old Sefidrud (Caspian Sea). Quat Int 408:78–92

Haghani S, Leroy SAG (2020) Recent avulsion history of Sefidrud, South West of the Caspian Sea. Quat Int 540:97–110

Haghani S, Leroy SAG, Wesselingh FP, Rose NL (2016) Rapid evolution of a Ramsar site in response to human interference under rapid sea level change: a south Caspian Sea case study. Quat Int 408:93–112

Hampton SE, McGowan S, Ozersky T, Virdis SGP, Vu TT, Spanbauer TL, Kraemer BM, Swann G, Mackay AW, Powers SM, Meyer MF, Labou SG, O'Reilly CM, DiCarlo M, Galloway AWE, Fritz SC (2018) Recent ecological change in ancient lakes. Limnol Oceanog 63:2277–2304

Herzfeld E (1947) Zoroaster and his world. Princeton University Press 2:411–851

Hinds DJ, Aliyeva E, Allen MB, Davies CE, Kroonenberg SB, Simmons MD, Vincent SJ (2004) Sedimentation in a discharge dominated fluvial-lacustrine system: the Neogene Productive Series of the South Caspian Basin, Azerbaijan. Mar Pet Geol 21:613–638

Hoogendoorn RM, Boels JF, Kroonenberg SB, Simmons MD, Aliyeva E, Babazadeh AD, Huseynov D (2005) Development of the Kura delta, Azerbaijan; a record of Holocene Caspian sea-level changes. Mar Geol 222–223:359–380

Hoogendoorn RM, Levchenko O, Missiaen T, Lychagin M, Richards K, Gorbunov A, Kasimov N, Kroonenberg SB (2010) High resolution seismic stratigraphy of the modern Volga delta, Russia. In: Proceedings of the International Conference on the Caspian region: environmental consequences of the climate change. Moscow, pp 32–37

Hoyle TM, Leroy SAG, López-Merino L, Miggins D, Koppers A (2020) Vegetation succession and climate change across the Plio-Pleistocene transition in eastern Azerbaijan, central Eurasia (2.77–2.45 Ma). Palaeogeogr Palaeoclimatol Palaeoecol 538. https://doi.org/10.1016/j.palaeo.2019.109386

Huseynov DA, Guliyev SI (2004) Mud volcanic natural phenomena in the South Caspian Basin: geology, fluid dynamics and environmental impact. Environm Geol 46:1012–1023

Hydroweb, no date. Time series of water levels in the rivers and lakes around the world. http://hydroweb.theia-land.fr/hydroweb. Last accessed 18 June 2018

Ibrayev RA, Özsoy E, Svhrum C, Sur Hİ (2010) Seasonal variability of the Caspian Sea three-dimensional circulation, sea-level and air-sea interaction. Ocean Sci 6:311–329

Ivanov TA (1986) Caspian Sea: hydrometeorological conditions on shelf zone of the soviet seas, vol 2. Gidrometeoizdat, Leningrad (in Russian)

Ivanov VP, Kamakin AM, Ushivtzev VB, Shiganiva T, Zhukova O, Aladin N, Wilson SI, Harbison RG, Dumont H (2000) Invasion of the Caspian Sea by the comb jellyfish *Mnemiopsis leidyi* (Ctenophora). Biol Inv 2:255–258

Kakroodi AA, Kroonenberg SB, Hoogendoorn RM, Mohammed Khani H, Yamani M, Ghassemi MR, Lahijani HAK (2012) Rapid Holocene sea-level changes along the Iranian Caspian coast. Quat Int 263:93–103

Kaplin PA, Selivanov AO (1995) Recent coastal evolution of the Caspian Sea as a natural model for coastal responses to the possible acceleration of global sea-level rise. Mar Geol 124:161–175

Karpevich AF (1975) Theory and practice of aquatic organism acclimatization. Pischevaya pronyshlennost, Moscow (in Russian)

Karpinsky MG (2010) On peculiarities of introduction of marine species into the Caspian Sea. Russian J Biol Inv 1(1):7–10

Karpinsky MG, Katunin DN, Goryunova VB, Shiganova TA (2005) Biological features and resources. In: Kostianoy A, Kosarev A (eds) The Caspian Sea environment. Springer, Berlin

Kasimov AG (1987) Wildlife of the Caspian Sea. Elm, Baku (in Russian)
Kasimov AG (1994) Ecology of the Caspian Lake. Azerbaijan, Baku (in Russian)
Kazancheev EN (1981) Fishes of the Caspian Sea. Lyogkaya i pishchevaya promyshlennost, Moscow (in Russian)
Kholodov VN, Lisitsina NA (1989) The Caspian Sea: Sedimentology. Nauka, Moscow (in Russian)
Khoshakhlagh F, Katigari AS, Saboori SH, Mojtahedi NF, Pour FM, Oskuee EA (2016) Trend of the Caspian Sea surface temperature changes. Nat Env Change 2:57–66
Klenova MV, Solovev VF, Aleksina IA, Vikhrenko NM, Kulakova LS, Maev EG, Rikhter VG, Skornyakava NS (1962) Geological structure of the Caspian bottom. Acad Sci USSR, Moscow
Klige RK (1990) Historical changes of the regional and global hydrological cycles. GeoJournal 20:129–136
Kosarev AN (1975) Hydrology of the Caspian and Aral seas. Moscow State University, 372 pp
Kosarev AN (ed) (1990) The Caspian Sea. Water structure and dynamics. Nauka, Moscow (in Russian)
Kosarev AN, Tuzhilkin VS (1995) Climatic thermohaline fields of the Caspian Sea. Sorbis, Moscow (in Russian)
Kosarev AN, Yablonskaya EA (1994) The Caspian Sea. The Hague
Kosarev AN, Kostianoy AG, Zonn IS (2009) Kara-Bogaz-Gol Bay: physical and chemical evolution. Aquat Geochem 15:223–236
Kostianoy A, Kosarev A (2005) The Caspian Sea environment. Springer, Berlin
Kostianoy AG, Ginzburg AI, Lavrova OY, Lebedev SA, Mityagina MI, Sheremet NA, Soloviev DM (2019) Comprehensive satellite monitoring of Caspian Sea conditions. In: Barale V, Gade M (eds) Remote sensing of the Asian Seas. Springer, Berlin, pp 505–521
Kouraev AV, Papa F, Buharizin PI, Cazenave A, Crétaux J-F, Dozortseva J, Remy F (2003) Ice cover variability in the Caspian and Aral seas from active and passive microwave satellite data. Polar Res 22(1):43–50
Kouraev AV, Papa F, Mognard NM, Buharizin PI, Cazenave A, Crétaux J-F, Dozortseva J, Remy F (2004) Synergy of active and passive satellite microwave data for the study of first-year sea ice in the Caspian and Aral Seas. IEEE Trans Geosci Remote Sens (TGARS) 42(10):2170–2176
Kouraev AV, Crétaux J-F, Lebedev SA, Kostianoy AG, Ginzburg AI, Sheremet NA, Mamedov R, Zhakharova EA, Roblou L, Lyard F, Calmant S, Bergé-Nguyen M (2011) The Caspian Sea. In: Vignudelli S, Kostianoy AG, Cipollini P, Benveniste J (eds) Handbook on Coastal altimetry, vol 19. Springer, pp 331–366
Krivonogov KS, Burr GS, Kuzmin YV, Gusskov SA, Kurmanbaev RK, Kenshinbay TI, Voyakin DA (2014) The fluctuating Aral Sea: a multidisciplinary-based history of the last two thousand years. Gondwana Res 26:284–300
Kroonenberg SB, Rusakov GV, Svitoch AA (1997) The wandering of the Volga delta: a response to rapid Caspian sea-level change. Sed Geol 107:189–209
Kroonenberg SB, Badyukova EN, Storms JEA, Ignatov EI, Kasimov NS (2000) A full sea-level cycle in 65 years: barrier dynamics along Caspian shores. Sed Geol 134:257–274
Kroonenberg SB, Simmons MD, Alekseevski NI, Aliyeva E, Allen MB, Aybulatov DN, Baba-Zadeh A, Badyukova EN, Davies CE, Hinds DJ, Hoogendoorn RM, Huseynov D, Ibrahimov B, Mamedov P, Overeem I, Rusakov GV, Suleymanova S, Svitoch AA, Vincent SJ (2005) Two deltas, two basins, one river, one sea: the modern Volga delta as an analogue of the Neogene Productive Series, South Caspian Basin. In: Giosan L, Bhattacharya J (eds) River deltas—concepts, models and examples, SEPM Spec Publ 83, pp 231–256
Kroonenberg SB, Abdurakhmanov GM, Badyukova EN, van der Borg K, Kalashnikov A, Kasimov NS, Rychagov GI, Svitoch AA, Vonhof HB, Wesselingh FP (2007) Solar-forced 2600 BP and Little Ice Age highstands of the Caspian Sea. Quat Int 173–174:137–143
Kuprin PN (2002) Apsheron threshold and its role in the processes of sedimentation and formation of hydrological regimes in the Southern and Middle Caspian basins. Water Res 29(5):473–484

Kuprin PN, Ferronsky VI, Popovchak VP, Shlykov VG, Zolotaya LA, Kalisheva MV (2003) Bottom sediments of the Caspian Sea as an indicator of changes in its water regime. Water Res 30(2):136–153

Lahijani HAK, Tavakoli V, Amini AH (2008) South Caspian river mouth configuration under human impact and sea level fluctuation. Env Sci 5(2):65–86

Lahijani H, Abbasian H, Naderi–Beni A, Leroy SAG, Haghani S, Habibi P, Hosseindust M, Shahkarami S, Yeganeh S, Zandinasab Z, Tavakoli V, Vahabi-Asil F, Azizpour J, Sayed-Valizadeh M, Pourkerman M, Shah-Hosseini M (2018a) Sediment distribution pattern of South Caspian Sea: possible hydroclimatic implications. Can J Earth Sci 56, https://doi.org/10.1139/cjes-2017-0239

Lahijani H, Naderi Beni M, Tavakoli V (2018b) Heavy metals in coastal sediments of South Caspian Sea: natural or anthropogenic source? Caspian J Envir Sci 16:35–43

Lebedev SA, Kostianoy AG (2008) Integrated using of satellite altimetry in investigation of meteorological, hydrological and hydrodynamic regime of the Caspian Sea. J Terr Atmos Oceanic Sci 19(1–2):71–82

Leontiev OK, Maev NG, Richagov GI (1977) Geomorphology of the Caspian coast and sea. Moscow State University Moscow (In Russian)

Leroy SAG, Marret F, Giralt S, Bulatov SA (2006) Natural and anthropogenic rapid changes in the Kara-Bogaz Gol over the last two centuries by palynological analyses. Quat Int 150:52–70

Leroy SAG, Marret F, Gibert E, Chalié F, Reyss J-L, Arpe K (2007) River inflow and salinity changes in the Caspian Sea during the last 5500 years. Quat Sci Rev 26:3359–3383

Leroy SAG, Kakroodi AA, Kroonenberg SB, Lahijani HAK, Alimohammadian H, Nigarov A (2013a) Holocene vegetation history and sea level changes in the SE corner of the Caspian Sea: relevance to SW Asia climate. Quat Sci Rev 70:28–47

Leroy SAG, Lahijani HAK, Reyss J-L, Chalié F, Haghani S, Shah-Hosseini M, Shahkarami S, Tudryn A, Arpe K, Habibi P, Nasrollahzadeh HS, Makhlough A (2013b) A two-step expansion of the dinocyst *Lingulodinium machaerophorum* in the Caspian Sea: the role of changing environment. Quat Sci Rev 77:31–45

Leroy SAG, Tudryn A, Chalié F, López-Merino L, Gasse F (2013c) From the Allerød to the mid-Holocene: palynological evidence from the south basin of the Caspian Sea. Quat Sci Rev 78:77–97

Leroy SAG, López-Merino L, Tudryn A, Chalié F, Gasse F (2014) Late Pleistocene and Holocene palaeoenvironments in and around the Middle Caspian Basin as reconstructed from a deep-sea core. Quat Sci Rev 101:91–110

Leroy SAG, Chalié F, Wesselingh F, Sanjani S, Lahijani HAK, Athersuch J, Struck U, Plunkett G, Reimer PJ, Habibi P, Kabiri K, Haghani S, Naderi Beni A, Arpe K (2018) Multiproxy indicators in a Pontocaspian system: a depth transect of surface sediment in the S-E Caspian Sea. Geol Belg 21:143–165

Mamedov AV (1997) The late Pleistocene-Holocene history of the Caspian Sea. Quat Int 41–42:161–166

Mangerud J, Jakobsson M, Alexanderson H, Astakhov V, Clarke GKC, Henriksen M, Hjort C, Krinner G, Lunkka J-P, Moller P, Murray A, Nikolskaya O, Saarnisto M, Svendsen JI (2004) Ice-dammed lakes and rerouting of the drainage of northern Eurasia during the last glaciation. Quat Sci Rev 23:1313–1332

Marret F, Leroy S, Chalié F, Gasse F (2004) New organic-walled dinoflagellate cysts from recent sediments of Central Asian seas. Rev Palaeobot Palynol 129(1–2):1–20

Mason IM, Guzkowska MAJ, Rapley CG, Street-Perrot FA (1994) The response of lake levels and areas to climate change. Clim Change 27:161–197

Mayev EG (2010) Mangyshlak regression of the Caspian Sea: relationship with climate. In: Proceedings of the International Conference on the Caspian region: environmental consequences of the climate change. Moscow, pp 107–109

Messager ML, Lehner B, Grill G, Nedeva I, Schmitt O (2016) Estimating the volume and age of water stored in global lakes using a geo-statistical approach. Nat Commun 13603

Mordukhai-Boltovskoy FD (1960) Caspian Fauna in the Azov and Black Sea Basin. Moscow-Leningrad: Izdatelstvo AN SSSR (in Russian)

Mordukhai-Boltovskoy PD (1979) Composition and distribution of Caspian fauna in the light of modern data. Int Rev Gesamten Hydrobiol 64:383–392

Mudie P, Marret F, Mertens K, Shumilovikh L, Leroy SAG (2017) Atlas of modern dinoflagellate cyst distributions in the Black Sea Corridor, including Caspian and Aral Seas. Mar Micropaleont 134:1–152

Naderi Beni A, Lahijani H, Mousavi Harami R, Arpe K, Leroy SAG, Marriner N, Berberian M, Ponel VA, Djamali M, Mahboubi A, Reimer PJ (2013) Caspian sea level changes during the last millennium: historical and geological evidences from the south Caspian Sea. Clim Past 9:1645–1665

Naghdi K, Moradi M, Kabiri K, Rahimzadegan M (2018) The effects of cyanobacterial blooms on MODIS-L2 data products in the southern Caspian Sea. Oceanologia 60:367–377

Nasrollahzadeh SH, Makhlough A, Eslami F, Leroy SAG (2014) Features of the phytoplankton community in the southern Caspian Sea, a decade after the invasion of *Mnemiopsis leidyi*. Iran J Fish Sci 13(1):145–167

Nazemosadat MJ, Ghasemi AR (2005) The effect of surface temperature fluctuations of the Caspian Sea in winter and spring seasons precipitation in northern and southwestern areas of Iran. J Sci Techn Agric Nat Resour 4:1–14 (in Persian)

Nejatkhah-Manavi P, Mazumder A (2018) Potential risk of mercury to human health in three species of fish from the southern Caspian Sea. Mar Pollut Bull 130:1–5

Nordyke MD (2000) The Soviet program for peaceful uses of nuclear explosions. Lawrence Livermore National Laboratory, 1 September 2000. https://e-reports-ext.llnl.gov/pdf/238468.pdf. Last accessed 29 July 2018

Osmakov A (2009) Caspian Sea level change. observation network and methods, quality and use of data. M.Sc. thesis, Brunel University, UK

Ozyavas A, Khan DS (2008) Assessment of recent short-term water level fluctuations of Caspian Sea using Topex/Poseidon. IEEE Geosc Remote Sens L 5:720–724

Ozyavas A, Shuhab DK, Casey JF (2010) A possible connection of Caspian Sea level fluctuations with meteorological factors and seismicity. Earth Planet Sci Lett 299:150–158

Palo JU, Väïnolä R (2006) The enigma of the landlocked Baikal and Caspian seals addressed through phylogeny of phocine mitochondrial sequences. Biol J Linn Soc 88:61–72

Plotnikov I, Aladin N, Crétaux J-F, Micklin P, Chuikov Yu, Smurov A (2006) Biodiversity and recent exotic invasions of the Caspian Sea. Verh Int Ver Limnol 29(5):2259–2262

Pobedonostsev SV, Abuzyarov ZK, Kopeikina TN (2005) On the quality of the Caspian Sea level observations. In: Proceedings of the State Hydrometeorological Center of the Russian Federation, 339, Sea and river hydrological calculations and forecasts (in Russian)

Proshkina-Lavrenko AI, Makarova IV (1968) Algae of the Caspian Sea plankton. Nauka, Leningrad (in Russian)

Putans VA, Merklin LR, Levchenko OV (2010) Sediment waves and other forms as evidence of geohazards in Caspian Sea. Int J Offshore Pol Eng 20(4):241–246

Ranjbaran M, Sotohian F (2015) Environmental impact and sedimentary structures of mud volcanoes in southeast of the Caspian Sea basin Golestan, Province Iran. Caspian J Env Sci 13:391–405

Rekacewicz P (2007) Population by administrative region, Caspian Sea region. UNEP/GRID-Arendal. https://www.grida.no/resources/6129. Last accessed 14 June 2018

Richards K, Bolikhovskaya NS, Hoogendoorn RM, Kroonenberg SB, Leroy SAG, Athersuch J (2014) Reconstructions of deltaic environments from Holocene palynological records in the Volga delta, northern Caspian Sea. Holocene 24(10):1226–1252

Richards K, van Baak CGC, Athersuch J, Hoyle TM, Stoica M, Austin WEN, Cage AG, Wonders AAH, Marret F, Pinnington CA (2018) Palynology and micropalaeontology of the Pliocene-Pleistocene transition in outcrop from the western Caspian Sea, Azerbaijan: potential links with the Mediterranean, Black Sea and the Arctic Ocean? Palaeogeogr Palaeoclimatol Palaeoecol 511:119–143

Rodionov SN (1994) Global and regional climate interaction: the Caspian Sea experience. Water Sci Technol Libr, Springer

Rychagov GI (1997) Holocene oscillations of the Caspian Sea, and forecasts based on palaeogeographical reconstructions. Quat Int 41(42):167–172

Sapozhnikov VV, Mordasova NV, Metreveli MP (2010) Transformations in the Caspian Sea ecosystem under the fall and rise of the sea-level. Oceanology 50:488–497

Sauer EW, Wilkinson TJ, Nokandeh J, Omrani Rekavandi H (2013) Persia's imperial power in late Antiquity. The great wall of Gorgan and frontier landscapes of Sasanian Iran. British Institute of Persian Studies. Archaeological Monographs series ii. Oxbow books. Oxford and Oakville

Shiklomanov IA, Georgievsky V, Kopaliani ZD (1995) Water balance of the Caspian Sea and reasons of water level rise in the Caspian Sea, IOC Workshop report No 108, Paris, pp 1–28

Starobogatov YI (1994) Systematics and paleontology. In: Starobogatov YI (ed) Species of fauna of Russia and neighboring countries. Zebra mussel *Dreissena polymorpha* (Pall.) (Bivalvia, Dreissenidae). Nauka, Moscow (in Russian)

Stepanjants SD, Khlebovich VV, Aleksees VR, Daneliya ME, Petryshev VV (2015) Identification keys for fish and invertebrates, vol 2. KMK, Moscow (in Russian)

Svitoch AA (2008) The Khvalynian transgression of the Caspian Sea and the New-Euxinian basin of the Black Sea. Water Res 35(2):165–170

Svitoch AA (2009) Khvalynian transgression of the Caspian Sea was not a result of water overflow from the Siberian Proglacial lakes, nor a prototype of the Noachian flood. Quat Int 197:115–125

Svitoch AA (2012) The Caspian Sea shelf during the Pleistocene regressive epochs. Oceanology 52(4):526–539

Tarasov AG (1996) Biological consequences of the Caspian basin pollution. Water Res 23(4):448–456

Tarasov AG, Kazantseva SZ (1994) Post-mortem transport of freshwater mollusc shells in the northern Caspian Sea: a cautionary note on the implications for palaeoecological reconstructions. Int J Salt Lake Res 3:49–52

Terziev FS, Kosarev AN, Kerimov AA (eds) (1992) Hydrometeorology and hydrochemistry of seas, vol 6, Caspian Sea, no. 1: Hydrometeorological conditions. Gidrometeoizdat, St. Petersburg (in Russian)

Tolosa I, de Mora S, Sheikholeslami MR, Villeneuve JP, Bartocci J, Cattini C (2004) Aliphatic and aromatic hydrocarbons in coastal Caspian Sea sediments. Mar Poll Bull 48:44–60

Toropov PA, Morozova PA (2010) Evaluation of Caspian Sea level at Late Pleistocene period (on the base of numeral simulation adjusted for Scandinavian glacier melting). In: Proceedings of the International Conference on the Caspian region: environmental consequences of the climate change. Moscow, pp 134–137

Tudryn A, Leroy SAG, Toucanne S, Gibert-Brunet E, Tucholka P, Lavrushin YA, Dufaure O, Miska S, Bayon G (2016) The Ponto-Caspian basin as a final trap for southeastern Scandinavian ice-sheet meltwater. Quat Sci Rev 148:29–43

Ulmishek GF (2001) Petroleum geology and resources of the North Caspian Basin, Kazakhstan and Russia. USGS Bull 2201-B:25

UNEP (2004) Freshwater in Europe. http://www.grid.unep.ch/products/3_Reports/freshwater_atlas.pdf. Last accessed 9 Aug 2018

Van Bavel CHM (1966) Potential evaporation: the combination concept and its experimental verification. Water Res 2:455–467

Varushchenko S, Varushchenko A, Klige R (1987) Changes in the regime of the Caspian Sea and closed basins in time. Nauka, Moscow

Vassilieva LM (2004) http://web.worldbank.org/archive/website00983A/WEB/OTHER/9E807689.HTM?Opendocument. Last accessed 9 August 2018

Vinogradov LG (1968) Order Decapoda. In: Birstein YaA, Vinogradov LG, Kondakov NN, Kun MS, Astahova TV, Romanova NN (eds) Atlas of invertebrates of the Caspian Sea. Pischevaya pronyshlennost, Moscow, pp 291–300 (in Russian)

Voropayev GV (1997) The problem of the Caspian Sea level forecast and its control for the purpose of management optimization. In: Glantz MH, Zonn IS (eds) Scientific, environmental, and political issues in the circum-Caspian region. Cambridge University Press, Cambridge, pp 105–118

Yanina Y, Sorokin V, Bezrodnykh Yu, Romanyuk B (2018) Late Pleistocene climatic events reflected in the Caspian Sea geological history (based on drilling data). Quat Int 465(Part A):130–141
Zaberzhinskaya EB (1968) Flora of algae-macrophytes of the Caspian Sea. Ph.D. thesis Fac. Biol., Baku State University (in Russian)
Zalasiewicz J, Waters CN, Summerhayes CP, Wolfe AP, Barnosky AD, Cearreta A, Crutzen P, Ellis E, Fairchild IJ, Gałuszka A, Haff P, Hajdas I, Head MJ, Assunção JA, Sul I, Jeandel C, Leinfelder R, McNeill JR, Neal C, Odada E, Oreskes N, Steffen W, Syvitski J, Vidas D, Wagreich M, Williams M (2017) The Working Group on the Anthropocene: summary of evidence and interim recommendations. Anthropocene 19:55–60
Zenkevich LA (1963) Biology of the seas of the USSR. Interscience Publishers, New York
Zonn IS (2001) Three hundred years in the Caspian (Chronology of the main historical events of the XVIII-XX centuries). Edel-M, Moscow (in Russian)
Zonn IS (2005) Environmental issues of the Caspian Sea. In: Kostianoy AG, Kosarev AN (eds) Handbook of Environmental Chemistry. Springer, Berlin, pp 223–242
Zonn IS, Kostianoy AG, Kosarev AN, Glantz M (2010) The Caspian Sea Encyclopedia. Springer, Berlin

Chapter 4
The Aral Sea: A Story of Devastation and Partial Recovery of a Large Lake

Philip Micklin, Nikolai V. Aladin, Tetsuro Chida, Nikolaus Boroffka, Igor S. Plotnikov, Sergey Krivonogov and Kristopher White

Abstract The Aral Sea was a huge brackish-water lake lying in a tectonic depression amidst the deserts of Central Asia. Water bodies of various dimensions have repeatedly filled this depression over the past several million years. Its modern incarnation is thought to be somewhat more than 20,000 years in age. In modern times, the sea supported a major fishery and functioned as a key regional transportation route. But since the 1960s, the Aral has undergone rapid desiccation and salinization, overwhelmingly the result of unsustainable expansion of irrigation that largely dried up its two tributary rivers, the Amu Dar'ya and Syr Dar'ya (dar'ya in the Turkic languages of Central Asia means river) before they reached the Aral Sea. The desiccation of the Aral Sea has had severe negative impacts, including, among others,

P. Micklin (✉)
Department of Geography, Western Michigan University, Kalamazoo, MI 49008 5424, USA
e-mail: Micklin@wmich.edu

N. V. Aladin · I. S. Plotnikov
Laboratory of Brackish Water Hydrobiology, Zoological Institute, Russian Academy of Sciences, Universitetskaya nab. 1, St. Petersburg 199034, Russia
e-mail: Nikolai.Aladin@zin.ru

I. S. Plotnikov
e-mail: Igor.Plotnikov@zin.ru

T. Chida
School of Global Governance and Collaboration, Nagoya University of Foreign Studies, 57 Takenoyama, Iwasaki, Nissin Aichi 470-0197, Japan
e-mail: tetsuroch@gmail.com

N. Boroffka
Eurasia Department, German Archaeological Institute, Im Dol 2-6, 14195 Berlin, Germany
e-mail: Nikolaus.Boroffka@dainst.de

S. Krivonogov
Institute of Geology and Mineralogy, Siberian Branch of Russian Academy of Sciences, Novosibirsk State University, 3 Koptyug Ave., 630090 Novosibirsk, Russia
e-mail: carpos@igm.nsc.ru

K. White
KIMEP University, 4 Abai Ave., Almaty 050010, Kazakhstan
e-mail: kwhite@kimep.kz

S. Mischke (ed.), *Large Asian Lakes in a Changing World*, Springer Water,
https://doi.org/10.1007/978-3-030-42254-7_4

the demise of commercial fishing, devastation of the floral and faunal biodiversity of the native ecosystems of the Syr and Amu Deltas, and increased frequency and strength of salt/dust storms. However, efforts have been and are being made to partially restore the sea's hydrology along with its biodiversity, and economic value. The northern part of the Aral has been separated from the southern part by a dike and dam, leading to a level rise and lower salinity. This has allowed native fishes to return from the rivers and revitalized the fishing industry. Partial preservation of the Western Basin of the southern Aral Sea may be possible, but these plans need much further environmental and economic analysis.

Keywords Lake history · Lake level · Irrigation · Water diversion · Dust · Salt

4.1 Introduction

The Aral Sea, in Russian "Aralskoye more" and in the Turkic languages of Central Asia "Aral Tengizi" (Kazak) or "Arol Dengizi" (Uzbek) is a terminal or closed-basin (endorheic) lake, lying amidst the vast deserts of Central Asia (Fig. 4.1). From the mid-seventeenth century until the 1960s, lake level variations were less than 4.5 m (Micklin 2016). During the first six decades of the twentieth century, the sea's water balance was remarkably stable and its annual average level fluctuated less

Fig. 4.1 Location of the Aral Sea in Central Asia. *Source* Micklin (2007)

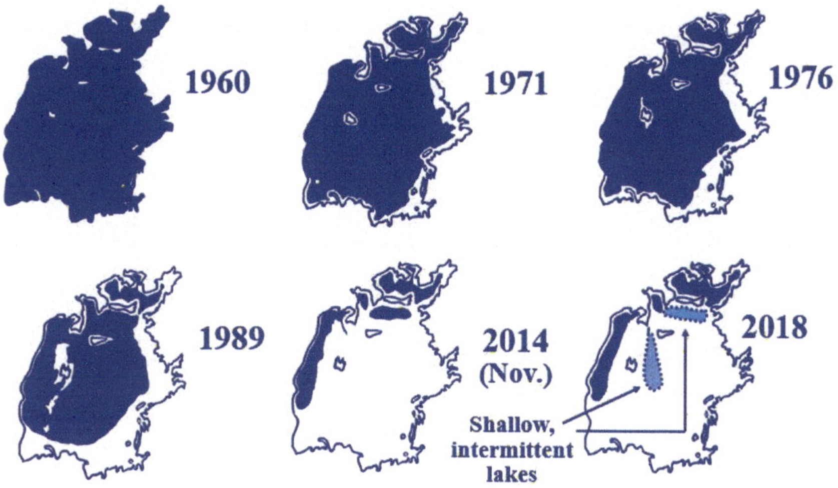

Fig. 4.2 The changing Aral Sea: 1960–2018

than a meter. At 67,500 km^2 in 1960, the Aral Sea was the world's fourth largest lake in surface area (Micklin 2010; Zonn et al. 2009). A brackish lake with salinity averaging near 10 grams/liter (g/l), less than a third of the ocean, it was inhabited by both freshwater and brackish-water fish species (Kostianoy and Kosarev 2010). The sea supported a major fishery and functioned as a key regional transportation route. The extensive deltas of the Syr Dar'ya and Amu Dar'ya sustained a diversity of flora and fauna, including endangered species. The deltas also had considerable economic importance supporting irrigated agriculture, animal husbandry, hunting and trapping, fishing, and harvesting of reeds, which served as fodder for livestock as well as building materials.

Post 1960, the Aral has undergone rapid desiccation and salinization, overwhelmingly the result of unsustainable expansion of irrigation that largely drained the two influent rivers (Fig. 4.2; Table 4.1). By June 2018 the Aral Sea consisted of five separate water bodies that at times are connected (Fig. 4.3). In summer 2014 the Eastern Basin of the Aral entirely dried. Its aggregate area and volume at that time were only 10% and 4%, respectively, of 1960 (Micklin 2016). Subsequently, the Eastern Basin has expanded and shrunk on a seasonal rhythm depending on inflow from tributary rivers. By June 2018, the level of the deeper Western Basin of the Aral Sea had fallen to a record low of a bit more than 23 m above the Kronstadt gauge (situated on the Gulf of Finland near St. Petersburg, Russia which has a zero 20 cm above ocean level), while owing to significant winter and spring inflow from the Amu Dar'ya and Syr Dar'ya, the very shallow Eastern Basin was somewhat higher (Table 4.1).

In the sections that follow we give a basic characterization of the Aral Sea in terms of basin history, natural state prior to the modern desiccation, human impacts on the lake, and future of the water body.

Table 4.1 Hydrological and salinity characteristics of the Aral Sea, 1960–2018

Year and portion of sea	Level (meters above Baltic Sea)	Area (km^2)	% 1960 area	Volume (km^3)	% 1960 volume	Average depth (meters)	Avg. salinity (g/l)	% 1960 salinity
1960 (all)	53.4	67,499	100	1089	100	16.1	10	100
Large	53.4	61,381	100	1007	100	16.4	10	100
Small	53.4	6118	100	82	100	13.4	10	100
1971 (all)	51.1	60,200	89	925	85	15.4	12	120
1976 (all)	48.3	55,700	83	763	70	13.7	14	140
1989 (all)		39,734	59	364	33	9.2		
Large	39.1	36,930	60	341	34	9.2	30	300
Small	40.2	2804	46	23	28	8.2	30	300
Sept 22, 2009 (all)		7146	10.6	83	7.7	10.8		
W. Basin Large	27	3588	26.2	56	17.9	15.1	>100	>1000
E. Basin Large	27	516	1.1	0.64	0.07	0.7	>150	>1500
Tshche-bas Gulf	28	292		0.51	7.1	1.4	~85	850
Small	42	3200	52	27	33	8.4	8	100–130
8/29 and 11/25, 2014 (all)		6990	10.4	81.7	4.4	6.9		
W. Basin Large	25.0	3120	22.8	54	17.2	15.4	>150	>1000

(continued)

Table 4.1 (continued)

E. Basin Large		0	0	0	0	0	0	0
Tshche-bas Gulf	28.5	372	NA	0.72	NA	1.4	89	890
Small	41.9	3197	52.3	27	33.2	8.5	8	80
6/20 and 6/21, 2018		9668	14.3	78.9	7.2	8.2		
W. Basin Large	23	2894	21	48	15.5	14.6	>150	>1500
E. Basin Large	27	2537	5.4	1.3	1.9	0.5	NA	NA
Tshche bas Gulf	28	403	NA	0.78	NA	1.9	85?	850?
Central Aral	29	422	NA	NA	NA	NA	NA	NA
Small	42+	3412	55.7	28.8	35.1	8.4	7	60

Sources (1) Values for 1960–1989 from Micklin (2010). (2) Area data for 2009, 2014 and 2018 calculated from MODIS 250-m resolution natural color images and Landsat 8 natural color images (30-m resolution) using ImageJ software (freeware developed by U.S. National Institutes of Health). (3) Volume data for 2009, 2014 and 2018 estimated from area changes. (4) Salinity data for 2009 are estimates based on measurements taken with a YSI-85 electronic meter during an expedition to the Aral Sea in September 2007 and in 2008. (5) Salinity data for 2013 based on measurements taken with the YSI-85 meter and an optical refractometer during an expedition to the Aral Sea in August and September 2011. (6) Salinity data for 2014 based on own data and data provided by Z. Ermakhanov, director of the Aral'sk Affiliate of the Kazakhstan Fisheries Institute. (7) Salinity data for 2018 based on data gathered at the Berg Strait dike and at Tastubek on the North Aral shore on May 31 and June 3, 2017

NA Not Available

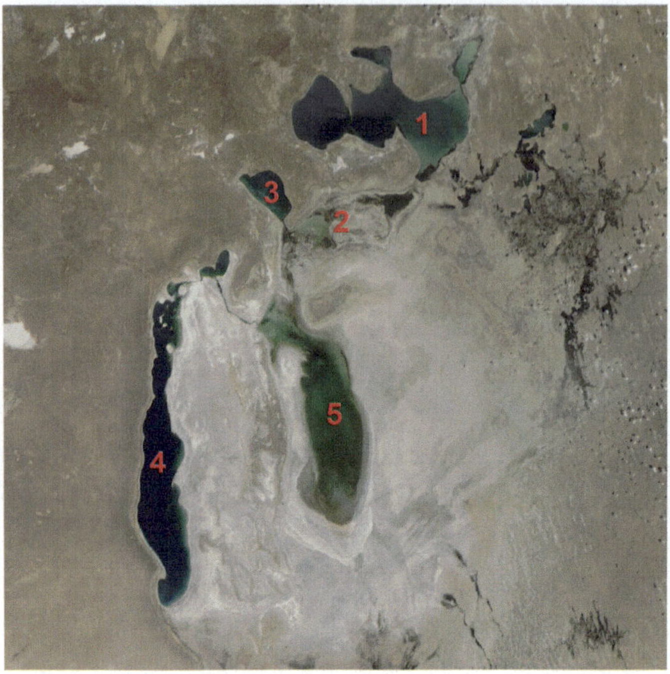

Fig. 4.3 MODIS 250-m resolution natural color image of the Aral Sea on June 20, 2018. Numbers indicate: 1—North Aral; 2—Central Aral; 3—Tshche-Bas Gulf; 4—Western Basin of Large Aral; 5—Eastern Basin of Large Aral. *Source* MODIS Rapid Response System (https://lance-modis.eosdis.nasa.gov/cgi-bin/imagery/realtime.cgi)

4.2 Basin History

Over the past three million years the Aral region was part of the Neogene Ponto-Caspian ancient seas (Akchagyl and Apsheron) and became a separate depression in the Pleistocene. Most experts believe that it began as a small depression, which collected local surface water. This runoff was slightly saline due to the dissolution of local salt deposits. When the water evaporated, it left behind a thin veneer of salts. The surface layer was highly sensitive to wind erosion. The process eventually deepened and enlarged the depression. The area was dry during a major part of the Pleistocene but geological data (Kes 1969, 1995; Rubanov et al. 1987) suggest large rivers reached the depression by the end of the middle to the beginning of the late Pleistocene, i.e., approximately 140,000 years before present (BP).

There have been several age estimates for the Aral Sea initiation. Lopatin (1957) calculated 18,000 years for the age of the Amu Dar'ya delta from its total volume and sedimentation rate. Chalov (1968), based on Uranium-isotope-ratio dating, concluded that the Aral Sea is as old as 139,000 ± 12,000 years, and that the Syr Dar'ya was the sole water source at that time. The Amu Dar'ya started to discharge water into the Aral Sea not earlier than 22,000 years ago. Kes (1995) believed that the highest level of the Aral Sea, up to 72 m above Kronstadt gauge occurred during the Last Glacial Maximum (LGM), and she referred to a terrace-derived ^{14}C date of 24,820 ± 820 years BP (Pshenin et al. 1984). Other experts, based on the bottom-sediment records, concluded the Aral is very young, up to 10,000 years, and the sediments reflect two shallow and deep-water stages in the middle Holocene (Nikolaev 1991, 1995). The most recent drilling data imply age of the deepest subbase layers of the lake sediments coarsely extrapolated to 19,000 years BP (Boomer 2012). The non-lacustrine substratum was dated to 23,800 years BP (Krivonogov 2014; Krivonogov et al. 2010). These ages were obtained from a cluster of boreholes to the south and north from Barsakelmes Island. The farthest to the south and the deepest borehole B-05-2009 is 15 m deep and includes 11 m of the Aral Sea sediments with a basal age of ca. 17,600 years BP (Burr et al. 2019). The previously published borehole depth (B-2008-01) east of Barsakelmes, where the sediments have been dated to 24,000 years BP, includes only 7 m of lacustrine sediments.

In general, researchers have placed the original filling stage of the Aral to 20,000 years BP (Oreshkin 1990, pp. 3–4). At this time and for a considerable period afterward, the Amu Dar'ya flowed westward into the Caspian Sea rather than northward into the Aral. The lake did not attain significant size until the Amu Dar'ya switched its course northward into the Aral Sea. This increased inflow to the lake by some threefold. It is believed to have occurred 10,000–20,000 years BP and was most likely due to a wetter climate that increased river discharge (Aladin et al. 1996). The Small (North) Aral only filled after the addition of the Amu's flow.

Approximately the last 10 millennia (corresponding with the Holocene Epoch) constitute the modern geological history of the Aral Sea (Micklin 2014a). Soviet scientists during the post-World War II era (from the late 1940s to 1991) intensively studied the evolution of the Aral over this time-period. Dating of relict shore terraces, fossils and deposits of various salts precipitating from the sea contained in sediment cores from the sea bottom, and of archeological sites, along with historical records point to repeated major recessions and transgressions of the sea.

Kes (1978), based on studies of terraces, believed lake level could have been as high as 57–58, 62–63 or even 70–73 m, measured above the Kronstadt gauge. The highest reliable standings of the sea discussed in early literature (e.g., Lymarev 1967; Rubanov et al. 1987) are the "Ancient Aral Transgression" that reached an estimated 57–58 m and lasted from approximately 2800–2000 years BP, "New Aral" at 54–55 m reached around 1000 years ago and the pre-1960 level around 53 m that dates from around the middle seventeenth century (350 years BP). More recent investigations using GIS and GPS techniques indicate the highest level the Aral reached over the last 10,000 years was no more than 54–55 m above sea level (Boomer et al. 2009; Boroffka et al. 2006; Reinhardt et al. 2008) and confirm the idea of Berg (1908) that the Aral never overflowed into the Sarykamysh Lake and Uzboy channel leading to the Caspian Sea. Feeding of the Sarykamysh and Uzboy was from direct diversions of the Amu Dar'ya. Diversions of the Amu Dar'ya westward toward the Caspian Sea caused regressions of the Aral. The change from a wetter to dryer climate leading to less flow into the Aral from both the Syr Dar'ya and Amu Dar'ya no doubt also affected lake levels.

Researchers from Moscow State University (Mayev et al. 1983, 1991; Mayeva and Mayev 1991) were the first who investigated and dated the Aral Sea sediment cores. They believed there were nine major regression/transgression cycles during the Holocene and showed that deep regressions were typical features in the Aral history. The deepest regression recorded by these authors was the so-called stage of the "Oxus swamp" that occurred about 1600 years BP (the radiocarbon dates were calibrated by Krivonogov et al. 2014), when the delta of the Syr Dar'ya was situated near the central part of the Large Aral Basin. The early transgressions and regressions of the sea are still not as well-known as later events. The level history for the second half of the Holocene (6000 years BP to the beginning of the modern drying in the 1960s) is better understood and the history of the last 2000 years is even more fully comprehended (Boroffka et al. 2006; Oberhänsli et al. 2007).

Regressions of the sea are related to the partial or full diversion of the Amu Dar'ya westward into the Sarykamysh Depression and from there via the Uzboy channel to the Caspian Sea. The change from a wetter to dryer climate leading to less flow into the Aral from both the Syr Dar'ya and Amu Dar'ya no doubt also played a role but cannot account for the size and rapidity of the most significant level declines.

Ancient civilizations also affected Aral levels. Human impacts included sizable irrigation withdrawals and periodic diversions of the Amu Dar'ya westward into the Sarykamysh Depression and Uzboy Channel. The first evidence of irrigation along the Amu dates to 3000 years ago (Kes 1978; Lunezheva et al. 1987, 1988) and irrigation may have covered five million hectares at times during the fourth century B.C. to fourth century A.D. (Micklin 2014b). However, the impact of ancient irrigation on river inflow to the sea was probably not as significant as it might seem (Kes 1978). Fields were small and withdrawals per hectare irrigated were much less than modern. Also, a much larger percentage of water withdrawn was returned via drainage flows to the rivers rather than being "lost" to evaporation in the arid surrounding deserts. Finally, canals were built and abandoned over time so that the area irrigated in a specific year was far smaller than the area covered by canal systems.

Human-caused diversions were by far the most important influence on levels over the past millennium. Some of these were accidental, caused by breaching of dikes and dams constructed for irrigation purposes during heavy flows of the river. Others occurred during wars and were purposeful with the intent to deprive an enemy of both water for drinking and irrigating crops. Thus, in 1221 the forces of Genghis Khan wrecked irrigation systems in Khorezem Khanate (Bartold 1902; Berg 1908). This caused the Amu to turn its course from northward to the Aral to westward into the Sarykamysh Depression and the Caspian Sea. A similar story is reported about Timur (Tamerlane) who is reputed to have diverted the Amu Dar'ya westward in 1406 to flood the city of Urgench in order to force its surrender. There is ample archeological and historical evidence of repeated settlement and agriculture around the Sarykamysh Depression and along the Uzboy, which would only be possible when the former was flooded and the latter contained a river (Vainberg 1999).

Research on historic level fluctuations of the Aral diminished greatly after the collapse of the USSR in 1991. The lake was no longer of great interest to research institutions in Moscow and Leningrad that had studied it both during Tsarist and Soviet times. Since the late 1990s, however, there has been resurgent interest in the topic. Motivating factors have been the need to better understand the modern regression by delving into past drying events and the fact that the receding sea is uncovering shoreline terraces, former river beds, archeological finds, and other evidence whose analysis provides a much clearer picture of past regressions than had hitherto been possible.

The most ambitious effort was developed within the CLIMAN Project (Holocene Climatic Variability and Evolution of Human Settlement in the Aral Sea Basin) in 2002–2005, funded by the European Union's INTAS Project (1993–1993–2007). The Aral-related program was intended as an interdisciplinary study to help distinguish between climatic variations and anthropogenically controlled environmental changes in the past (Boroffka et al. 2006; Oberhänsli et al. 2007). The focus was on previous lake levels and the evolution of human settlement and agriculture in the Aral Sea Basin. Differential GPS elevation measurements of shorelines around the sea convincingly argued against the Aral's level standing any higher than about 55 m for at least the past 35,000 years (Reinhardt et al. 2008). Based on archeological evidence, relict shorelines, and sediment core analyses, the CLIMAN group delineated seven transgressions and six regressions over the past 5,000 years. However, the best documented of these (by sediment cores analyses and shoreline traces) are four regressions dated to 350–450, 700–780, around 1400, and 1600–2000 years BP (Austin et al. 2007; Huang et al. 2011; Oberhänsli et al. 2011; Sorrel et al. 2006, 2007a, b).

The second post-Soviet effort to reveal changes of the Aral Sea level was the joint US CRDF—Russian RFBR project "Environmental history of the Aral Sea in the last 10,000 years: natural and anthropogenic components" in 2008–2010. The key activity was drilling and dating the sediments in the dry bottom of the Large Aral Basin. A series of high and low stands were identified through the Holocene (Krivonogov 2014; Krivonogov et al. 2010). Special attention was paid to the last 2000 years, for which two large regressions and two transgressions (prior to the modern technogenic regression) were substantiated by multiproxy data both original and published by predecessors (Krivonogov et al. 2014). Special attention was paid to the extent and timing of the last major desiccation of the Aral prior to the modern drying (Krivonogov 2009; Krivonogov et al. 2014).

The regression occurred from the 11th to sixteenth centuries when the level may have fallen below 29 m above the zero level of the Kronstadt gauge. Historical records as well as archeological sites, preserved tree stumps, and relict river channels on the dried bottom of the Aral attest to this event. The most convincing evidence was the discovery by Kazakh hunters at the end of the twentieth century of a mazar (Islamic holy gravesite) on the dried bottom of the Eastern Large Aral, northeast of the former Island of Barsakelmes, which in the early 1960s was 19 m below the surface of the Aral (Boroffka et al. 2005, 2006; Micklin 2007; Smagulov 2001, 2002). The gravesite is known as Kerdery #1. Consequently, two other archeological sites were found at about ten kilometers from Kerdery #1: Aral-Asar settlement and Kerdery #2 gravesite (Catalogue 2007). Archeologists date the sites in the range from late 13th to early fifteenth centuries. Radiocarbon dating of wood and bones from the sites (Krivonogov et al. 2010, 2014) gave a range of 500–1000 years BP.

The major cause of this recession very likely was an anthropogenic diversion of the Amu westward toward the Caspian Sea prior to the Mongol invasion of Central Asia in the thirteenth century. The Mongol intrusion probably increased this effect. The Amu returned (or was returned) to the Aral and the sea recovered by the mid-1600s. The sea was generally in a relatively stable "high" phase until the modern regression began in the early 1960s. Level fluctuations were no more than 4–4.5 m and were chiefly related to climatic variation with, perhaps, some effects from irrigation.

4.3 Natural State of the Aral Sea Prior to the Modern Desiccation

The Aral Sea, as nearly all the earth's large lakes, has suffered significantly from human actions for some time. So, to speak of its "natural state" we would need to go back several thousand years as made clear in the section above on the lake's history. The discussion below is focused primarily on the condition of the lake during the first six decades of the twentieth century prior to the modern desiccation that started in the 1960s.

4.3.1 Geographical Setting of the Aral Sea

The Aral Sea is in the heart of Central Asia on the Eurasian continent (Fig. 4.1). Its drainage basin covers 2.2 million km^2 (World Bank 1998, p. 1). The basin is mainly lowland desert (Micklin 2014a). The climate is desert and semi-desert with cold winters and hot summers in the north and central parts and very hot summers and cool winters in the south (Goode's World Atlas 1982, pp. 8–9). High mountains ring the basin on the east and south (Tian Shan, Pamir, Kopet-Dag), with peaks in the Pamirs over 7000 m.

Annual precipitation in the lowland deserts ranges from less than 100 mm to the south and east of the Aral Sea to near 200 mm approaching the foothills of the southeastern mountains (Atlas of the USSR 1983, p. 102) The foothills and valleys of the mountainous south and southeast are substantially more humid with precipitation ranging from 200 to over 500 mm. The high Pamir and Tian Shan ranges are wet with average annual precipitation from 800 to 1600 mm giving this zone a marked surplus of moisture. This, in turn, has created large permanent snow-fields and glaciers that feed the two major rivers, the Amu Dar'ya and Syr Dar'ya, flowing across the deserts to the Aral Sea.

4.3.2 Hydrology of the Aral Sea Basin

Although the majority of the Aral Sea Basin is desert, it has substantial water resources. The mountains on the south and southeast capture the plentiful precipitation, storing most of it in snowfields and glaciers (Micklin 2014a, c). Runoff from these, heaviest during the spring-early summer thaw, feeds the region's rivers. Estimated average annual river flow in the Aral Sea Basin is 116 km^3, including flow of the drainage basins of the Amu Dar'ya and Syr Dar'ya.

The Amu is the most important river within the Aral Sea Basin. Originating primarily among the glaciers and snowfields of the Pamir Mountains of Tajikistan, its drainage basin covers 465,000 km^2. The river flows 2620 km from the mountains across the Kara-Kum Desert and into the Aral Sea. During this journey, the river flows along the borders and across four Central Asian nations: Tajikistan, Afghanistan, Turkmenistan, and Uzbekistan, entering, leaving, and reentering the last two states several times (Fig. 4.1).

Average annual flow from the drainage basin of the Amu is around 79 km^3. This includes not only the flow of the Amu Dar'ya and its tributaries but several "terminal" rivers that disappear in the deserts (Micklin 2000, pp. 6–7). The Amu is "exotic," which, hydrologically means that essentially all its flow originates in the well-watered Pamir Mountains, but that this flow is substantially diminished by evaporation, transpiration from vegetation growing along its banks, and bed exfiltration as the river passes across the Kara–Kum Desert to the Aral Sea. The Amu Delta accounted for very large flow losses owing to evaporation and transpiration. Prior to the Aral Sea's modern desiccation, average annual inflow of the river to it decreased to 40 km^3 from the 62 km^3 coming out of the Pamir Mountains on to the desert plain of Turkmenistan.

The Syr Dar'ya flows from the Tian Shan Mountains, located to the north of the Pamirs. The melt of glaciers and snowfields are its main source of water. Its drainage basin covers 462,000 km^2. With a length of 3078 km, it is longer than the Amu (Micklin 2014a). Average annual flow of the Syr at 37 km^3, is considerably less than that of the Amu. Prior to the 1960s, flow diminution was substantial during its long journey across the Kyzyl-Kum Desert with less than half (around 15 km^3 on an average annual basis) of the water coming from the mountains reaching the Aral Sea.

4.3.3 Physical Characteristics of the Aral Sea Before the Modern Desiccation

The Aral Sea lies at the bottom of the Turan Depression by the eastern edge of the Ust-Urt Plateau (Bortnik and Chistyayeva 1990, p. 6). Its name derives from the word Aral, which means "island" in the Turkic languages of Central Asia. It may have been thusly named because it was an "island of water" in the vastness of the

Central Asian deserts. The name may also be connected to the many islands present in the sea prior to its modern desiccation (Ashirbekov and Zonn 2003, p. 6).

The Aral Sea occupies the lowest part of a vast erosional-tectonic hollow of middle Cenozoic age (Micklin 2016). It is geologically young, having arisen at the end of the Quaternary period, coincident with the last glacial epoch about 20,000 years BP (Boomer 2012; Burr et al. 2019; Krivonogov 2014). The lake's level in 1960 was 53.4 m with an area of 67,499 km^2, making it the world's fourth largest lake in extent at that time after the Caspian Sea in Eurasia (371,000 km^2), Lake Superior in North America (82,414 km^2) and Lake Victoria in Africa (69,485 km^2; Micklin 2014a; Table 4.1). The Aral in 1960 had a maximum depth of 69 m, average depth of 16 m, volume of 1089 km^3, and shoreline stretching for more than 4430 km. More than 1100 islands, with an aggregate area of 2235 km^2 dotted the sea. The largest were Kok-Aral (311 km^2), Barsakelmes (170 km^2) and Vozrozhdeniya (170 km^2; Kosarev 1975, p. 23).

The Aral was divided into a so-called "Small Sea" (or "North Aral") on the north and "Large Sea" (or "South Aral") to the south, which were connected by the Berg Strait. The Small Aral had an area of 6118 km^2, volume of 82 km^3, maximum depth of 29 m and average depth of 13.4 m (Table 4.1). It consisted of a deeper central basin and several shallower gulfs. The largest town and most important port and fishing center (Aral'sk) was situated at the northern end of the Gulf of Saryshaganak.

The Large Aral had a considerably greater surface area and volume (61,381 km^2 and 1007 km^3). It was divided into two basins by a north-south stretching underwater ridge that protruded through the surface to form a chain of small islands, the largest of which was named Vozrozhdeniye ("Resurrection"). This island became famous, perhaps better to say "infamous" as the location of the USSR's most important, super-secret testing grounds for biological weapons. The Eastern Basin had an area of 47,461 km^2 and the Western Basin 13,920 km^2. However, the former was shallow (maximum depth of 28.4 m and average depth of 14.7 m) whereas the Western Basin was considerably deeper with a maximum depth of 69 m and average depth of 22.2 m (Micklin 2014a).

The estimated average annual water balance for the Aral Sea for 1911–1960 (considered the quasi-stationary period for the Aral's level) is below (Bortnik and Chistyayeva 1990, Table 4.1, p. 36, Fig. 2.5, p. 20, pp. 34–39).

1. **Gain**: river inflow (56 km^3) + sea-surface precipitation (9.1 km^3) = 65.1 km^3
2. **Loss**: sea-surface evaporation = 66.1 km^3
3. **Volume change** = (−1.0 km^3)

The main elements determining the Aral's level, area, and volume were river inflow and surface evaporation, with sea-surface precipitation playing a secondary role on the gain side of the balance. There was also a net groundwater inflow, but it was believed small (up to 3.4 km^3) and ignored in calculating the sea's water budget.

The Aral Sea was brackish with an average salinity around 10 g/l, slightly less than one-third that of the open ocean. Salinity was lower than the average near the entrance of the two main rivers, particularly during peak-river inflow in spring/early summer when it could fall below 4 g/l near the mouth of the Amu. High salinity levels (17–18 g/l) were reached during summer and winter in the gulfs of the east and southeast part of the Large Aral owing to high rates of evaporation during summer and ice formation (which concentrates salts in the remaining water volume thus raising salinity) in winter (Kosarev 1975, p. 228). Levels of salinity in isolated portions of the Gulf of Saryshaganak could reach 80–150 g/l.

Researchers considered Aral water exceptionally transparent (Zenkevich 1963, p. 510). On average, a Secchi disk, used to determine this, could be seen at 8.2 m, with maximum readings of 23.5 m in the central part of the Large Aral, 24 m in the Small Aral, and 27 m in Chernishov Gulf at the northern end of the Western Basin of the Large Sea (Bortnik and Chistyayeva 1990, p. 95).

Maximum water temperatures were reached in July and August, when the surface layer along the shoreline could reach 29 °C and 24–26 °C in the open sea (Bortnik and Chistyayeva 1990, pp. 43–49; Zenkevich 1963, Table 236, p. 510). As heating of the water mass progressed, a significant thermocline and temperature discontinuity formed in the deep Western Basin of the Large Sea, where the surface temperature would average around 24 °C while at depths below 30 m it would range from 2 to 6 °C. The shallower Eastern Basin of the Large Sea, on the other hand, had relatively uniform temperatures throughout the water column, with a difference of only a few degrees between the surface and bottom.

Vertical stability was primarily determined by temperature and only in the southern parts of the sea by both temperature and salinity (Bortnik and Chistyayeva 1990, pp. 82–85; Kosarev 1975, pp. 237–240, 247–260). Intensive heating of the Aral's surface waters in spring and summer led to the formation of a stable surface layer (down to the temperature discontinuity) and a stable bottom layer below that. Hence mixing between the surface and bottom layers was prevented. With the onset of cooling in fall, the surface to bottom temperature gradient weakened considerably, sometimes turning negative, leading to greatly diminished stability and convective mixing. During winter, ice formation and the resulting salt concentration increased surface water layer density and further enhanced convective mixing. The fall-winter convective mixing, which affected all parts of the sea and encompassed all water layers was considered the most important process determining the hydrologic structure of the Aral waters, particularly for the deeper parts of the sea.

4.3.4 Biology of the Aral Sea Prior to the Modern Desiccation

The aboriginal fauna of the Aral Sea was represented by more than 200 species of free-living invertebrates (Mordukhai-Boltovskoi 1974; Plotnikov 2016), over 200 species of parasitic invertebrates (Osmanov et al. 1976) and 20 species of fish (Nikolsky 1940). Among the species of free-living invertebrates, the inhabitants of freshwater, brackish-water and saline continental waterbodies composed 78% Ponto-Caspian species (species originating from the Black and Caspian seas that are relict fauna of the ancient Paratethys Ocean—these seas are its remnants). In modern times, most

Ponto-Caspian species are endemics living primarily in the Black Sea, Sea of Azov and Caspian Sea; only a few lived in the Aral prior to its modern drying.

The abundance of only a few species of free-living invertebrates in the Aral was high. Among the zooplankton, the most numerous copepod *Arctodiaptomus salinus* represented fauna of continental saline water bodies. Ponto-Caspian cladocerans *Cercopagis pengoi aralensis, Evadne anonyx, Podonevadne camptonyx* and *P. angusta* were also numerous. Among the freshwater euryhaline Cyclopoida, the most common was *Mesocyclops leuckarti*. The highest diversity of zooplankton was in the freshened parts of the sea due to freshwater species (Andreev 1989; Kortunova 1975).

Freshwater and Caspian species prevailed in the aboriginal benthic fauna of the Aral Sea. Its basis was mollusks, oligochaetes, higher crustaceans and larvae of Chironomidae (Mordukhai-Boltovskoi 1974). Bivalve mollusks were numerous, including *Dreissena* spp. and *Hypanis* spp., oligochaetes *Nais elingius* and *Paranais simplex*, ostracod *Cyprideis torosa*, amphipod *Dikerogammarus aralensis*, larval chironomids *Chironomus behningi* and caddis flies *Oecetis intima.* Mollusks accounted for 63% of zoobenthos biomass, and chironomid larvae for 33% (Karpevich 1975).

Almost all aboriginal ichthyofauna of the Aral Sea consisted of generatively freshwater (usually breeding in freshwater) species. In it, 60% were cyprinids (Ermakhanov et al. 2012; Nikolsky 1940). The best places for spawning were freshened bayside deltas, rivers and deltaic lakes. All aboriginal fish, except for the stickleback, *Pungitius platygaster aralensis*, whose main food was zooplankton, were benthophagous (feeding on bottom-dwelling organisms) or predators (Nikolsky 1940).

Vegetation and flora of aquatic and coastal-aquatic plants were monotonous and poor by species. Only two species and two plant communities dominated—reeds in coastal shallow waters and eelgrass (*Zostera*) at depths of up to 11 m on silty sands. In the central part on muds charophytes were found at depths of 11–22 m, and in the shallows—watermilfoil *Myriophyllum* sp. and pondweed *Potamogeton perfoliatus*. By the 1960s in the Aral Sea flora 24 species of higher plants, six species of charophytes and about 40 other species of macroalgae were known. Aquatic vegetation formed zones of helophytes (plants rooted in the bottom, but with leaves above the waterline) and zones of hydrophytes (plants that complete their entire life cycle submerged, or with only their flowers above the waterline). Along the shore reed-beds of *Phragmites australis* dominated. In the northern part of the sea beyond the reed zone often was a zone of bulrush *Scirpus kasachstanicum*. Other helophytes did not form large thickets (Plotnikov et al. 2014a).

Communities of hydrophytes presented diverse associations that formed vast underwater meadows. Extensive deep-water thickets of Charophyta existed at the beginning of the twentieth century but were absent by the 1950s. In their place, yellow-green algae *Vaucheria dichotoma* were found. In freshened bays, the basis of macrophytobenthos was higher flowering plants. In closed saline bays and inlets charophytes dominated (Plotnikov et al. 2014a).

4.4 Human Impact on the Aral in the Modern Era (Mainly Post 1960)

Humans have affected the Aral Sea in important ways for millennia as has been discussed above in Sect. 4.2 on Basin History. However, the human touch has been especially dramatic, powerful and far reaching since the 1960s. The subsections that follow provide a description and discussion of the major changes that have and are unfolding.

4.4.1 The Changing Physical Character of the Sea and Surrounding Region

The physical character of the Aral has undergone unprecedented changes since 1960. Sadly, these have not been as well studied, documented and analyzed as one would wish as the well-developed and pursued research and monitoring effort on the Aral Sea in the years 1925–1941 and after World-War II faltered in the 1970s and 1980s as the sea shrank and shallowed at a rapid pace, hydrometeorological stations closed and cruises by research ships became more difficult and infrequent (Bortnik and Chistyayeva 1990). The situation worsened after the Soviet Union collapsed at the end of 1991, and the Aral Sea became part of the two new riparian countries Kazakhstan and Uzbekistan. But since the mid-1990s matters have somewhat improved as research on the Aral has been renewed and revitalized by funding from national, regional, and international organizations and conducted by both national and international research teams.

Since the early 1960s, the sea has steadily shrunk and salinized (Table 4.1). The main causative factor has been expanding irrigation that greatly diminished discharge from the two tributary rivers Amu Dar'ya and Syr Dar'ya (Micklin 2014c, 2016). Irrigation has been practiced in the Aral Sea Basin for at least three millennia. Until the 1960s this did not substantially diminish inflow to the sea, owing to substantial return flows from irrigated fields to the Amu Dar'ya and Syr Dar'ya and other compensatory factors such as reduced losses to transpiration from phreatophytes (water-loving plants) along the lower courses of the rivers and in the deltas as well as lowered evaporation from reduced spring flooding in the deltas of these rivers. However, growth in the irrigated area from around 5 million ha in 1960 to 8.2 million by 2010 pushed irrigation development beyond the point of sustainability reducing or eliminating these compensatory effects and leading to a marked reduction of river discharge to the Aral.

River inflow to the Aral began declining in the 1960s and accelerated in the 1970s and 1980s. More precipitation in the mountains and some reduction in irrigation withdrawals increased river discharge during the 1990s and reduced the water balance deficit, slowing the sea's recession (Cretaux et al. 2019). There was a severe drought in 2000–2001 and, consequently, river inflow was very low (Micklin 2014c). Higher inflows on the Amu Dar'ya characterized the period 2002 through 2005 and water balance deficits for the Large Aral were significantly lessened. However, dry conditions returned from 2006 to the middle of 2009 resulting in rapid drop of lake level, reduction of surface area and volume accompanied by rising salinity (Table 4.1).

The Aral separated into two water bodies in 1987–1989—a "Small" Aral Sea in the north, also known as the North Aral Sea and a "Large" Aral Sea in the south (Micklin 2014e). The Syr Dar'ya flows into the Small Aral, and the Amu Dar'ya into the Large Aral. A channel formed connecting the two lakes, allowing water to flow from the former to the latter. Local authorities constructed an earthen dike in 1992 to block outflow to raise the level of the Small Sea, lower salinity, and improve ecological and fishery conditions. This makeshift construction breached and was repaired several times. In April 1999 after the level of the Small Aral had risen well over 43 m, the dike was overtopped, breached and destroyed during a wind storm, with the death of two people.

Under study and design since 1993, the World Bank and the Government of Kazakhstan funded construction of an engineeringly sound 13-km earthen dike with a concrete, gated outflow-control structure to regulate the flow from the Small to Large seas. Construction was completed from 2003 to 2005 (Aladin 2014; Dam of the North Aral Sea and the hydrocomplex Aklak 2017; Micklin 2014e, 2016; World Bank 2001, 2014). The structure raised and stabilized the level of the Small Aral at near 42 m above the Baltic Sea in early 2006. The total cost of this project was 23.2 million USD, but other related infrastructure projects along the Syr Dar'ya added another 62.6 million USD to project costs for a total of 85.8 million. The World Bank provided a loan of 64.5 million and the Government of Kazakhstan funded the remaining 21.3 million.

The desiccation of the Aral Sea has had severe negative impacts (Micklin 2007, 2014d, 2014f, 2016; Micklin and Aladin 2008). The vibrant commercial fishing industry ended in the early 1980s as the brackish-water indigenous species that provided the basis for the fishery disappeared owing to their inability to adjust to rising salinity. The more salinity-tolerant Black Sea flounder (*Platichthys flesus lulscus*) was introduced to the Aral in the 1970s. It flourished in the Small Aral and provided a sizable non-commercial catch. But it disappeared from the Large Aral as salinity rose. Tens-of-thousands of people were thrown out of work because of the loss of the commercial fishery and associated activities. Employment in these occupations today, although rising owing to the partial recovery of the North Aral Sea (discussed below), remains only a fraction of what it was.

The level-stabilization project reinvigorated the fishery in the North Aral by lowering average salinity below the 10 g/l level of the early 1960s (Ermakhanov et al. 2012; Plotnikov et al. 2014a, b; Toman et al. 2015; White 2016). This has allowed the return and flourishing of commercially valuable indigenous species such as the Sudak or Pike-perch (*Lucioperca lucioperca*), Sazan (*Cyprinus carpio*) and Lyeshch or Bream (*Abramis brama orientalis*), types of carp, Plotva or Roach (*Rutilis rutilis aralensis*) as well as several other species. The North Aral catch rose from 695 metric tons in 2005 to 6000 metric tons in 2016 (Micklin et al. 2018). Unfortunately, millions of fish are being carried over the discharge gates of the Kok-Aral dam and ending up in the Central Aral where they perish from higher salinity and temperatures and lower dissolved oxygen than in the North Aral.

The formerly biologically diverse and rich ecosystems of the deltas of the Amu Dar'ya and Syr Dar'ya have suffered considerable harm from reduced river flows, elimination of spring floods, and declining groundwater levels leading to spreading desertification (Micklin 2000, pp. 13–23; 2014d; Novikova 1999). Salts have formed pans (solonchak) on the surface where practically nothing will grow. Expanses of unique Tugay forests along the main and secondary watercourses have drastically shrunk. Desiccation of the deltas has significantly diminished the area of lakes, wetlands, and their associated reed communities. These changes caused the number of species of mammals and birds to drop precipitously. Strong winds blow sand, salt and dust from the dried bottom of the Aral Sea onto surrounding lands causing harm to natural vegetation, crops, and wild and domestic animals (Indoitu et al. 2015; Novikova 1996, 1999). As most of the sea has dried and more of the bottom has been exposed, dust storms with entrained salts in particulate and aerosol (a colloid of fine solid particles or liquid droplets in air) form have become more frequent and intense, covering at times more than 100,000 km^2 and extending downwind more than 500 km.

Owing to the sea's shrinkage, climate has significantly changed in a band up to 100 km wide along the former shoreline in Kazakhstan and Uzbekistan (Micklin 2014a). Summers have warmed and winters cooled, spring frosts are later and fall frosts earlier, humidity is lower, and the growing season shorter. The population living around the sea suffers acute health problems (Micklin 2007, 2014d). Some of these are direct consequences of the sea's recession such as respiratory and digestive afflictions from inhalation of blowing salt and dust and poorer diets from the loss of Aral fish as a major protein source.

The darkest consequence of the Aral's modern shrinkage is what has happened to Vozrozhdeniya Island (Micklin 2007, 2014d). The Soviet military in the early 1950s selected this, at the time, tiny, isolated island in the middle of the Aral Sea, as the primary testing ground for its super-secret bioweapons program. This program stopped with the collapse of the USSR in 1991. As the sea shrank, Vozrozhdeniya grew and in 2001 united with the mainland to the south as a peninsula. There was concern that weaponized organisms survived decontamination measures by the departing Russian military and could escape to the mainland via infected rodents or that terrorists might gain access to them. The U.S. government worked with the Government of Uzbekistan to ensure the destruction of any surviving weaponized pathogens in 2000.

By September 2009, the Aral had shrunk to a small remnant of its 1960 size and separated into four parts (Table 4.1). The dike and dam constructed to regulate flow from the Small to Large Aral had raised and stabilized the level of the former leading to greatly improved ecological conditions and a revitalized fishery. The Large Sea in the south was not so fortunate. The level of the deeper Western Basin (max. depth 69 m in 1960) had fallen 26 m and salinities reached over 100 g/l, creating conditions where fishes could not survive. The Eastern Basin became a shallow pond with salinity likely above 150 g/l. It appeared that it would dry up completely during the summer of 2010. However, a heavy flow year on the Amu in 2010 partially revitalized the basin. In summer 2014 the Eastern Basin dried completely for probably the first time in 600 years. Since then, the Eastern Basin has expanded and shrunk on a seasonal rhythm related to the annual hydrologic flow pattern combined with longer-term cycles of wet and dry years in the Aral Sea Basin. Table 4.1 and Fig. 4.2 show the physical characteristics of the Aral in June 2018.

The physical and chemical character of the Aral Sea has also changed dramatically as the unitary water body shrunk, salinized and separated into multiple lakes with different hydrologic parameters. A dearth of primary field work has made a comprehensive picture of changes impossible. But according to scientists from the Shirshov Institute of Oceanology in Moscow who along with an international team studied the Aral (mainly the deeper Western Basin, of the Large Aral) in 12 expeditions between 2002 and 2010 (Zavialov et al. 2012, p. 212.):

> The shallowing of the sea and accompanying changes of its morphometric characteristics, mainly the catastrophic salinization led to deep transformations of its physical and chemical regimes, all processes determining its condition and dynamics – from large-scale circulation to turbulent mixing and from the variability of the ionic-salt composition to the energy exchange with the atmosphere.

Among other findings, instrumental measurements found a strong vertical stratification in temperature, salinity and density that resulted in hydrogen sulfide contamination of the deeper portions of the basin owing to lack of exchange with upper layers.

Other international teams of investigators concentrated mainly on the North Aral in expeditions in 2005, 2007 and 2011, chiefly focusing on measuring salinity, dissolved oxygen, and temperature at shoreline and shallow water locations (Micklin 2014e). These showed that salinities around this water body were steadily falling and were in the range favorable for the native brackish-water fish species. High dissolved oxygen levels were also found – another positive sign for fish and other desirable aquatic species.

4.4.2 Changes in the Biological Character of the Sea

By the 1950s, it became clear to fishery experts that given irrigation-expansion plans for the basin of the Aral Sea, river runoff would be significantly reduced, and the salinity of the Aral Sea would increase. It was anticipated the freshwater and brackish-water species that formed the basis of the biota would gradually disappear and the sea would lose its fishery importance. Therefore, it was necessary in advance to form a salt-tolerant biota by acclimating suitable species. First, it was necessary to create a phytoplankton, then zooplankton and then zoobenthos food base, and only then introduce fish, but only benthophagous and predators. However, this was not done carefully, and several unwanted species of invertebrates and fish were introduced (Karpevich 1975).

In 1954–1956 during unsuccessful introduction of the mullets *Liza aurata* and *L. saliens* from the Caspian Sea, unwanted non-commercial fishes were introduced:

atherine *Atherina boyeri caspia*, pipefish *Syngnatus abaster caspius* and six species of gobies: *Knipowitschia caucasicus, Neogobius fluviatilis, N. melanostomus, N. syrman* and *N. kessleri*, and *Proterorhinus marmoratus* (Ermakhanov et al. 2012). In addition, the shrimp *Palaemon elegans* was brought to the sea (Karpevich 1975). This naturalized shrimp became a competitor with the aboriginal amphipod *Dikerogammarus aralensis* and even ate it. This, but not salinization, caused the disappearance of the amphipod by 1973 (Aladin and Kotov 1989; Aladin and Potts 1992; Andreeva 1989; Mordukhai-Boltovskoi 1972).

The most profound consequences were caused by the planktophage Baltic herring *Clupea harengus membras* introduced in 1954–1956 (Karpevich 1975). Because of the introduction of this alien species, as well as atherine and gobies, the abundance and biomass of zooplankton crustaceans sharply decreased. This led to the mass death of herring and atherine from starvation in ensuing years (Kortunova 1975; Kortunova and Lukonina 1970; Osmanov 1961).

The first purposely introduced invertebrates in the Aral Sea (1958–1960) were Ponto-Caspian misyds—a valuable food for fish and able to tolerate salinity of 17–20 g/l (Karpevich 1960). Of the three species—*Paramysis lacustris, P. intermedia* and *P. baeri*—taken from the delta of the Don, only the first two were naturalized. A fourth species, *Paramysis ullskyi*, migrated independently from reservoirs on the Syr Dar'ya where it was introduced earlier (Kortunova 1970). The next planned invertebrate species to be introduced (1960–1961) was the euryhaline polychaete worm *Hediste diversicolor* from the Sea of Azov. This worm quickly naturalized and settled (1973–1974) the whole Aral (Karpevich 1975). The euryhaline bivalve *Syndosmya segmentum* was introduced from the Azov Sea in 1960–1963. By 1973, it settled all over the sea and became the main component of zoobenthos (Karpevich 1975).

The euryhaline marine planktonic copepod *Calanipeda aquaedulcis* was introduced from the Sea of Azov in 1965 and 1970. By 1971 this crustacean became one of the species dominating the zooplankton (Andreev 1989; Karpevich 1975). By 1974, *C. aquaedulcis* displaced *Arctodiaptomus salinus* (Mordukhai-Boltovskoi 1972) and the latter species became extinct in the Aral Sea.

In the deltaic areas of the Syr Dar'ya and Amu Dar'ya, commercial freshwater fishes were acclimatized in 1958–1960: macro-phytophage grass carp *Ctenopharyngodon idella*, phyto-planktophague silver carp *Hypophtalmichthys molifrix*, zooplanktophage bighead carp *Aristichtys nobilis* and introduced inadvertently benthophage black carp *Mylopharyngodon piceus*. Except for bighead carp, all were successfully naturalized and became commercially important (Karpevich 1975).

Another cause of changes in the composition of the biota has been anthropogenic change in the hydrological regime of the Aral Sea, its desiccation and salinization. During the period 1961–1970 the Aral Sea desiccation and increase of its salinity occurred very slowly. Over these 10 years salinity increased only by 1.5 g/l, and by 1971 it reached 11.5 g/l. At this early stage of the Aral Sea's modern regression, changes in the species composition of its fauna were mostly the result of the introduction of new fishes and invertebrate species and to a lesser extent were the result of increasing salinity.

Throughout the period 1961–1971 the species composition of larval Chironomidae fauna in the Aral Sea remained unchanged. A small increase in salinity of the Aral Sea caused a very significant reduction in the total number of bivalves *Dreissena* after 1964. It should be noted that this slight salinization was unfavorable only for *D. polymorpha aralensis* and *D. p. obtusecarinata* but not for *D. caspia pallasi*, more resistant to salinity and not numerous (Andreeva 1989).

In the 1970s the rate of Aral Sea desiccation and salinity rise increased. Since that time, the main factor influencing the fauna of the Aral Sea has been continued increase in salinity of its waters. In 1971–1976 invertebrate fauna of the Aral Sea passed through the first crisis period caused by salinization over 12–13 g/l (Plotnikov et al. 1991). Increasing salinity became an obstacle for further existence of freshwater species.

During this first crisis period the most species-rich, freshwater component of fauna disappeared. Only eight species of rotifers remained. From them only a few species of the genus *Synchaeta* were common and numerous. With increasing salinity *Ceriodaphnia reticulata* and *Alona rectangula* disappeared by 1974. By 1975, of Cladocera species in the fauna only representatives of the Ponto-Caspian fauna *Evadne anonyx, Podonevadne camptonyx, P. angusta* and *Cercopagis pengoi aralensis* remained. Instead of freshwater *Mesocyclops leuckarti* the most numerous species became euryhaline marine *Halicyclops rotundipes aralensis* (Andreev 1989).

All mollusks from the genus *Hypanis* – *H. vitrea bergi, H. minima minima* and *H. m. sidorovi* disappeared after 1977. Further increases in salinity affected the sea forms of *Dreissena* inhabiting the sea differently. It was unfavorable for *Dreissena polymorpha aralensis* and *D. p. obtusicarinata*, but favorable for *D. caspia pallasi* that tolerates higher salinities. The growth of salinity led to the reduction in the area and number of the bivalve *Cerastoderma rhomboides rhomboides*, and conversely was favorable for *C. isthmicum*. After 1978 *C. rhomboides rhomboides* was no longer found and *C. isthmicum* took its place. Rising of salinity above 12–14 g/l favored the mollusk *Syndosmya segmentum*. The abundance of the halophilic gastropods *Caspiohydrobia* spp. began to grow. Since 1973, when the salinity of the sea reached 12 g/l, Oligochaeta were no longer found. By 1974 most of larval Chironomidae species had disappeared and only *Chironomus salinarius* and *Ch. halophilus* remained in the salinized bays (Andreeva 1989). By 1980, the leading forms of zoobenthos were *Syndosmya segmentum, Cerastoderma isthmicum, Hediste diversicolor* and *Caspiohydrobia* spp. After 1977 when the salinity had reached 15 g/l, all mysids were absent from the sea but were preserved in the rivers and their deltas (Andreeva 1989).

As a result of this first crisis, freshwater and brackish-water species of freshwater origin disappeared from the free-living invertebrate fauna of the Aral Sea. This provided an advantage to Caspian and marine euryhaline species and halophilic species (Andreev 1989). Despite the continuing salinity growth, the first crisis period for the free-living invertebrate fauna of the Aral Sea transitioned into a period of relative stability between 1977 and 1985.

By 1987 salinity of the Aral Sea rose to 27 g/l. Crossing this boundary meant free-living invertebrate fauna of the Aral Sea entered the period of the second crisis during which occurred the next reduction of species diversity (Plotnikov et al. 1991).

Because of this all the Ponto-Caspian cladocerans of the family Podonidae disappeared by 1990. After the second crisis period, of the native zooplankton species in the sea, remained only the rotifers *Synchaeta* spp., *Notholca squamula, N. acuminata, Keratella quadrata, Brachionus plicatilis, B. quadridentatus* and perhaps a few species of copepods (*Calanipeda aquaedulcis* and *Halicyclops rotundipes aralensis*), as well as several species of Harpacticoida. Among aboriginal species in the benthic fauna only the mollusks *Cerastoderma isthmicum, Caspiohydrobia* spp. and ostracod *Cyprideis torosa* survived. Among introduced species only the polychaete *Hediste diversicolor*, mollusk *Syndosmya segmentum*, crab *Rhithropanopeus harrisii tridentata* and the shrimp *Palaemon elegans* remained. After this crisis period, in the free-living invertebrate fauna of the Aral Sea were marine species and euryhaline species of marine origin as well as representatives of euryhaline halophilic fauna of inland saline waters. This crisis was followed by a new period of relative stability (Plotnikov 2016).

Soon after the separation of the Aral Sea into two parts (North or Small Aral and South or Large Aral) in 1987–1989, when the decrease in salinity in the Small Aral began, reappeared *Podonevadne camptonyx* from dormant eggs. In 1999 were larvae of Chironomidae found in the benthos again (Aladin et al. 2000).

Salinity in the separated Large Aral grew and it was transformed into a hypersaline water body. In the mid-1990s when salinity exceeded 47–52 g/l came another period of crisis. A rapid change in the composition of all the Large Aral biota occurred. By 2004, when salinity became 100–105 g/l, most invertebrates disappeared. Only rotifers *Hexarthra fennica* and *Brachionus plicatilis*, ostracod *Cyprideis torosa*, Turbellaria *Mecynostomum agile*, and some species of Foraminifera, Nematoda and Harpacticoida remained. But by this time in the Large Aral appeared some halophilous invertebrates such as ciliates *Fabrea salina* and *Frontonia marina*, copepod *Apocyclops dengizicus*, ostracod *Eucypris mareotica*, brine shrimp *Artemia parthenogenetica*, and larval chironomids *Beotendipes noctivaga* (Aladin and Plotnikov 2008; Mokievsky and Miljutina 2011).

4.4.2.1 Aral Fishery

Freshened deltaic bays and lakes were the best places for fish spawning (Bervald 1964). Decline in the Aral Sea water level, salinization and drying of deltas significantly altered the living conditions for fishes, especially for their reproduction (Ermakhanov et al. 2012). This sharply affected the state of commercial fish populations. The first signs of the negative impacts of salinization on the ichthyofauna occurred in the mid-1960s, as salinity reached 12–14 g/l. Salinity increased faster in shallow spawning areas than in the open sea, exceeding 14 g/l in 1965–1967. At the end of the 1960s, spawning conditions for semi-anadromous fishes significantly worsened.

Beginning in 1971, when average salinity in the open sea reached 12 g/l, the first signs of negative effects of salinity on adult fishes appeared. The rate of growth slowed for many fish species, with their numbers falling sharply. By the middle of the

1970s, when the average salinity of the sea exceeded 14 g/l, the natural reproduction of Aral fishes was destroyed. As a result, in the second half of the 1970s, recruitment of new members was absent for the populations of many fish species. By 1981, when salinity exceeded 18 g/l, the Aral Sea had completely lost its fishery. The remaining ichthyofauna consisted of nine-spined stickle-back, as well as gobies, atherine and Baltic herring. Aboriginal commercial fishes survived only in the Syr Dar'ya and Amu Dar'ya rivers and deltaic lakes (Ermakhanov et al. 2012).

To restore the Aral Sea fishery, the flounder-gloss *Platichthys flesus luscus* from the Sea of Azov was successfully introduced in 1979–1987 (Lim 1986). This marine fish can reproduce at salinities from 17 to 60 g/l. A fishery was established by the early 1990s. Acclimatized flounder-gloss remained the only commercial fish species in the Aral Sea from 1991 to 2000 (Ermakhanov et al. 2012).

After construction of the Kok-Aral dike, freshening of the Small Aral Sea began, and the zone with lower salinity began increasing. Aboriginal fish, including Aral roach, bream, carp, zander, and asp began to be found again in the Small Aral Sea after many years. The fish fauna expanded their spawning and feeding zones to almost the entire area of the Small Aral Sea, except Butakov Bay, where the salinity remained too high (Ermakhanov et al. 2012). By the end of the 1990s, the salinity of the Large Aral reached 60–70 g/l, resulting in the complete disappearance of all fish. It has become a lake without fishes since that time (Ermakhanov et al. 2012).

4.4.2.2 Aral Aquatic Vegetation

Regression and salinization of the Aral Sea caused the destruction of the majority of vegetational complexes. Freshwater and freshwater-brackish water submerged higher plants were not able to survive. During the 1970s, the species composition was depleted and a few euryhaline species became dominant. Reed-beds in the 1980s disappeared completely owing to rising salinity. By the end of the 1980s there was only *Ruppia* spp. tolerating salinity of 50 g/l (Plotnikov et al. 2014b).

In the Small Aral in the 1990s, the bulk of macrophytobenthos production belonged to the macroalgae *Chaetomorpha linum, Cladophora glomerata* and *Cl. fracta*. Macrophyte communities were formed of flowering plants *Phragmites australis, Ruppia cirrhosa, Ruppia maritima, Zostera noltii*, and charophytes *Lamprothamnium papulosum* and *Chara aculeolata*. Near the Syr Dar'ya delta reed-beds began to form. At present, the salinity of the Small Aral Sea continues to gradually decrease, and this water body is being settled widely by species of hydrophytes and helophytes coming from other continental brackish water bodies (Plotnikov et al. 2014b).

In the hypersaline Large Aral microphytobenthos (diatoms and blue-green algae) dominates. Among macrophytobenthos only *Cladophora* and *Vaucheria* were found. From higher plants sterile specimens of *Ruppia* sp. were found (Plotnikov et al. 2014b; Zavialov et al. 2012).

4.5 Future of the Aral Sea

The view by some that the Aral Sea is destined to dry up completely in the twenty-first century is false (Micklin 2010, 2014f). Even if river inflow from the Amu Dar'ya and Syr Dar'ya were reduced to zero, a very improbable scenario, there would still be residual input of irrigation drainage water, groundwater, and snow melt and rain that would maintain at least two substantial lakes: the deeper western (Shevchenko Gulf) and deeper parts of the central Small Aral Sea in the north, and the Western Basin of the Large Sea in the south. These lakes would be hypersaline and of little ecological or economic value, except, perhaps for the commercial production of brine shrimp (*Artemia*) eggs.

Return of the sea to its early 1960s state is possible but very unlikely in the foreseeable future. Based on a spreadsheet fill-time model developed by Micklin (2016), it would necessitate restoring average annual river inflow to 56 km^3 and take more than 100 years. Restoration would follow a logistic curve: rapid at first as inflow greatly exceeded net evaporation, then slowing and approaching zero as net evaporation grew and approached inflow. However, the sea would reach an area of 60,000 km^2 (91% of stability area) and level of 50 m in 45 years. Estimated average annual inflow to the entire sea from 1992 to 2011 is 8.8 km^3 − 16% of what would be needed for realization of this scenario. The only realistic means for substantially increasing inflow to the Aral is reducing the use of Aral Sea Basin river flow for irrigation as it accounts for 92% of water withdrawals (Micklin 2014b, c). Irrigation efficiency in this region is low and certainly could be raised to free more water for the Aral. But the cost would be huge and the time to implement long. Hence, to free the large amount of water needed would also require a substantial reduction in the area irrigated. Given the dependence of Aral Sea Basin nations' economies on this activity, such a reduction is improbable, if not impossible, any time soon.

It is engineeringly possible to bring water to the Aral Sea from outside Central Asia (Micklin 2014g). The Soviet government developed plans in the 1960s and 1970s to divert up to 60 km^3 from the Siberian rivers Irtysh and Ob' to the Aral Sea Basin as the best means to solve regional water problems for the long-term. The initial phase (27 km^3) was near implementation when stopped in 1986 by Gorbachev, then head of the USSR Government and Communist Party. Attempts have been made to revitalize the project, but they appear futile.

4.5.1 Further Restoration of the North (Small) Aral

On the other hand, various partial rehabilitation scenarios for the Aral Sea hold considerable promise (Micklin 2014f, 2016). About 3.24 km^3/year, on average, is the inflow needed from the Syr Dar'ya to maintain the current nominal level of the Small Aral (42 m above the Kronstadt gauge) with an area of 3200–3300 km^2 and allow sufficient outflow through the Berg Strait (Kok-Aral) Dam to regulate salinity.

For 1992–2016 the average annual inflow was around 5.9 km^3 indicating there is more than enough water available to maintain the current stabilized hydrologic status of the Small Aral Sea.

The Kazakhstan Government, with World Bank support, is planning a second phase of the Small Aral restoration project (Micklin 2016; World Bank 2014). One of the two alternatives is to raise the level of water only in the Gulf of Saryshaganak, which extends northeast off the eastern part of the Small Sea, to 50 m above the Kronstadt gauge (Figs. 4.4 and 4.5). For this a new dike and dam, with an outflow structure and navigation locks for ingress and egress, would be necessary at the Gulf's mouth. Part of the flow of the Syr Dar'ya would be diverted northward via a canal into Saryshaganak to maintain its level. The gulf now converted into a reservoir with an area of 825 km^2, volume of 6.3 km^3, and average depth of 7.6 m, would be brought back very near the town of Aral'sk the former main port at the northern end of the Aral Sea. This would allow fishing vessels via a short canal direct access to the new and rebuilt fish processing plants in that town. Cost of this project is estimated at 150 million USD (World Bank 2014).

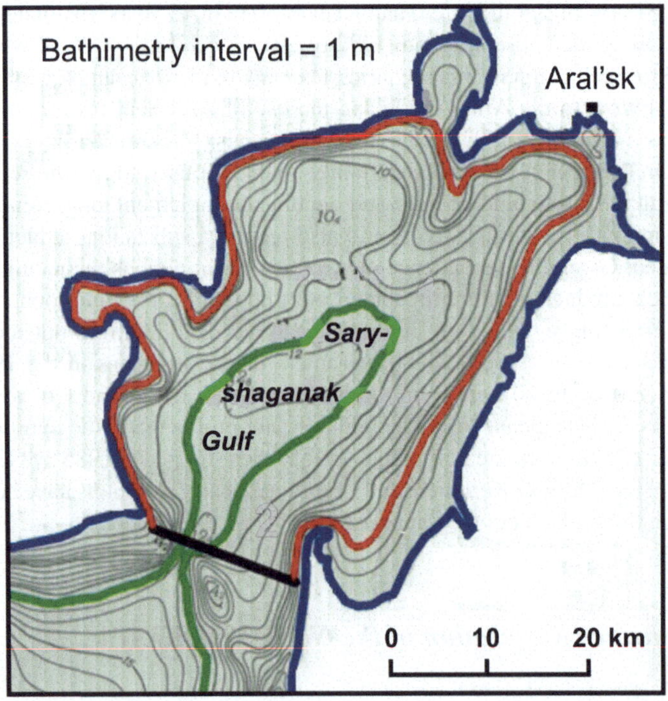

Fig. 4.4 Saryshaganak Gulf Reservoir Plan. Black line shows position of proposed dike and dam. Blue line is 1960 level (53 meters above Kronstadt gauge), red line is level at 50 meters as proposed in 2nd stage of Saryshaganak Plan, green line is 2019 level (42 m). Base map is digitized version of 1:500,000 bathymetric map of the Aral Sea produced by the Institute of Water Problems, Soviet Academy of Sciences in 1981. *Source* Micklin (2016)

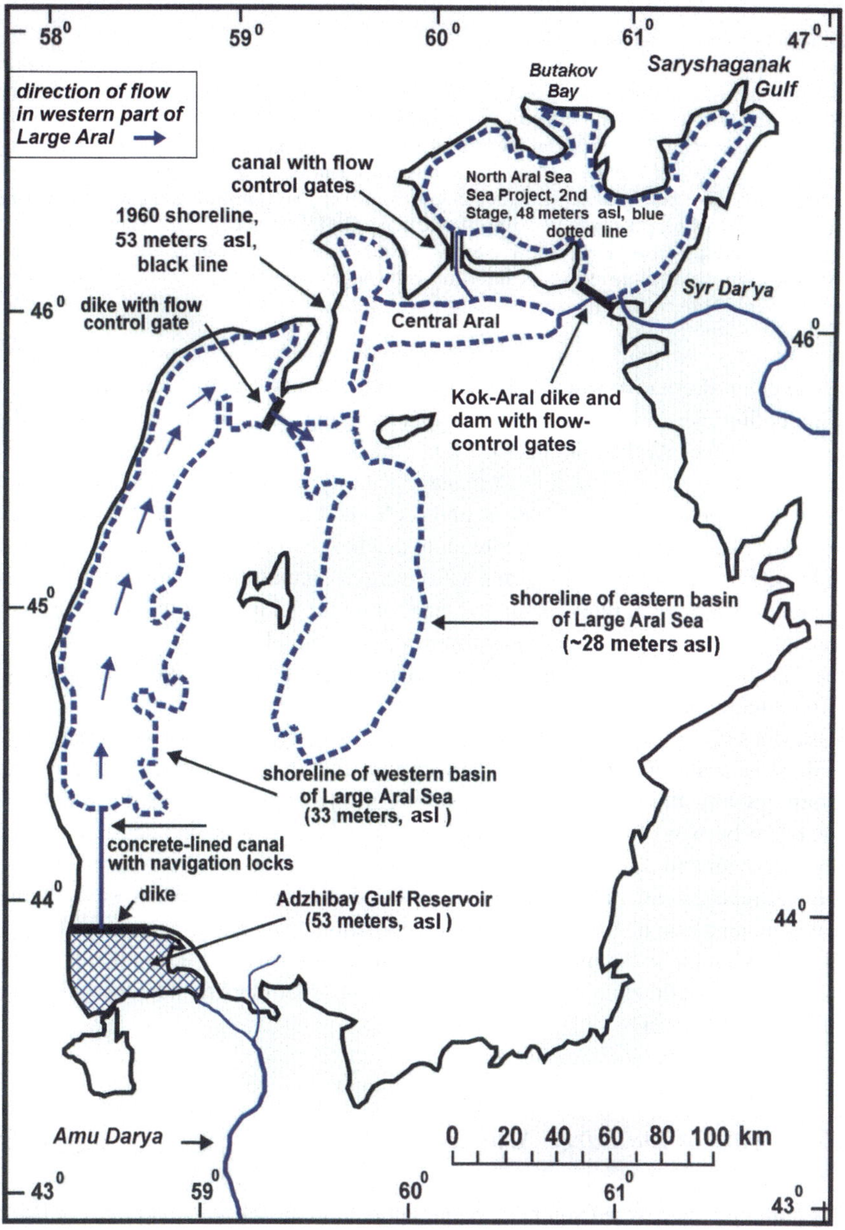

58⁰
59⁰
60⁰
61⁰
47⁰
Butakov Bay
Saryshaganak Gulf
direction of flow in western part of Large Aral
canal with flow control gates
North Aral Sea Sea Project, 2nd Stage, 48 meters asl, blue dotted line
1960 shoreline, 53 meters asl, black line
Syr Dar'ya
46⁰
dike with flow control gate
Central Aral
Kok-Aral dike and dam with flow-control gates
45⁰
shoreline of eastern basin of Large Aral Sea (~28 meters asl)
shoreline of western basin of Large Aral Sea (33 meters, asl)
concrete-lined canal with navigation locks
44⁰
dike
Adzhibay Gulf Reservoir (53 meters, asl)
Amu Darya
0 20 40 60 80 100 km
43⁰

◂**Fig. 4.5** Optimistic Scenario of the Future Aral Sea (after 2030). Legend (figures are average annual values). Small Aral Sea: level – 8 m, surface area – 4927 km^2, volume 54 km^3, river inflow – 5.0 km^3, net groundwater inflow – 0.1 km^3, outflow – 1.0 km^3, salinity – 6–7 g/l. Western Basin of Large Aral Sea: level – 3 m, surface area – 6200 km^2, volume – 85 km^3, river inflow – 6.4 km^3, net groundwater influx 2.0 km^3, outflow to Eastern Basin – 3.6 km^3, salinity steadily decreasing reaching 42 g/l by 2055 and 15 g/l by 2110. Eastern Basin of Large Aral Sea: level ~28.0 m, surface area ~3800 km^2, volume ~7.6 km^3, inflow from Western Basin Aral 3.6 km^3, inflow from Central Aral highly variable, salinity >200 g/l. Adzhibay Gulf Reservoir: level – 53 m, surface area – 1147 km^2, volume – 6.43 km^3, inflow – 8 km^3, outflow to Western Basin of Aral Sea – 6.6 km^3, salinity – 2 g/l. Level in meters above the Kronstadt gauge on the Baltic Sea near St. Petersburg, Russia. *Source* Micklin (2016)

The other alternative would rebuild the Kok-Aral dike and dam, raising the level of the entire lake to 48 m above the Kronstadt gauge (Fig. 4.5). At this level, the area of the Small Aral would be 4927 km^2. Inflow from the Syr Dar'ya to maintain such a level, including outflow to manage salinity, is 4.85 km^3. This alternative would likely provide more economic and ecological benefits than the Saryshaganak Reservoir plan but would also require more discharge from the Syr Dar'ya. The lake could be filled to 48 m in 17 years with average annual inflow of 5.0 km^3. After reaching design level, releases, on average, of about 1 km^3/year would maintain a relatively stable level. Salinity, over time, would reach about 6 g/l, which would be ideal for the 11 most important commercially caught fish in the North Aral Sea in 2016 (Micklin et al. 2018; White 2016).

As the salinity of the North Aral decreases, the species diversity will rise as a result of natural reintroduction of many species that disappeared due to the earlier salinity growth. Reintroduction of planktonic invertebrates can occur by waterfowl transfer or by wind from fresh or brackish water bodies. It is also possible to take these organisms for reintroduction from the Syr Dar'ya and from associated lakes in its lower reaches, which act as refugia. Further decrease in the salinity of the North Aral Sea can cause new changes in the composition of its fauna. Strong freshening will be unfavorable for marine species and representatives of the fauna of saline water bodies of arid zone, favored by the salinization of the Aral Sea, as well as for brackish-water species.

4.5.2 Fate of the Large Aral

The future for the Large (Southern) Aral Sea is more problematic (Micklin 2014f). The Eastern Basin, depending on inflow from the outflow over the Berg Strait Dam and inflow from the Amu Dar'ya, is at times an extensive, very shallow lake or a dry playa basin contributing to salt/dust storms arising from the dried Aral Sea bottom (also often called simply Aralkum which means Aral Desert). The lake when present has high salinity and limited ecological value except to mitigate salt/dust storms by reducing the area of dry bottom subject to wind deflation. The Eastern Basin also has some potential for raising brine shrimp and harvesting their eggs, but the commercial promise of this is limited. The deeper Western Basin depends largely on net groundwater inflow, direct runoff from rain and snowmelt, and some input from the Central Aral (via the connecting channel) when that water body is sufficiently filled by outflow from the Small Aral. The Western Basin also receives inflow via the connecting channel from the Eastern Basin when that water body is sufficiently filled by inflow from the Amu Dar'ya. On June 20–21, 2018, its level was around 23 m above the Kronstadt gauge and its area was about 2894 km^2 (Table 4.1). The Eastern Basin on those dates was a few meters higher and had an area about 2537 km^2.

Given a continuation of present trends, the level of the Western Basin will continue to decrease for some time, perhaps stabilizing around 21 m above the Kronstadt gauge. At that level its area would be 2560 km^2. It would continue toward hyper-salinization. As salinity of the Western Basin grows, a new reduction in the species diversity of its fauna will begin. Ostracod *Cyprideis torosa* and all rotifers will disappear. When salinity exceeds 250 g/l, only *Artemia*, which can tolerate up to 350 g/l, would remain, If the salinity exceeds this limit, the remainder of the Large Aral will become like the Dead Sea (Oren et al. 2010; Plotnikov 2016).

But there are more optimistic scenarios for the Western Basin of the Large Aral (Micklin 2016). Figure 4.5 shows a more hopeful future and is based on earlier work by two Soviet experts (Lvovich and Tsigelnaya 1978). It would require an average annual inflow in the lowest reaches of the Amu Dar'ya of around 12.5 km^3, a bit more than double recent estimated average annual flow (5.4 km^3/year for 1992–2011), which could be accomplished via feasible improvements in irrigation efficiency in the Amu Dar'ya River Basin. This alternative would likely cost more than the 85 million USD expended on the first stage of the Small Aral restoration. The greatest obstacles to implementation of this plan are political and economic related to the fact that the plan would complicate the ongoing exploration for and exploitation of oil and gas deposits from parts of the now dried bottom of the southern part of the Western Basin of the Aral Sea.

Rehabilitation and preservation of the lower Amu Dar'ya delta through creation of artificial ponds and wetlands and rehabilitation of former lakes and wetlands in the delta and on the dry bed of the Aral Sea has been a priority since the late 1980s (Micklin 2016; Novikova 1999). Benefits are enhanced biodiversity, improved fisheries, greater forage production, treatment of wastewater by aquatic vegetation, and some reduction in salt and dust transfer from the dried sea bottom. Efforts to improve wetlands and lakes in the lower Syr Dar'ya delta have also been made.

Since the early 1990s efforts have been made to stabilize the dried bottom of the Large Aral in Uzbekistan and to lower the deflation potential with planting of salt-tolerant shrubs, grasses and trees. The largest scale project in this regard is the "Stabilization of the desiccated Aral Sea bottom in Central Asia" (Dukhovny et al. 2007). This program has been managed by the German foreign aid agency (GTZ) and the Forestry Research Institute of Uzbekistan. Between 1995 and 2007 drought-resistant tree and shrub species, primarily Black Saksaul were planted on 300 km^2. Since 2005 the project has also included an integrated remote sensing/GIS and sea bed-based monitoring component to assess surficial dynamics, desertification risks, and other negative impacts of the continuing drying.

4.6 Conclusions

The Aral Sea is geologically a relatively young large lake, no more than 20 thousand years old. As a terminal water body, strongly impacted by human actions, its history has been turbulent. Levels have risen and fallen significantly accompanied by major transgressions and recessions of the shoreline, changes in salinity and accompanying alterations in biotic communities. Early fluctuations owed to natural forces of climate change and diversions of the Amu Dar'ya away from the sea. But for several millennia man has had a growing influence through irrigation related reductions of inflow and inadvertent and purposeful diversions of the Amu westward toward the Caspian Sea and away from the Aral.

The most dramatic human-caused impacts occurred in the twentieth century, particularly after 1960, and have continued into the twenty-first century. Invertebrates and fishes (and their parasites) were introduced both consciously and inadvertently beginning early in the 1920s. Some of these substantially and negatively affected native species. But by far the most damaging action was the major expansion of irrigation beyond the point of hydrologic sustainability that began in the 1950s and that led to the modern desiccation of the Aral after 1960. This ongoing process, by 2018, has likely led to the greatest level drop, most dramatic shoreline retreat and highest salinities experienced by the Aral in the past several millennia. These physical changes have devasted and simplified the biologic communities of most of what was formerly the Large Aral and destroyed their ecologic and economic value.

But is all lost? The answer is clearly "no". The partial restoration project for the North Aral has, so far, been a resounding success with major positive ecologic and economic impacts. A further stage of restoration for this water body is contemplated that could bring even greater benefits. The picture for the South Aral is much gloomier. Major restoration would require additional flow from the Amu Dar'ya that is not in the cards in the short or medium term but might be possible in the long perspective. Somewhat raising, stabilizing, and bringing back into ecologic and economic use the deeper Western Basin is possible by realizable improvements in irrigation efficiency to increase the Amu's inflow, but economic justification and political will for this seems absent.

Acknowledgements The work of SK was supported by a State assignment of the IGM SB RAS, and those of NVA and ISP by the theme of the State assignment for 2019-2021 "AAAA-A19-119020690091-0. Studies of biological diversity and the mechanisms of the impact of anthropogenic and natural factors on the structural and functional organization of ecosystems of continental water bodies. Systematization of the biodiversity of salt lakes and brackish-water inland seas in the zone of critical salinity, study of the role of brackish water species in ecosystems".

References

Aladin N (2014) The dam of life or dam lifelong. The Aral Sea and the construction of the dam in Berg Strait. Part one (1988–1992). El Alfoli Boletin Semestral de IPAISAL IPAISAL's Biyearly J 15:3–17
Aladin NV, Kotov SV (1989) The Aral Sea ecosystem original state and its changes under anthropogenic influences. Proc Zool Inst USSR Acad Sci 199:4–25 (in Russian)
Aladin N, Plotnikov I (2008) Modern fauna of residual water bodies formed on the place of the former Aral Sea. Proc Zool Inst Russ Acad Sci 312(1/2):145–154 (in Russian)
Aladin NV, Potts WTW (1992) Changes in the Aral Sea ecosystem during the period 1960–1990. Hydrobiologia 237:67–79
Aladin NV, Plotnikov IS, Orlova MI, Filippov AA, Smurov AO, Piriulin DD, Rusakova OM, Zhakova LV (1996) Changes in the form and biota of the Aral Sea over time. In: Micklin P, Williams W (eds) The Aral Sea Basin (Proceedings of an Advanced Research Workshop, May 2–5, 1994, Tashkent, Uzbekistan), NATO ASI series, vol 12. Springer, Heidelberg, pp 33–55
Aladin NV, Filippov AA, Plotnikov IS, Egorov AN (2000) Modern ecological state of the Small Aral Sea In: Ecological research and monitoring of the Aral Sea deltas: a basis for restoration, Book 2. UNESCO Aral Sea Project 1997–1999: Final Scientific Report, pp 73–81
Andreev NI (1989) Zooplankton of the Aral Sea in the initial period of its salinization. Proc Zool Inst USSR Acad Sci 199:26–52 (in Russian)
Andreeva SI (1989) Zoobenthos of the Aral Sea in the initial period of its salinization. Proc Zool Inst Acad Sci USSR 199:53–82 (in Russian)
Ashirbekov U, Zonn I (2003) History of a disappearing sea. Dushanbe, Tadjikistan: Executive Committee of the International Fund for Saving the Aral Sea (in Russian). Available at http://www.cawater-info.net/library/icc.htm
Atlas of the USSR (1983) Moscow, Soviet Union: GUGK (in Russian)
Austin P, Mackay A, Palagushkina O, Leng M (2007) A high-resolution diatom-inferred palaeoconductivity and lake level record of the Aral Sea for the last 1600 yr. Quat Res 67:383–393
Bartold VV (1902) Records on the Aral Sea and the lower stream of the Amudarya since the earliest times to the XVIIth Century. Turkestan Branch of the Russian Geographic Society, Tashkent (in Russian)
Berg LS (1908) The Aral Sea. Attempt at a physical–geographical monograph. Stasyulevich Publisher, St. Petersburg (in Russian)
Bervald EA (1964) Ways of organizing a rational fish economy on inland water bodies. Rostov State University, Rostov-on-Don, p 148 (in Russian)
Boomer I (2012) Ostracoda as indicators of climatic and human-influenced changes in the Late Quaternary of the Ponto-Caspian region (Aral, Caspian and Black Seas). In: Horne DJ, Holmes J, Rodriguez-Lazaro J, Viehberg FA (eds) Ostracoda as proxies for Quaternary climate change. Elsevier Science, Amsterdam, pp 205–215
Boomer I, Wünnemann B, Mackay A, Sustin P, Sorrel P, Reinhardt C, Keyser D, Guichard F, Fontugne M (2009) Advances in understanding the late Holocene history of the Aral Sea region. Quat Int 194:79–90
Boroffka N, Oberhänsli H, Achatov G, Aladin N, Baipakov K, Erzhanova A, Hörnig A, Krivonogov S, Lobas D, Savel'eva T, Wünnemann B (2005) Human settlements on the northern shores of Lake Aral and water level changes. Mitig Adapt Strat Glob Change 10:71–85
Boroffka N, Oberhänsli H, Sorrel P, Demory F, Reinhardt C, Wünnemann B, Alimov K, Baratov S, Rakhimov K, Saparov N, Shirinov T, Krivonogov S (2006) Archaeology and climate: settlement and lake-level changes at the Aral Sea. Geoarchaeol Int J 21:721–734
Bortnik V, Chistyaevaya S (eds) (1990) Hydrometeorology and Hydrochemistry of the Seas of the USSR, vol 7. Gidrometeoizdat, Aral Sea, Leningrad (in Russian)
Burr GS, Kuzmin YV, Krivonogov SK, Gusskov SA, Cruz RJ (2019) A history of the modern Aral Sea (Central Asia) since the Late Pleistocene. Quat Sci Rev 206:141–149

Catalogue of monuments of the Kazakhstan Republic history and culture (2007). Aruna, Almaty (in Russian)
Chalov PI (1968) Non-equilibrium uranium dating. Ilim, Frunze (in Russian)
Cretaux J-F, Kostianoy A, Bergé-Nyuyen M, Kouraev A (2019) Present-day water balance of the Aral Sea seen from satellite. In: Barale V, Gade M (eds) Remote Sensing of the Asian Seas. Springer, Berlin, pp 523–539
Dam of the North Aral Sea and hydrocomplex Aklak (2017) Brochure distributed at the International Forum on Sustainable development of the Aral Region, Kyzyl-Orda, Kazakhstan, May 30–31, 2017 (in Russian)
Dukhovny V, Navratil P, Rusiev I, Stulina G, Rushenko Y (eds) (2007) Comprehensive remote and ground studies of the dried Aral Sea bed. SIC ICWC, Tashkent
Ermakhanov Z, Plotnikov I, Aladin N, Micklin P (2012) Changes in the Aral Sea ichthyofauna and fishery during the period of ecological crisis. Lakes Reservoirs Res Manag 17:3–9
Goode's World Atlas (1982) Edited by E. Espenshade Jr. Rand McNally and Company, Chicago
Huang X, Oberhänsli H, von Suchodoletz H, Sorrel P (2011) Dust deposition in the Aral Sea: implications for changes in atmospheric circulation in Central Asia during the past 2000 years. Quat Sci Rev 30:3661–3674
Indoitu R, Kozhoridze K, Batyrbaeva M, Vitkovskaya I, Orlovsky N, Blumberg D, Orlovsky L (2015) Dust emission and environmental changes on the dried bottom of the Aral Sea. Aeol Res 17:101–115
Karpevich AF (1960) Biological bases of aquatic organisms' acclimatization in the Aral Sea. Trudy VNIRO 43(1):76–115 (in Russian)
Karpevich AF (1975) Theory and practice of aquatic organisms acclimatization. Pischevaya Promyshlennost, Moscow (in Russian)
Kes AS (1969) The main stages in the development of the Aral Sea. In: Geller SY (ed) Problem of the Aral Sea. Nauka, Moscow, pp 160–172 (in Russian)
Kes AS (1978) Reasons for the level changes of the Aral during the Holocene. Izv Akad Nauk SSSR Ser Geog 4:8–16 (in Russian)
Kes AS (1995) Chronicle of the Aral Sea and the Sub-Aral region. GeoJournal 35:7–10
Kortunova TA (1970) Some data on food organisms introduced to the Aral Sea. Proc. VNIRO 76(3):178–184 (in Russian)
Kortunova TA (1975) On the changes in the Aral Sea zooplankton in 1959–1968. Zool Zhurnal 54(5):567–669 (in Russian)
Kortunova TA, Lukonina NK (1970) Quantitative characteristic of the Aral Sea zooplankton. In: Rybnye resursy vodoemov Kazakhstana i ikh ispolzovanie, vol 6. Nauka, Alma-Ata, pp 52–60 (in Russian)
Kosarev AN (1975) Hydrology of the Caspian and Aral seas. Moscow University, Moscow (in Russian)
Kostianoy AG, Kosarev AN (eds) (2010) The Aral Sea Environment. The Handbook of Environmental Chemistry. Springer, Berlin
Krivonogov S (2009) Extent of the Aral Sea drop in the Middle Age. Dokl Earth Sci 428(1):1146–1150
Krivonogov S (2014): Changes of the Aral Sea level. In: Micklin P, Aladin N, Plotnikov I (eds) The Aral Sea: the Devastation and Partial Rehabilitation of a Great Lake. Springer, Heidelberg, pp 77–111
Krivonogov S, Burr G, Gusskov S, Khazin L, Kuzmin Y, Kurmanbaev R, Kenshinbay T, Nurgizarinov A, Zhakov E (2010) Environmental changes of the Aral Sea (Central Asia) in the Holocene — major trends. Radiocarbon 52:555–568
Krivonogov S, Burr G, Gusskov S, Kenshinbay T, Kurmanbaev R, Kuzmin Y, Voyakin D (2014) The fluctuating Aral Sea: a multidisciplinary-based history of the last two thousand years. Gondwana Res 25:284–300
Lim RM (1986) About acclimatization of flounder-gloss in the Aral Sea. In: Biologicheskie osnovy rybnogo khozyaistva vodoemov Sredney Azii i Kazakhstana. Ashhabad, pp 249–250 (in Russian)

Lopatin GV (1957) Structure of the Amu Darya delta and its formation history. Proc Lake Res Lab USSR Acad Sci 4:5–34 (in Russian)
Lunezheva MS, Kiyatkin AK, Polischuk V (1987) Central Asia and Kazakhstan – ancient regions of irrigation. Reclam Water Manag 10:65–70 (in Russian)
Lunezheva MS, Kiyatkin AK, Polischuk V (1988) Central Asia and Kazakhstan - ancient regions of irrigation. Reclam Water Manag 8:60–63 (in Russian)
Lvovich M, Tsigelnaya I (1978) Management of the Aral Sea water balance. Izv Akad Nauk SSSR Ser Geog 1:42–54 (in Russian)
Lymarev VI (1967) Shores of the Aral Sea an inland basin of the arid zone. Nauka (in Russian), Leningrad
Mayev EG, Mayeva SA, Nikolaev SD, Parunin OB (1983) New data on the Holocene history of the Aral Sea. In: Mayev EG (ed) Paleogeography of the Caspian and Aral Seas in the Cenozoic. vol 2. Moscow, MGU, pp 119–133 (in Russian)
Mayev E, Mayeva S, Karpychev Y (1991) The Aral Sea in the Holocene. In: Aral crisis. Acad Sci USSR, Moscow, pp 76–86 (in Russian)
Mayeva SA, Mayev YG (1991) Changes of the level of the Aral Sea for the recent millennia. Geodesia i Aerofotosemka 1:124–132 (in Russian)
Micklin P (2000) Managing water in Central Asia. The Royal Institute of International Affairs, London
Micklin P (2007) The Aral Sea disaster. Annual Review of Earth and Planetary Sciences. Ann Rev 35:47–72
Micklin P (2010) The past, present, and future Aral Sea. Lakes Reservoirs Res Manag 15:193–213
Micklin P (2014a) Introduction to the Aral Sea and its Region. In: Micklin P, Aladin N, Plotnikov I (eds) The Aral Sea: the Devastation and Partial Rehabilitation of a Great Lake. Springer Earth System Sciences. Springer, Heidelberg, pp 15–40
Micklin P (2014b) Irrigation. In: Micklin P, Aladin N, Plotnikov I (eds) The Aral Sea: the Devastation and Partial Rehabilitation of a Great Lake. Springer Earth System Sciences. Springer, Heidelberg, pp 207–232
Micklin P (2014c) Aral Sea Basin water resources and the changing Aral water balance. In: Micklin P, Aladin N, Plotnikov I (eds) The Aral Sea: the Devastation and Partial Rehabilitation of a Great Lake. Springer Earth System Sciences. Springer, Heidelberg, pp 111–137
Micklin P (2014d) Introduction. In: Micklin P, Aladin N, Plotnikov I (eds) The Aral Sea: the Devastation and Partial Rehabilitation of a Great Lake. Springer Earth System Sciences. Springer, Heidelberg, pp 1–11
Micklin P (2014e) Efforts to revive the Aral Sea. In: Micklin P, Aladin N, Plotnikov I (eds) The Aral Sea: the Devastation and Partial Rehabilitation of a Great Lake. Springer Earth System Sciences. Springer, Heidelberg, pp 361–380
Micklin P (2014g) The Siberian water transfer schemes. In: Micklin P, Aladin N, Plotnikov I (eds) The Aral Sea: the Devastation and Partial Rehabilitation of a Great Lake. Springer Earth System Sciences. Springer, Heidelberg, pp 381–404
Micklin P (2016) The future Aral Sea: hope and despair. Environ Earth Sci 75 (9):1–15
Micklin P, Aladin N (2008) Reclaiming the Aral Sea. Sci AM 98(4):64–71
Micklin P, Aladin N, Plotnikov I (2014) An Expedition to the Northern Part of the Small Aral Sea. In: Micklin P, Aladin N, Plotnikov I (eds) The Aral Sea: the Devastation and Partial Rehabilitation of a Great Lake. Springer Earth System Sciences. Springer, Heidelberg, pp 337–351
Micklin P, White K, Alimbetova Z, Ermakhanov Z (2018) Partial Recovery of the North Aral Sea: a Water Management Success Story in Central Asia. Presentation at the IGU Thematic Meeting, Moscow, Russia, June 4–7, 2018 (www.researchgate.net/profile/Philip_Micklin2/publications)
Mokievsky VO, Miljutina MA (2011) Nematodes in meiofauna of the Large Aral Sea during the desiccation phase: taxonomic composition and redescription of common species. Russ J Nematology 19(1):31–43 (in Russian)
Mordukhai-Boltovskoi FD (1972) Modern state of the Aral Sea fauna. Gidrobiol Zhurnal 3:14–20 (in Russian)

Mordukhai-Boltovskoi FD (ed) (1974) Atlas of the Aral Sea invertebrates. Fishing Industry, Moscow, p 272 (in Russian)
Nikolsky GV (1940) Fishes of the Aral Sea. MOIP, Moscow, p 216 (in Russian)
Nikolaev SD (1991) The development of the Aral Sea from oxygen isotope data. In: Sevastyanov DV, Mamedov ED, Rumyantsev VA (eds) History of the Lakes Sevan, Issyk Kul, Balkhash, Zaysan and Aral. Nauka, Leningrad, pp 246–249 (in Russian)
Nikolaev SD (1995) Isotope paleogeography of intracontinental seas. VNIRO Publishers, Moscow (in Russian)
Novikova N (1996) The Tugai of the Aral Sea is dying: can it be restored? Russ Conserv News 6:22–23
Novikova N (1999) Priaralye ecosystems and creeping environmental changes in the Aral Sea. In: Glantz M (ed) Creeping environmental problems and sustainable development in the Aral Sea Basin. Cambridge University Press, Cambridge, pp 100–127
Oberhänsli H, Boroffka N, Sorrel P, Krivonogov S (2007) Climate variability during the past 2,000 years and past economic and irrigation activities in the Aral Sea Basin. Irrig Drain Syst 21:167–183
Oberhänsli H, Novotna K, Piskova A, Chabrillat S, Nourgaliev DK, Kurbaniyazov AK, Grygar TM (2011) Variability in precipitation, temperature and river runoff in W Central Asia during the past ~2000 years. Glob Planet Change 76:95–104
Oren A, Plotnikov IS, Sokolov SB, Aladin NV (2010) The Aral Sea and the Dead Sea: disparate lakes with similar histories. Lakes Reservoirs Res Manag 15:223–236
Oreshkin DB (1990) Aral catastrophe. EarthScience, 2. Znanie, Moscow (in Russian)
Osmanov SO (1961) On the death of atherine in the Aral Sea. Annals of the Karakalpak Branch. Uzbekistan Rep Acad Sci 3(5):95–96 (in Russian)
Osmanov SO, Arystanov EA, Ubaidullayev K, Yusupov OY, Teremuratov TA (1976) Problems of Aral Sea parasitology. Fan, Tashkent (in Russian)
Plotnikov IS (2016) Long-term changes of the free-living aquatic invertebrate fauna of the Aral Sea. Zoological Institute, Russ Acad Sci, St. Petersburg, Russia (in Russian)
Plotnikov IS, Aladin NV, Filippov AA (1991) The past and present of the Aral Sea fauna. Zool J 70(4):5–15 (in Russian)
Plotnikov I, Aladin N, Ermakhanov Z, Zhakova L (2014a) Biological Dynamics of the Aral Sea Before its Modern Decline (1900–1960). In: Micklin P, Aladin N, Plotnikov I (eds) The Aral Sea: the devastation and partial rehabilitation of a Great Lake. Springer Earth System Sciences. Springer, Heidelberg, pp 131–171
Plotnikov I, Aladin V, Ermakhanov Z, Zhakova L (2014b) The new aquatic biology of the Aral Sea. In: Micklin P, Aladin N, Plotnikov I (eds) The Aral Sea: the Devastation and Partial Rehabilitation of a Great Lake. Springer Earth System Sciences. Springer, Heidelberg, pp 131–171
Pshenin GN, Steklenkov AP, Cherkinskii AE (1984) Origin and age of pre-Holocene terraces of the Aral Sea. Reports Acad Sci USSR 276:675–677 (in Russian)
Reinhardt C, Wünnemann B, Krivonogov SK (2008) Geomorphological evidence for the Late Holocene evolution and the Holocene lake level maximum of the Aral Sea. Geomorphology 93(3–4):302–315
Rubanov IV, Ischniyazov DP, Baskakova MA (1987) The geology of the Aral Sea. Fan, Tashkent (in Russian)
Smagulov E (2001) The discovery and study of the mazar on the Aral Sea bottom. Otan Tarikhy 4:77–81 (in Russian)
Smagulov E (2002) A mausoleum on the sea bottom. Qumbez 4:50–52 (in Russian)
Sorrel P, Popescu S-M, Head MJ, Suc JP, Klotz S, Oberhänsli H (2006) Hydrographic development of the Aral Sea during the last 2000 years based on a quantitative analysis of dinoflagellate cysts. Palaeogeogr Palaeoclimatol Palaeoecol 234:304–327
Sorrel P, Oberhänsli H, Boroffka N, Nourgaliev D, Dulski P, Röhl U (2007a) Control of wind strength and frequency in the Aral Sea basin during the late Holocene. Quat Res 67(3):371–382

Sorrel P, Popescu S-M, Klotz S, Suc J-P, Oberhänsli H (2007b) Climatic variability in the Aral Sea basin (Central Asia) during the late Holocene based on vegetation changes. Quat Res 67:357–370
Toman M, Plotnikov I, Aladin N, Micklin P, Ermakhanov Z (2015) Biodiversity, the present ecological state of the Aral Sea and its impact on future development. Ac Biol Slovinica 58:45–59
Vainberg BI (1999) Ethnogeography of Turan in the past. VII c. BC–VIII c. CE. "Eastern Literature" Russ Acad Sci Publ House, Moscow (in Russian)
White K (2016) Kazakhstan's Northern Aral Sea today: partial ecosystem restoration and economic recovery. In: Friedman E, Neuzil M (eds) Environmental Crises in Central Asia (from steppes to seas, from deserts to glaciers), New York, Routledge, pp 127–138
World Bank (1998, May) Aral Sea Basin Program (Kazakhstan, Kyrgyz Republic, Tajikistan, Turkmenistan, and Uzbekistan). Water and Environmental Management Project. Washington, DC, World Bank
World Bank (2001) Project Appraisal Document on a Proposed Loan in the Amount of U.S. $64.5 Million to the Republic of Kazakhstan for the Syr Dar'ya Control and Northern Aral Sea Phase-1 Project, Report No. 22190-KZ, 98 pages including annexes and maps (PDF format)
World Bank (2014) Project Information Document (PID), Concept Stage, Syr Dar'ya Control and Northern Aral Sea Project, Phase 2 (P15200), Report No.: PIDC12697, September 24, 2014, pp 1-9
Zavialov P, Arashkevich E, Bastida I, Ginsburg A, Dikarev S, Zhitina A, Izhitskiy A et al (2012) The Large Aral Sea at the beginning of the XXI Century: physics, biology, chemistry. Shirshov Oceanology Institute, Nauka, Moscow (in Russian)
Zenkevich LA (1963) Biology of the Seas of the USSR. Acad Sci USSR, Moscow (in Russian)
Zonn IS, Glantz M, Kostianoy AG, Kosarev AN (2009) The Aral Sea Encyclopedia. Springer, Berlin

Chapter 5
Geological History and Present Conditions of Lake Balkhash

Renato Sala, Jean-Marc Deom, Nikolai V. Aladin, Igor S. Plotnikov and Sabyr Nurtazin

Abstract Lake Balkhash is a large endorheic water body, the third largest by size in Eurasia and the second largest salt lake of the world. With its half-moon elongated morphology and 78% of inflows provided by the Ili River from the West, the lake has a freshwater basin in the West and saline water basin in the East, separated by the 4-km narrow and 6-m deep Uzunaral Strait. The average bathymetry is shallow, with a maximum water depth of 11 m in the West and 26 m in the East. According to investigations of the geological history of the lake in Soviet times and in international projects during the last 15 years, a large lake was formed by the Ili River in the Balkhash region encompassing the present area of the Kapchagai Reservoir during the Middle Pleistocene. The large lake basin was subsequently transformed by a series of tectonic deformations. Around 300 kiloyears before present (ka BP) the Ili River was diverted to the North where it formed a large megalake, the Ancient Balkhash, in the Balkhash-Alakol Depression. Around 110 ka BP, the lake became divided into two basins forming the Alakol Lake in the East and the modern Lake Balkhash in the West. Hydrological conditions mostly controlled by precipitation, evaporation and meltwater discharge caused three different lake-level stages in the Late Pleistocene and Holocene: lake levels between 349 and 355 m above sea level (asl) prevailed during glacial periods, between 341 and 348 m asl during the pluvial early and middle Holocene, and between 335 and 348 m asl during the arid late

R. Sala (✉) · J.-M. Deom
Laboratory of GeoArchaeology, Faculty of History Archaeology Ethnology, Al-Farabi Kazakh National University KazNU, 71 Ave. Al-Farabi, 050040 Almaty, Kazakhstan
e-mail: ispkz@yahoo.com

N. V. Aladin · I. S. Plotnikov
Laboratory of Brackish Water Hydrobiology, Zoological Institute, Russian Academy of Sciences, Universitetskaya nab. 1, 199034 St. Petersburg, Russia
e-mail: Nikolai.Aladin@zin.ru

I. S. Plotnikov
e-mail: Igor.Plotnikov@zin.ru

S. Nurtazin
Department of Biodiversity and Bioresources, Faculty of Biology, Al-Farabi Kazakh National University, 71 Ave. Al-Farabi, 050040 Almaty, Kazakhstan
e-mail: Sabyr.Nurtazin@kaznu.kz

S. Mischke (ed.), *Large Asian Lakes in a Changing World*, Springer Water,
https://doi.org/10.1007/978-3-030-42254-7_5

Holocene when extreme regressions at ca. 5.0, 1.2 and 0.8 ka BP divided the lake into more than one basin. The present lake water balance results from a major regression due to a recent phase of aridization and the filling of the Kapchagai Reservoir in the 1970s and 1980s and the compensation of lower precipitation by increased meltwater discharge from glaciers. However, meltwater runoff will diminish with rapidly shrinking glaciers in the next 50 years. This alarming perspective requires careful water-basin management which was not yet implemented. Lake Balkhash is exposed to the threat of exaggerated anthropogenic water subtraction due to an accelerated infrastructural and demographic boost that doubled the irrigated farmlands in the Chinese part of the catchment in less than 20 years. Due to the lake's hydrology, catchment rock and hydrochemical conditions, the water of the Eastern Balkhash has high concentrations of potassium and magnesium, unfavorable for hydrobionts. Any further increase in salinity will soon cause a considerable diminution of the lake's biomass. The ichthyofauna of the lake has been intensively manipulated during the twentieth century, with the introduction of new species and the decline of the original ones. The substitution of the native fish fauna by introduced species caused a decrease of valuable commercial fish in the lake and a decrease of the total fish catch. Thus, Lake Balkhash faces serious environmental risks today and its near future depends on the collective will and decisions of the responsible agencies.

Keywords Balkhash-Alakol depression · Ili River · Kapchagai Reservoir · Water balance · Salinity

5.1 Geographical and Hydrological Features of Modern Balkhash

The Balkhash is a permanent, endorheic (terminal, closed, athalassic), large, shallow, and slightly saline lake, 588–614 km long and 9–74 km wide, covering an area of 18,210 km^2 between 44°57′–49°19′ N and 73°24′–79°14′ E. The lake represents an oasis in the Saryesik-Atyrau Desert and lowest part of the Balkhash-Alakol Depression. Lake Balkhash drains waters from a 413,000 km^2 large region of which 153,000 km^2 actively contribute water to the lake. Salinity of the terminal lake is in a range from 0.2 to 5.0 g/L (Fig. 5.1).

The climate of the Balkhash region is arid continental (BWk) with a mean annual temperature of 5.8 °C (mean January temperature −15.2 °C, mean July temperature 24.3 °C), continentality index of 39.5, diurnality index of 11.1, mean annual precipitation of 142 mm (maximum in May 17 mm, minimum in September 4 mm), annual evapotranspiration of 668 mm, aridity index of 4.7, and an arid season from 15 April to 30 October (Algazi meteorological station, data series 1936–1960). Lake Balkhash receives the highest river inflow between May and August, peaking in July, and its water surface is frozen from November-December to March-April. Predominating winds are from the East and Northeast, inducing waves up to 3.0–3.5 m in the East and no more than 2.5 m in the West, and currents along the lake's longitudinal axis.

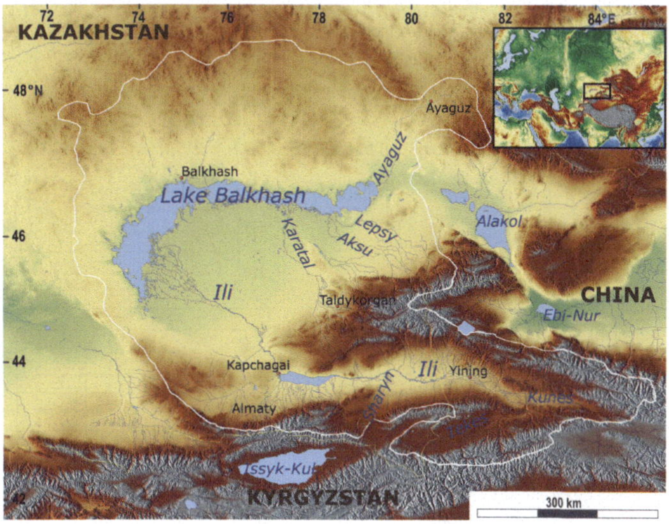

Fig. 5.1 Hydrological basin of the modern Lake Balkhash (outlined in white)

The mean hydrological characteristics of Lake Balkhash in the years 1939–1969 and similarly during 2000-2020 are: water volume of 106 km^3, water surface of 18,210 km^2, water level at 342 m asl (with seasonal variation between 340.8 and 343.1 m asl) and average depth of 5.7 m (Fig. 5.2a).

The equation of the average lake-water balance includes a total yearly water input of 18 km^3 which represents river inflow (15.1 km^3, 83.9%), local precipitation (2.1 km^3, 11.7%), and shore-spring runoff (0.8 km^3, 4.4%), balanced by a corresponding water output as evaporation (−17.2 km^3, 95.6%) and groundwater infiltration (−0.8 km^3, 4.4%; Fig. 5.2b, c). The annual river inflow volume would be 3.0 km^3 higher without water withdrawal from tributaries of Lake Balkhash for irrigation farming (Shnitnikov 1973; Kezer and Matsuyama 2006; Dostay 2009).

The lake is characterized by high and accelerated variability under changing climate conditions.[1] The sedimentary deposits carried by rivers into the lake do not account for more than 0.005 km^3 per year. Ice-melt water is variable, depending from glaciological parameters. Annual meltwater discharge accounted for ca. 0.8 km^3, i.e., ca. 4.4% of the annual inflow, in the last decades. Of the 15.1 km^3 of river inflow, 11.8 km^3 come from the Ili River (78%) and 3.2 km^3 (21%) from the rivers Karatal, Aksu-Lepsy and Ayaguz (Figs. 5.1 and 5.2c).

The quantity of natural and anthropogenic water subtraction above the lake's basin is more or less equivalent to the amount of actual river inflow: 15.1 km^3 per year consisting of 7.0 km^3 of water subtraction by riparian vegetation and of 7.9 km^3 for human use (values of 2014). Without riparian and anthropogenic water subtraction,

[1]During the pluvial years 1958–1961 the total annual river inflow in the lake rose to 23.88 km^3, which with evaporation levels of 17.63 km^3, supported a sensible increase of water level to 343 m asl (Shnitnikov 1973, 134).

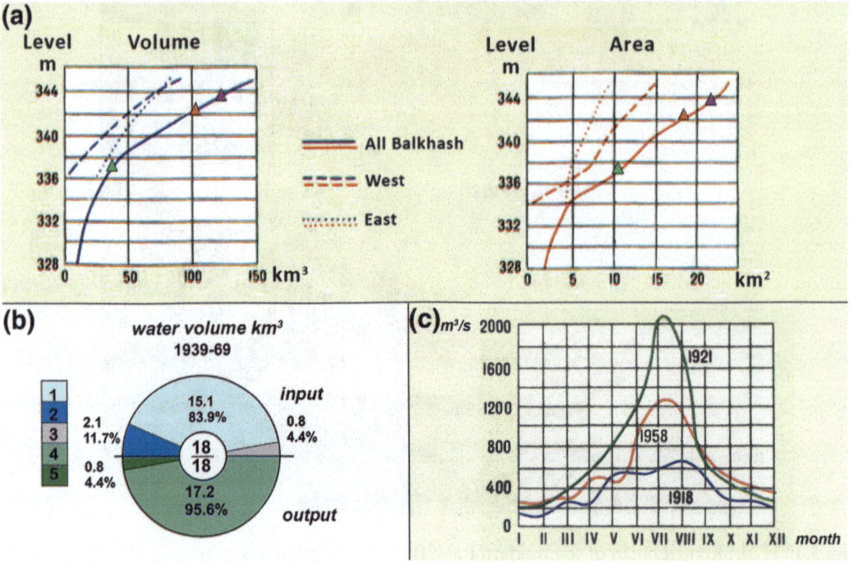

Fig. 5.2 Hydrological parameters of Lake Balkhash. **a** Relationship of water level (m asl) and water volume (km^3) at left and area (km^2) at right. Triangular marks indicate three water-level stages: average at 342 m in 1939–1969 and today (orange), high at 343.5 m in 1908 (magenta), critically low at 337 m at the end of the 18th century (green). **b** Water balance volumes (km^3) for 1939–1969: 1 river inflow, 2 precipitation, 3 groundwater and shore springs, 4 evaporation, 5 infiltration. **c** Ili River mean monthly discharge (m^3/s): low in 1918 (blue), average in 1958 (red), high in 1921 (green). Adapted from Atlas (1982)

the virtual water volume of Lake Balkhash would be 123 km^3, the surface area of the lake 22,000 km^2, and its water level at 343.5 m asl. This high lake level was reached for the last time during the transgressive phase in 1908 (Fig. 5.2a).

In order to discuss the hydrological balance of Lake Balkhash, it must be underlined that its average water volume of 106 km^3 in the year 2000 corresponds to less than seven years of water input (or evaporation, i.e., ca. 17.2 km^3 per year) and to an almost similar amount of water stored as ice in the mountain glaciers of the basin (ca. 90 km^3 per year). Therefore, the hydrological balance of the lake is very sensitive to temperature-driven changes of ice volumes in the uppermost reaches of the tributaries. Effects of climate change on lake-water volumes, levels, salinity, etc., might be delayed. At the start of an arid and warm phase as it occurs at present, the increase of ice melting will postpone the decrease of the lake level by a few tens of years until glaciers reached a new equilibrium or disappeared. In contrast, at the start of a cold pluvial phase, the additional accumulation of ice would postpone the rise of Lake Balkhash's level for more than one century until glaciers grew and reached a new equilibrium.[2]

---

[2]The high salinity levels of Lake Balkhash reconstructed for the period from 2.4 to 2.2 ka BP in spite of the cold and moist climate phase between 2.5 and 2.0 ka are attributed by Feng et al. (2013) to

The hydrological regime of Lake Balkhash is strongly determined by its shape: latitudinally outstretched as a narrow crescent of 600 km length, with a shallow bottom progressively sloping West to East from 333 to 316 m asl, and with its western and eastern parts separated by the 4-km wide and 6-m deep Uzunaral Strait (Fig. 5.3).

The Western Balkhash Basin is larger (10,600 km^2, 58% of the total lake area) and wider (74 km) but shallower (average depth of 4.8 m, maximum depth of 11 m at 331 m asl) and less voluminous (47.8 km^3). The Eastern Balkhash Basin is smaller (7500 km^2) and narrower (9–19 km) but significantly deeper (average depth 9.0 m, maximum depth 26 m at 316 m asl) and holding 56.6 km^3, i.e., 54% of the total water volume.

The Western Balkhash is mostly fed by the Ili River (78%), which carries sediments responsible for its turbidity and yellow-gray colour (visibility 5–10 m) and for its even silty-sandy lake bottom. The Ili River inflow causes the low mineralization rates (0.8–1.8 g/L) of the western basin. The Eastern Balkhash receives a significantly lower input from smaller rivers that all together constitute only 21.1% of the total river inflow. It experiences a yearly water deficit of 1.15–2.80 km^3 that requires an equivalent amount of water inflow from the western basin and makes its water five times saltier (3.5–4.2 g/L) with emerald-blue waters of higher turbidity (visibility 0.4–1.0 m) and an undulatory lake bottom covered by light-grey limy silts including authigenic dolomite.

Besides the Ili River, the other tributaries Karatal, Aksu and Lepsy were and are still discharging into the Eastern Balkhash. Their discharge alone does not support the actual water level of the eastern basin, which is co-fed by water inflow from the Ili and Western Balkhash. The Ayaguz River used to enter the lake from the North, but is has no surface inflow today. As a whole the sediments of all tributary deltas contributed to the partition of the Balkhash basin in five different subbasins of different depth (Fig. 5.3).[3]

---

reduced runoff from glacial meltwater. Similarly, the water-level rise (+8.1 cm per year) during the period 1992–2010 under increasing temperature (+0.08 °C per year) and precipitation (+11.6 mm per year) is largely attributed to the increasing contribution of glacial meltwater by Propastin (2012).

[3]The modern Balkhash is divided in five basins: West-South, West-North, Middle, Lepsinsky and Burly-Tyubin, with depth increasing from West to East, i.e., lake bottom at 334, 333, 327, 326 and 316 m asl, respectively (corresponding to water depths of 10, 11, 15, 16 and 26 m today).

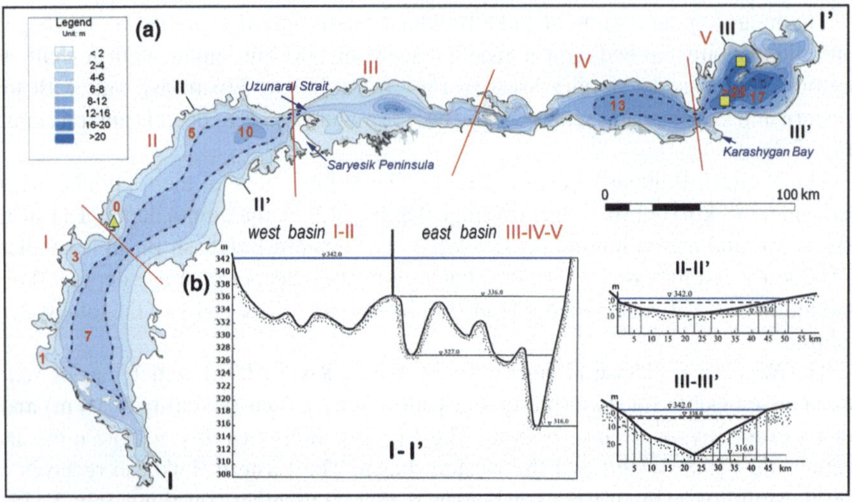

Fig. 5.3 Bathymetric map of Lake Balkhash. **a** Roman numerals I–V in red define the main morphometric partitions of the lake; Arabic numerals show average water depth in metres. Punctuated blue line: Balkhash reservoir under the worst forecasted scenario, with water levels at 337 m asl (Tursunov 2002). Yellow triangle: position of core BAL07; yellow squares: position of cores 0901 and 0902. **b** Morphology of bottom relief: line I–I′ is a W-E profile, and lines II–II′ and III–III′ are N-S profiles of the Western and Eastern basins, respectively. Adapted from Dostay (2009), Myrzakhmetov et al. (2017)

5.2 Geological History of the Lake Balkhash Region

The history of the research of Lake Balkhash during Soviet times can be summarized in three major campaigns of data acquisition and knowledge integration: the works of the 1940s led by D. G. Sapozhnikov who organized the collection and analyses of sediment cores from 160 sites (Sapozhnikov 1951); the studies of Quaternary deposits in the area of the Balkhash-Alakol Depression led by T. N. Dzhurkashev in the 1960s; and the collective researches of the 1970s and 1980s led by N. N. Verzilin who supervised new analyses and introduced the first absolute chronology for lake deposits (which, unfortunately, resulted in a misleading interpretation of the lake history still often used nowadays).[4]

[4]The first important studies of the geological history of the modern Balkhash, although focused on geological characters with scanty information about hydrological phases, had been published by Berg (1904), by Rusakov (1933), and Kostenko (1946), constituting for decennia the main references on the lake. Besides the fundamental studies of Sapozhnikov (1951), other important contributions have been those of Kurdyukov (1958) and Tarasov (1961). Dzhurkashev (1964, 1972) started to consider water-level fluctuations and found evidence for a short-term isolation of the Eastern Balkhash in the early eighteenth century, which decreased its water level by 6–7 m. Only later on, under the stimuli of the Balkhash regression provoked by the realization of the Kapchagai Reservoir, Venus (1985) studied buried peat bogs and inferred regression events, and N. N. Verzilin provided the first qualitative reconstruction of the Holocene lake-level fluctuations on the basis of

Concerning more recent international researches following the perestroika, significant investigations of the modern Lake Balkhash were conducted during the years 2007–2012 in the frame of the Kazakh-Japanese research project "Historical interaction between multi-cultural societies and the natural environment in a semi-arid region in Central Asia"[5] (hereafter shortened as "Ili Project"). The accumulated data enabled the quantitative evaluation of the historical scenario already drawn by N. N. Kostenko and T. N. Dzhurkashev. Among the main results of the fieldwork surveys and laboratory analyses are reconstructions of transgressive hydrological events during the last 35 ka BP based on dating of gravel bars exposed on the northern shore, and a detailed reconstruction of the Balkhash history for the last 8000 years decoded from three cores of bottom sediments (Watanabe and Kubota 2010; Sala et al. 2016).[6]

5.2.1 History of the Ancient and Modern Balkhash Basins Since 300 ka BP

Lake Balkhash did not exist until ca. 300 ka BP. The tectonic Balkhash-Alakol Depression was already established ca. 10–15 millions years ago as a lacustrine landscape made of ponds fed by little streams flowing down the Dzungarian Alatau and the northern Pre-Balkhash hillocks. The largest water body of the basin was the Ili Lake, which was extending West of the present Chinese border, occupying the bed

---

mineralogical, hydrochemical and biotic analyses of sediment cores (Sevastyanov 1991; Tursunov 2002; Dostay 2009). Y. P. Khrustalev and Y. G. Chernousov published an article in 1992 reproducing their historical reconstruction of the lake already included in Sevastyanov (1991) which constitutes today the main (and unfortunately partly confusing) reference concerning the Holocene lake history (see also note 8).

[5]The "Ili Project" was based on the cooperation between the "Research Institute for Humanity and Nature" (RIHN) of Kyoto and three organizations based in Almaty: the "Institute of Geological Sciences named after KI Satpaev", the "Kazakh State Research Institute of the Cultural Heritage of the Nomads" and the "Laboratory of Geoarchaeology". The researches of the Ili Project focused on the Ili-Balkhash basin (Semirechie) and included several fields: the reconstruction of palaeoclimate-environmental changes, historical fluctuations of Balkhash's lake levels and of ice deposits, archaeological traces of land and water use, Late Medieval historical accounts concerning the territory, and Soviet documents and post-Soviet interviews about the pastoralist and agricultural activities in the basin.

[6]The investigation of lake sediments was initiated in 2007 by two international teams working in cooperation with the Satpaev Institute of Geological Sciences of Kazakhstan (B. Aubekerov): a Chinese team led by Zhaodong Feng and Chengjun Zhang, and a Japanese team of the RHIN Institute led by Jumpei Kubota. That year, both teams drilled sediment cores at short distance from each other in the lake near the Tasaral Island in the western basin of Lake Balkhash. The preliminary analyses of the 'Japanese' core BAL07 were published in 2010 (Endo et al. 2010) and more detailed results based on fossil diatom assemblages were issued in 2016 (Chiba et al. 2016). The pollen-based study of the 'Chinese' core BK-A was published with incorrect location of the coring site by Feng et al. (2013) and an additional study of the ostracod record of the same core by Mischke et al. (2020). In 2009, the Japanese team drilled two additional cores (0901 and 0902) in the easternmost part of Lake Balkhash, and preliminary analyses of the cores were published by Endo et al. (2012; Sect. 5.2.2).

of the modern Kapchagai Reservoir and proceeding further West as Ili River until its confluence with the Chu River and reaching the Aral Sea.

Under the tectonic rise of the Zailisky Alatau, the water inflow increased progressively and a critical deformation uplifted the Karaoi Plateau (700 m asl), all together provoking a northward deviation of the Ili Lake that was displaced into the southwestern part of the Balkhash-Alakol Depression (at the time at 200–250 m asl, 200 m lower than today) and formed here its first Ili Delta (Akdala Delta; Dzhurkashev 1972).

As shown in Fig. 5.4b, the Paleozoic substratum of the Balkhash-Alakol Depression has a cup-like profile with a central depression at -100 m asl in the region

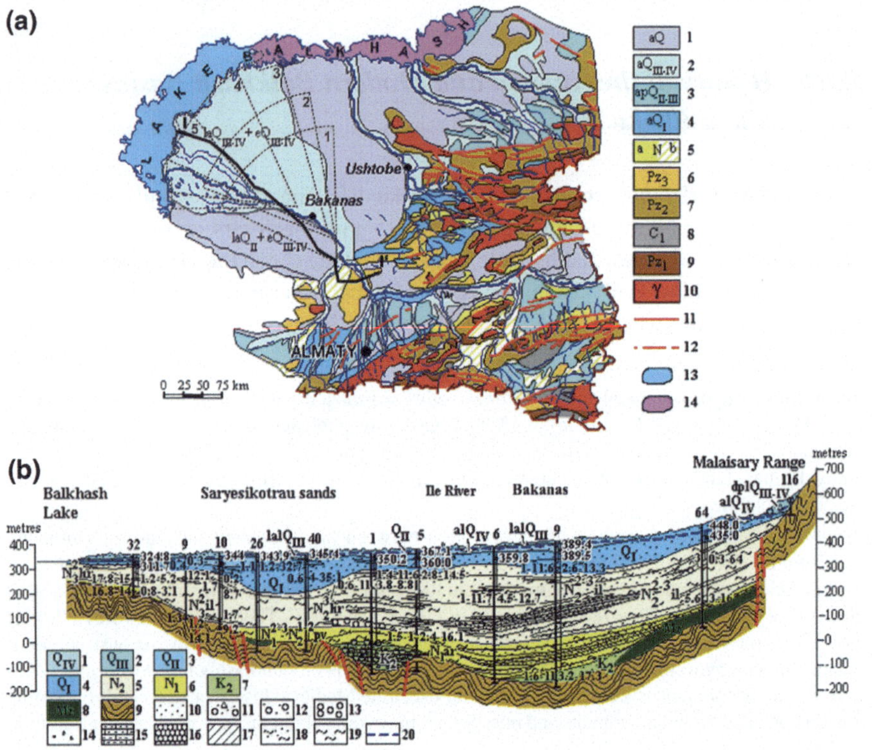

Fig. 5.4 Hydrogeological map and longitudinal section of the South Balkhash Depression. **a** Map of the region showing the successive historical deltas of the Ili River (dotted fans numbered 1–5) crossed by the transect I-I' (black line); 1–4 Quaternary deposits, 5 Neogene, 6–10 Paleozoic, 11–12 aquiferous fissures, 13 freshwater lake, 14 slightly saline lake. **b** Hydrogeological section along transect I-I' from the southern shore of the Western Balkhash to the Malaisary Range (Northwestern piedmonts of Dzungarian Alatau); 1–4 Quaternary deposits, 5–6 Neogene, 7 Upper Cretaceous, 8 Mesozoic, 9 Paleozoic, 10–19 soil and rock features, aquiferous faults, 20 shallow aquifer. Acronyms inside the section refer to locations: il Ili Suite, al alluvial lake, dp deluvial-proluvial deposits. (Adapted from Akhmedsafin et al. 1973; http://www.water.unesco.kz/bal_ch_5_e.htm; and Abdrasilov 1996)

of Bakanas village rising to +250 m asl on both sides against the Zhungarian and the northern Pre-Balkhash pre-mountain zones. Middle Pleistocene deposits (Q_{II}) between +230 to +330 m asl, suggest that, following the Ili diversion, at first a lake (Bakanas Lake) or a system of several lakes was formed in depressions around Bakanas, which then had been filled and displaced North-East by the accumulation of Ili River sediments and the formation of the Ili Delta. Gradually, between 300 and 100 ka BP, a huge lake called Ancient Balkhash was 'sediment-dammed' and seized the entire northeastern part of the depression unifying the present Balkhash, Sasykkol, Alakol and Aibi basins.

This mega-lake was reshaped at the very beginning of the last glacial period (ca. 110 ka BP) following intensive orogeny of the Dzungarian Alatau that hoisted the Arkarly Mountains at 780 m asl dividing the Ancient Balkhash in two different water bodies: the Alakol Lake in the East (today at 347 m asl) and the modern Lake Balkhash (inclusive of its eastern and western parts) in the West (today at 342 m asl; Kurdyukov 1958; Dzhurkashev 1972; Aubekerov et al. 2009a, 2010).[7,8]

The configuration of the shorelines of the modern Lake Balkhash is determined by a 20–30-m high, tectonically disturbed rocky terrain in the West and North and by flat sandy alluvial and aeolian deposits in the South. The lake as a whole is lying on the most elevated and faulted northern slopes of the Paleozoic depression which favors groundwater circulation between the lake bottom and the vertical overflow of pressure waters of the southern Pre-Balkhash artesian basin, across a few hundred metres of semi-permeable sediments. The lowest reaches of a huge accumulation of alluvial

[7]The average sediment thickness in the area of the modern Western Balkhash is 70 m, parted as 44 m of Neogene and 26 m of Quaternary deposits, the latter with distinct 2–3 horizons (the second horizon uneven and eroded). Total sediments' thickness decreases from 70 m in Western Balkhash to 50 m in Eastern Balkhash on the account of the lesser river and sediment input of the eastern basin (Sevastyanov 1991).

[8]The reconstruction of the history of the lake provided by Khrustalev and Chernousov (1992) is relatively different and confusing. Using a chronological framework established by Maksimov (1961) for the glacial stages of the Dzungarian Alatau, the authors attribute the appearance of an Ancient Balkhash ("Balkhash-Alakol") basin outrivaling the Aral Sea in size, even if not reached yet by the Ili River inflow to the post-glacial early Holocene period. The disappearance of the Ili Lake by Ili River diversion is assigned to the end of the Atlantic period (5.6 ka BP) and is quoted as responsible for the formation of the "Balkhash", so that, in the view of the authors, the entire formation and rotation of the Ili deltas occurred during the last 5 ka.

Then, ignoring geological reports and only referring to lithological and chemical data from core samples provided by N. N. Verzilin's fieldwork in 1979–1981, Khrustalev and Chernousov reconstructed the succession of four stages of the lake during the Holocene, each stage starting with a transgression (T) and ending with a regression (R). These fluctuations are named by four specific toponyms and approximate chronological boundaries (all in ka BP) were provided: (I)-Ancient Balkhash (T 10.3, R 8.3–5.6), (II)-Balkhash (T 5.6, R 4.4–3.5), (III)-New Balkhash (T 3.5, R 2.6–1.9), (IV)-Modern Balkhash (T 1.9, R intermittent). The present authors infer that three of these regressive events reduced the Eastern Balkhash into a series of "isolated or semi-isolated pools": between 8.3 and 5.6, at 3.8 and 0.75 ka BP (quoted from Venus 1985).

The described middle-late Holocene history of the modern Balkhash, by being the most recent and best furnished in terms of radiometric ages, has been diffused and adopted as reference by many scientists (e.g., Tursunov 2002; Solomina and Alverson 2004; Krylov et al. 2014; etc.) against the reliable geochronology of the basin established by geologists before.

sediments in its South favor the water input from rivers and from shallow aquifers. The resulting groundwater dynamics, accompanied by seepage of ion-enriched lake waters, contribute with other processes such as the dispersion of coastal salt deposits by wind, to the maintenance of a relatively low salinity of the endorheic lake (Smolyar and Mustafaev 2007).[9]

The Ili Delta has always been the principal feeder of the lake by freshwater and biomass during the late Pleistocene and Holocene, and it continuously evolved, changing its morphology, location and size. Arid phases, through diminishing the discharge of delta distributaries, favor the lowering and erosion of the existing river channels, the formation of terraces and, in general, the stabilization of the existing delta morphology. The establishment of a new delta typically coincides with pluvial phases that enhance the energy levels of the distributaries and promote diversions in various directions.

The Late Pleistocene, postglacial and Holocene evolution of the Ili Delta occurred in five stages (Fig. 5.4a), featuring as a whole a gradual northern and anticlockwise rotary displacement under the forcing of sedimentary and tectonic factors (Abdrasilov 1996; Deom et al. 2019):

(I) The first stage, barely detectable, refers to the Akdala Delta (possibly several superposed deltas): the head was around Bakbakty village, and the delta area was huge, with distributaries feeding the eastern part of the modern Lake Balkhash during the glacial period.

(II) The Bakbakty Delta was created during the postglacial period: it was fed by the so-called Palaeo-Ili River course with head at Birlyk, 30 km North of the former one and an anticlockwise rotation of the distributaries.

(III) With the start of the Holocene, ca. 10 ka BP, the Uzunaral Delta (or Older Bakanas Delta) was created, with head displaced further North-West (30 km southeast of Bakanas village, at 500 m asl) and the delta front reaching the present lake shore in correspondence with the Saryesik Peninsula. The last consists of alluvial sediments of 30–50 m thickness, shaping the connection between the Western and Eastern Balkhash into a narrow and shallow strait (the Uzunaral Strait, today 6 m deep).

(IV) Only ca. 2 ka BP, the new Ili course became activated, parallel to the Palaeo-Ili in the West, and opening a delta with head at Bereke (30 km North of Bakanas). It further rotated anticlockwise, and in that way, the Bakanas Delta (or Younger Bakanas Delta) was established, significantly reduced in size and discharging in the western part of the lake.

(V) Only recently, during the pluvial phase of the sixteenth to eighteenth centuries, the delta rotated further West to form the modern Ili Delta (Dzhurkashev 1964).

[9]Especially aeolian processes promote the relatively low salinity of the lake's water: "The stock of salt reserves in the lake's area amounts to only 260–300 million tons, which is not large since, in addition to their consumption in the process of carbonate formation, some of them are lost by entering the numerous bays of the winding coastline. During phases of water level recession, the bays separate from the lake, dry out, and their salt deposits are blown by the wind, thus reducing the salt's reserve of the lake" (Sevastyanov 1991).

5.2.2 Hydrological Stages of Lake Balkhash During the Late Pleistocene and Holocene

The most recent changes of the modern Balkhash across the Last Glacial Maximum up to now occurred mostly under the activity of rivers and changing climate, as a result of sediment accumulation and transgression-regression cycles, because no major tectonic events arose anymore. In that sense, the "Ili Project" focused on the analysis of Balkhash's water level changes during the last 40 ka; and significant study objects became the tracers of water-level variations, such as the absolute heights of sand bars on the shore and the physical, chemical and biotic composition of the lake sediments.

The study and optically stimulated luminescence (OSL) dating of exposed gravel bars at the shore of the Karashygan Bay in the southeastern part of the lake (Fig. 5.3) revealed the presence of very relevant transgressions, the oldest to 354 m asl (+13 m higher than today's water level), dated to 36 ka BP, ascribed to the interstadial stage of Marine Isotope Stage 3. Two successive transgressions of lesser entity were detected at ca. 10 m above the present lake and dated to the LGM between 25 and 17 ka BP. The early Holocene lake level had decreased significantly ca. 8.4 ka BP when the lake reached for the first time its present level at 342 m asl. These events attest a tendency to very high lake levels during the very cold phases with minimal evaporation of the Last Glacial Maximum and postglacial period, followed by lower water levels during the drier middle and late Holocene.

Analyses of the lake sediments of two cores drilled in the deepest part of the eastern basin (cores 0901 and 0902) and one core from the shallow western basin (BAL07; Fig. 5.3) allow the quantitative reconstruction of Balkhash's water-level fluctuations during the last 8 ka BP.

The sedimentary columns of cores 0901 and 0902 have been recovered at a water depth of 20 m in the easternmost Burlyutyubin Basin of the Eastern Balkhash, with a length of 6 m and a chronology of the last 8 ka BP (Fig. 5.5). The sediments were dated using the radiocarbon technique, and examined by complex analyses including lithology, geochemistry, magnetic susceptibility, fossil diatoms (siliceous algae) and ostracods (micro-crustaceans), pollen and spores which are partly still not completed. The recorded data indicate a transgressive mode during the pluvial Atlantic period (7.0–5.5 ka BP) and three main regressions at 5.5–5.0, 2.7–2.4 and 1.3–0.8 ka BP. If the second and third regressions were severe but presumably not falling below 336 m asl, the first regression at 5.5–5.0 ka BP definitely went below 336 m asl which is the level of the Uzunaral Strait floor at which the Western and Eastern Balkhash become separated water bodies. Thus, the two basins were disconnected and isolated lakes formed in the Eastern Balkhash in the middle Holocene.

At 3.7–3.0 m of their columns (the sediments at 2.9 m depth in core 0901 have an age of 4.7 ka BP), both cores revealed sediments characteristic of an arid or semiarid environment. Core 0902 contains a layer of abundant gypsum at 3.67 m, indicating strong evaporation effects on the lake waters. Core 0901 includes laminated sediments rich in Fe and Si at 3.50 m which are overlain by silt and fluvial sand (Sugai

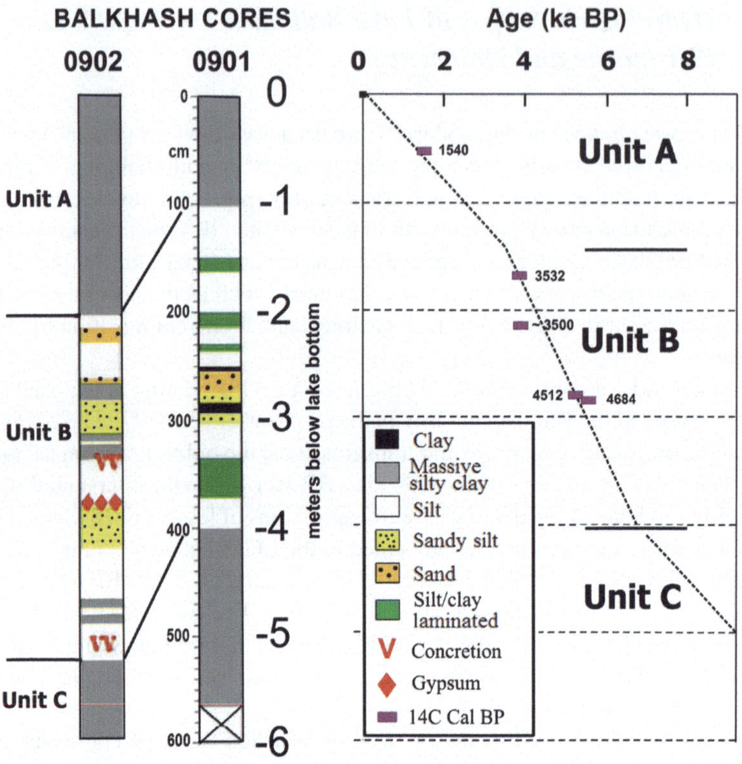

Fig. 5.5 Lithological profiles and age-depth plot (approximate accumulation rate 0.5 m/ka; Endo et al. 2012) of cores 0901 and 0902 from the Eastern Balkhash (original figure published in Endo et al. 2012, Fig. 5)

et al. 2011). The depth of the gypsum layer in core 0902 corresponds to ca. 318–319 m asl, i.e., 23–24 m below the present water surface of the lake, suggesting an extreme lake-level drop of 24 m in the Burlyutyubin Basin and the reduction of the lake to a small brackish pond ca. 5.5–5.0 ka BP.[10]

A similar inference is confirmed by the biotic proxies: palynological spectra, ratios of benthic saline and planktonic freshwater diatoms, and the contraction and diversity change of different ostracod species (Endo et al. 2012, 46).

Several peat layers have also been recorded by Soviet geologists (Sapozhnikov 1951; Dzhurkashev 1972; Venus 1985) but a significant lake-level drop as those in the middle Holocene was not detected previously, not even suspected, underlining the vulnerability of the Eastern Balkhash Basin and the importance of this discovery.

The core BAL07 was drilled in the shallow western part of the lake near the Tasaral Island at a water depth of 3 m. The core is 6 m long and covers the time span of the last 2000 years. Its lithology and biotic content disclosed two major

[10]Taking into account sedimentation rates, such pond could have been 4–5 m deep.

medieval regressions dated to 750 and 1150 AD, roughly the start and the end of the Medieval Warm Period (MWP).[11] The water level dropped by ca. 5 m during the earlier backset and the Uzunaral Strait might have desiccated (which at that time was supposedly crossed by a caravan road linking the delta to the northern shore). The later regression was less intense but longer (Fig. 5.6). Afterwards, the lake experienced a long transgressive phase under the pluvial climate of the Little Ice Age, between 1300–1800 AD, interrupted by three regressions at the start and the end of the sixteenth century and around the late eighteenth century.

Two main factors are regarded as the causes of the historical lake-level regressions: climatic aridification diminishing the river inflow and enhancing evaporation, and anomalous transmission losses, natural or anthropogenic.

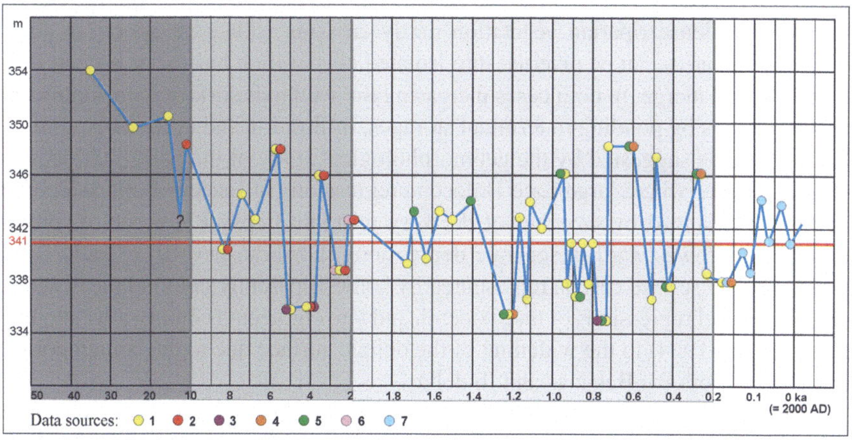

Fig. 5.6 Synoptic reconstruction of the water level of Lake Balkhash during the last 36 ka. *Sources* 1 OSL data of gravel bars of the Karashygan Bay (earliest transgressions dated to 36, 25 and 17 ka BP), and ^{14}C data of cores 0901, 0902 and BAL07 (Endo et al. 2012; Chiba et al. 2016); 2 ^{14}C data from Y. P. Khrustalev's lithological analyses (10.3, 8.3, 5.6, 4.4, 3.6, 2.7 ka BP) extended by the dating of an additional peat layer by N. N. Verzilin (3.97 ka BP; Sevastyanov 1991), 3 ^{14}C data from B. G. Venus' analyses of peat layers (5.05, 3.86, 0.75 ka BP; Venus 1985); 4 palynological data of core BAL07 (Aubekerov et al. 2009b); 5 inferred climatic phases: multi-secular long minimum (~1.25 ka BP) at 335.5 m, multi-secular long maximum (0.7–0.5 ka BP) at 348.5 m, secular maximum (0.27 ka BP) at 346 m, secular minimum (0.16 ka BP) at 338 m (Kurdin 1976; Shnitnikov 1957); 6 ^{14}C, pollen and ostracod data from the core BK-A (Feng et al. 2013; Mischke et al. 2020) 7 Gauge measurements of water levels for the period 1879–1967 (Semenov and Kurdin 1970). The decreasing intensity of gray shading refers to four different chronological scales, from left to right: 10 ka, 1 ka, 0.1 ka, 0.05 ka

[11]Using diatom assemblages correlated with Ca content in ostracods radiometrically dated, the Japanese team placed the two low-level stands in a series of seven regressive periods: 0–300, 330–360, 750–790, 1060–1260, 1560–1600, 1780–1840, 1950–1990 AD (Chiba et al. 2016). The pollen-based reconstruction by Feng and co-authors confines itself to the outline of three climatic stages in the basin: cool-wet (500 BC–200 AD), moderately warm-dry (200–1350 AD) and cool-wet for the past 650 years with a last warm-dry century (Feng et al. 2013).

The climate of the Balkhash region is under the influence of the northern Atlantic atmospheric circulation, which evolves over the northern Atlantic Ocean in response to sea surface temperature fluctuations driven by the Atlantic Multidecadal Oscillation (AMO), and to atmospheric pressure changes of the North Atlantic Oscillation (NAO). In western Central Asia, the late Holocene and in particular the Subatlantic period was characterized by the alternation of hot-dry and cold-wet climate phases. Arid phases were detected at 5.4–5.2, 4.6–3.9, 3.2–2.8, 2.0–1.6 and 1.3–0.8 ka BP, intercalated by moist phases (Chen et al. 2008; Hill 2018). Moreover, the effects of fluctuations between arid and humid phases on decreasing or increasing levels of Lake Balkhash are postponed by negative or positive changes in glacial ice volumes (Footnote 3; Feng et al. 2013).

The tributary rivers of Lake Balkhash are flowing on broad flat plains with discharges significantly decreasing downstream, particularly in the very variable deltaic courses with dense riparian vegetation partly crossing isolated evaporation pans. Delta distributaries can be diverted into isolated evaporation basins, or instead they can widen and merge, in both cases increasing the wetlands surface with significant transport losses by ponding in terminal storages, infiltration and evapotranspiration. The first case is suggested by the geomorphological study of the palaeo-terraces of the Lepsy River where large Late Holocene regressions of Eastern Balkhash could have been enhanced by switches of the very unstable Lepsy River bed, diverting its waters southwestward in separate depressions (Dzhurkashev 1972; Endo et al. 2012, 44). The second case is represented by some apparently not climatically triggered lake-level regressions of the sixteenth and late eighteenth century attributed by Dzhurkashev (1964) to the widening of the deltaic surface due to the contemporary activation of both the Bakanas and Ili deltas.

The Late Pleistocene and Holocene hydrological regime of Lake Balkhash can be summarized as three successive stages, each correlated to specific climate conditions and lake-level ranges (Fig. 5.6):

- the first stage, with high water levels between 355 and 349 m asl, is connected with glacial and postglacial conditions with very low evaporation;
- the second stage, with water levels between 348 and 341 m asl, occurred during the early and middle Holocene, and is characterized by warm and moderately wet conditions and still significant accumulation of glacier ice in the mountains of the catchment;
- the third and present stage, with water levels between 348 and 335 m asl, exists since the late Holocene, and is characterized by an arid climate and small volumes of glacier ice, and by relevant lake-level fluctuations between arid and humid phases. Henceforth, three main phases can be distinguished within this third stage: regressive (5.0, 1.2, 0.8 ka BP), transgressive (0.7–0.1 ka BP), and regressive (0.2, 0.1, 0 ka BP).

5.3 Modern Balkhash: Present Hydrological Conditions Under Increasing Climate Warming and Water Subtraction

The history of Lake Balkhash is significantly better known for the last 120 years thanks to gauge stations introduced in 1879. Today, quantified data concerning the lake level, water quality and biological environment are regularly collected, analyzed and used for monitoring by three main institutes based in Almaty: the Hydrometeorological Institute "Kazhydromet", the Department of Water Resources of the Institute of Geography, and the Akmedsafin Institute of Hydrogeology and Geoecology.

According to their annual measurements, Lake Balkhash, like most Central Asian lakes, manifests a tendency of progressive desiccation during the last century, with the exception of two high stands in 1908–1912 (343.7 m) and in 1960–1972 (343 m). The lake level increase after 2000 (342.5 m) is attributed to melting of glaciers (Fig. 5.7a).

During the period 1960–1986, the lake was confronted with three major regressive factors: the continuation of the secular dry climate trend, the filling of the Kapchagai Reservoir (1970–1986), and the increase of the irrigation area within the basin from 300,000 to 550,000 ha and of anthropogenic water withdrawal from 3.5 to 5.5 km^3. The total river inflow decreased by 3.8 km^3 per year (−23%), of which 62% were attributed to natural variability and 38% to anthropogenic subtraction; and Balkhash's water level fell from 343 m asl in 1970 to 340.6 m in 1986 when the Kapchagai crisis culminated. This lake-level decrease was accompanied by a very sensible decrease of the water surface by 25% with shorelines receding 2–8 km, and by a reduction of the water volume by 40% (Kezer and Matsuyama 2006).

Balkhash's lake level regained to 341.5 m asl between 1986 and 2008, following the reduction of the active capacity of the Kapchagai Reservoir to 6.64 km^3 (23.7% of the originally projected capacity)[12] and the post-perestroika economical crisis in Kazakhstan which was, anyhow, counterbalanced by an increasing use of Ili River waters on the territory of the People's Republic of China (Shaporenko 1995; Fig. 5.7).

During the period from 1955 to 2000, the accelerated melting of the glacier ice in Balkhash's catchment basin exceeded the global average ice melting by a factor of four. Thus, the meltwater surplus compensated the increased evaporation losses from the lake as a result of the warming trend. Ice volume in Balkhash's catchment decreased from 122 to 90 km^3, i.e., −26% between 1955 and 2000 (−0.8 km^3 on average per year; Bolch 2015; Severskiy et al. 2016). The corresponding surface decreased from 2000 to 1500 km^2, i.e., −25% between 1960 and 2007 (−0.6 km^2 per year; Xu et al 2015; Severskiy et al. 2016). The persisting glacier shrinkage provides a significant 5–10% excess of yearly river inflow that will disappear under ongoing climate trends and glacier retreat within less than 50 years.

The shallow bathymetry of Lake Balkhash with an average depth of 5.7 m makes it sensitive to inflow changes. The ratio between water surface (18,210 km^2) and

[12] A water volume of 6.6 km^3 in the Kapchagai Reservoir corresponds to a water surface of 1250 km^2 (7.1% of Lake Balkhash's surface) and an annual evaporation of 1.2 km^3.

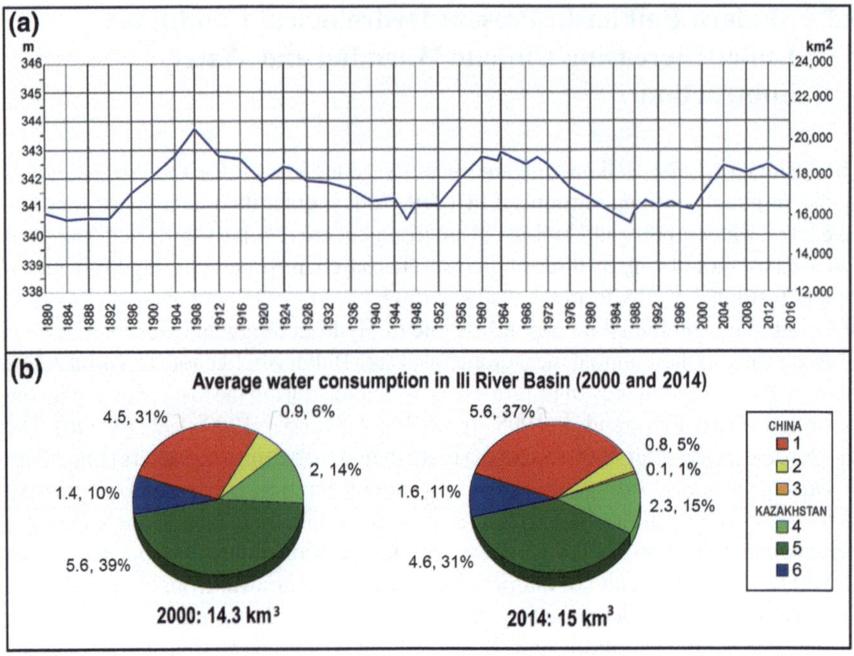

Fig. 5.7 Historical development of Lake Balkhash's level (left axis) and area (right axis) since 1880 and of water consumption in the Ili River Basin after 2000. **a** Water-level and area fluctuations between 1880 and 2016 (based on data from Semenov and Kurdin 1970; Propastin 2012; Myrzakhmetov et al. 2017). **b** Water consumption (km^3) in the Ili River Basin in 2000 (37% in China, 63% in Kazakhstan) and in 2014 (43% in China, 57% in Kazakhstan). Legend: China 1 Agriculture, 2 Riparian vegetation, 3 Lakes and reservoirs; Kazakhstan 4 Agriculture, 5 Riparian vegetation (including delta vegetation), 6 Lakes and reservoirs (Balkhash excluded; data from Thevs et al. 2017)

water volume (106 km^3) is very high and implies a correspondingly large volume of evaporated water (yearly 17.2 km^3). Thus Balkhash's resilience is very low; the lake would disappear without water inflows in less than seven years.

Today, a mean annual water input of 18 km^3 ensures a lake level at 342 m asl but future decreases of meltwater discharge and increasing aridification with higher evaporation will result in a lower lake level. Moreover, demands for anthropogenic water withdrawal are rising, mostly through the expansion of irrigated surfaces. Thus, Lake Balkhash requires a strict water management system for its catchment area including common rules for all water consumers because the current state of the lake can only be secured with a reduced and better use of the Ili River water.

The most recent change in the complex hydrological system of Lake Balkhash resulted from plans prepared in the 1990s to multiply the area of irrigated fields in the upper Ili Basin in Xinjiang by three to four times accompanied by a drastic demographic increase.[13] Today, the resulting irrigation farming and established infrastructure and population density reached the alarming capacity of a yearly subtraction of a few km^3 of water that may soon cause a drop of Balkhash's level below the critical height at 336 m asl and eventually the disappearance of the Eastern Balkhash.

The size of irrigated areas on the Chinese side grew between 2000 and 2014 from 400,000 to 500,000 ha and the water retention from 5.0 to 5.6 km^3, reaching the level of the implementations on the Kazakh side during Soviet times. On the Kazakh side, irrigated areas and water use were slowly increased after the collapse which followed perestroika. In 2014, an area of 250,000 ha was irrigated and 2.3 km^3 of water was used (Thevs et al. 2017).

The combined water withdrawal for irrigation purposes in Balkhash's catchment in Kazakhstan and China amounts to a total of 7.9 km^3 (52% of the river inflow) today which will further increase in the future (Fig. 5.7b). If the annual water consumption grows by additional 3 km^3, the lake level of Balkhash will drop to 340.2 m asl and the lake surface will shrink to 15,000 km^2. A higher increase of the consumed water by additional 6 km^3 would cause a lake-level fall to 338 m asl and a corresponding lake surface of 10,000 km^2. An even higher increase by additional 9 km^3 would result in a lake-level drop to less than 336 m asl and a division of the lake in three to six different basins of which some would completely desiccate (Propastin 2013).

5.4 Modern Balkhash: Lithology of the Sediments and Hydrochemistry of the Water Column

5.4.1 Types of Sediments

The geological features and hydrological conditions of the Balkhash Basin determine the character of the lake-bottom sediments and of the water (Sect. 5.4.2) and the character of its biotic components (Sect. 5.5).

The sediments of Lake Balkhash, suspended or dissolved[14] in the water column and precipitated to the lake bottom, consist of allochthonous materials transported by rivers, waves and winds, and of autochthonous materials produced within the lake itself by chemical processes and biotic activity. They consist of suspended inorganic and organic debris and of dissolved chemical particles differently distributed within the lake according to the location and discharge of the individual river deltas,

[13]Chinese authorities do not provide information about their hydraulic plans in the region and, up to now, did not respond to repeated requests for the establishment of an international consortium for the management of the trans-boundary hydrological system of the Ili-Balkhash Basin.

[14]A solution is the homogeneous mixture of a solute and a liquid where, in contrast to a suspension, the solute cannot be separated from the solvent by filtration.

the lithological and chemical characteristics of the water inflow, and the peculiar elongated morphometry and resulting water circulation in the basin (Fig. 5.8).

Within the total transported sediments, the clastic terrigenous materials, mainly in the form of sand, are by far the most abundant (ca. 4–6 million tons per year, 0.005 km^3), accounting for 92–93% of the annual total. They are made of mineral particles of quartz, feldspar, volcanic rock fragments, limestone and shale, sized between

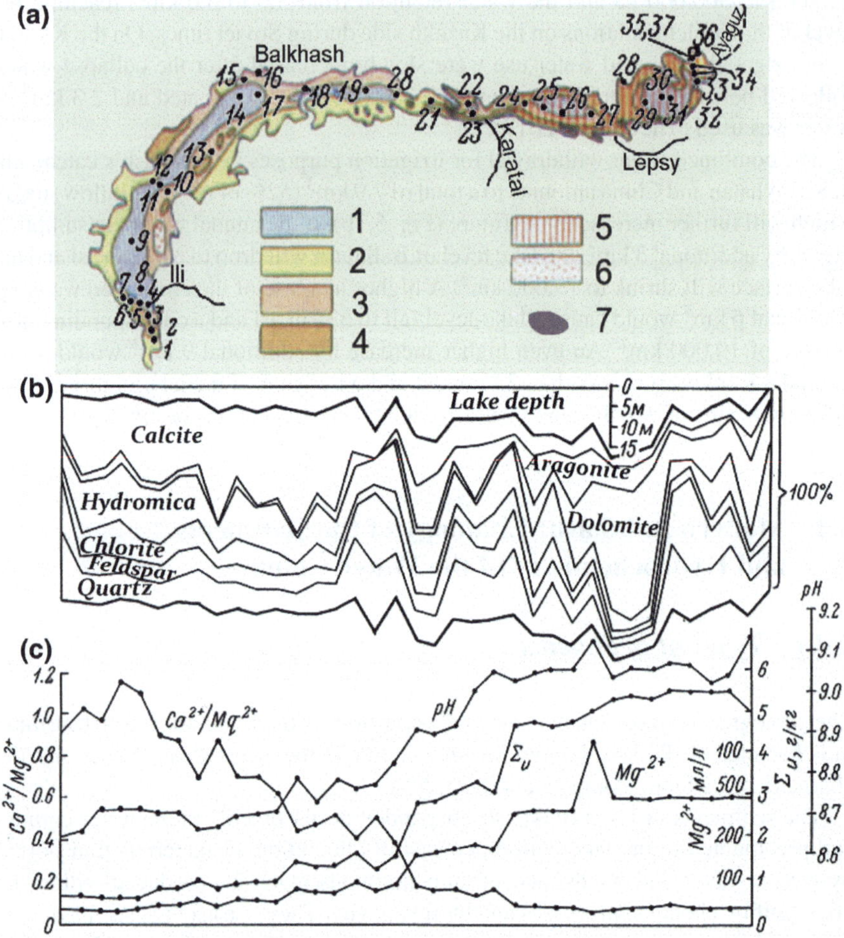

Fig. 5.8 Lithological and hydrochemical characteristics of Lake Balkhash. **a** Location of 36 sediment cores (6 m long) retrieved at regular intervals along the lake's long axis. Main granulometric concentration: 1 silt; 2 silty sand; 3 sandy silt; 4 pebbles, gravel; 5 dolomite; 6 sandbank; 7 Neogene clay. **b** Above: West-to-East profile of lake depth. Below: West-to-East profiles of ratios in percent of lithological composition of bottom sediments (calcite, aragonite, dolomite, etc.). **c** Hydrochemical composition of the water column (ratios of Ca^{2+}/Mg^{2+}, pH, salinity $\Sigma\upsilon$, Mg^{2+}; from Atlas Kazakhskoi SSR 1982: 59 and Sevastyanov 1991, Fig. 32)

clay and gravel. The mineral composition of the lake deposits highly depends on the transport mechanism and size of the sediment particles (Fig. 5.8a). Granulometric analyses of the bottom sediments (Fig. 5.8a) show that sands, mainly carbonaceous, constitute the absolute majority around delta mouths and account for more than 50% of the deposits in the middle part of the lake. The mineral composition of the lake deposits highly depends on the size of the sediment particles. Wind-transported aeolian silts, mainly siliceous, abound all along the dunes near the southern coast. Clays of different color and composition (mainly green chlorites and hydromicas) accumulate abundantly in the western basin and in the bays along the northern coast where they constitute more than 70% of the surficial lake-bottom sediments (Sapozhnikov 1951).

The mineralogy of the bottom sediments is dominated by rock-forming components from the same terrigenous geochemical province, i.e. mainly as a result of suspended materials carried by the Ili River to the lake (Fig. 5.8b). Carbonate minerals [calcite and aragonite $CaCO_3$, dolomite $CaMg(CO_3)_2$] are the most abundant components and also those reacting fastest and precipitating earliest from the waters, so that their portion is >50% in bottom sediments around the delta mouths (Ili, Aksu, Lepsy) and between 10 and 20% in the remote central regions of the basin, balanced here by an increase of more stable silica and sodium oxide. Iron concentration varies between 3 and 9% and is evenly distributed in high concentration along the northern coast. In contrast, sediments in the easternmost basin fed by the Lepsy and Ayaguz tributaries represent an anomalous mineralogical composition with a tenfold higher concentration of magnesium and manganese oxides, and a decrease of iron, resulting in a significantly different province of authigenic geochemical sediments there (Sevastyanov 1991).

Chemogenic sediments consist of solid and gaseous undissociated compounds, molecules and ions dissolved in water, from which a part precipitates by chemical reactions. Their yearly input from rivers represents a small fraction (less than 2–3%) of the yearly total sedimentary budget but they represent by far the main sedimentary components dissolved in the water column. A total mass of 13 million tons (i.e., a total volume of 0.025 km^3) precipitates per year from the water column. Their dissolved mass and composition result from thousands of years of chemogenic processes and, together with water inflow and evaporation, control the hydrochemical conditions of Balkhash's biotic system.

Biogenic materials consist of terrestrial macrofossils, phytoliths, pollen and charcoal from fires. They are suspended in low quantity in the water and practically absent in most bottom deposits. A larger fraction of the sediments is represented by calcareous and siliceous skeletal elements of aquatic organisms in the lake. These are mostly ostracod valves and diatom tests, and to a lesser degree also bivalve and gastropod shells. Organic carbon concentration of bottom sediments is ca. 1% in the proximity of delta mouths and less than 0.5% in the central parts of the lake (Sapozhnikov 1951).

5.4.2 Composition of the Water Column

Lake Balkhash's water is almost isothermal and isochemical without vertical stratification due to the shallow depth of the lake and the action of wind and currents.

Mean annual surface-water temperature is +1.1 °C, mean December surface-water temperature is −3.3 °C, and mean July surface-water temperature is +23.0 °C. Summer-water temperatures are close to ambient air temperatures (Sect. 5.1) but winter surface-water temperatures are significantly higher than air temperatures.

The solid particles suspended in the water are clay- and silt-sized, carried by flowing waters until they settle out when flow is insufficient to keep them in suspension. Together with dissolved chemical particles, they determine the moderate turbidity in the western basin (visibility 5–10 m) and the increasing turbidity to the East enhanced by winds, currents and salinity until a visibility of 1 m or less.

Among gases, dissolved oxygen (O_2) as the most critical indicator of a lake's environment and of a healthy aquatic ecosystem, is in a range from 6 to 10 mg/L in Lake Balkhash. Dissolved oxygen saturation is typically 100% at the surface and 90% at the lake bottom with variations due to local water temperature, salinity and depth.

Total Dissolved Solids (TDS) is defined as all substances contained in water (metals, minerals, salts, calcium and other compounds which can be both inorganic and organic) that can pass through a 2 micron filter. Thus, TDS is a measure of the sum of the amount of dissolved ions (as mg/L).[15] The relatively larger and most abundant ions or macro-components in Lake Balkhash are carbonates (hydrocarbon-bicarbonate HCO_3^- and carbonate CO_3^{2-}) and calcium (Ca^{2+}) in equal proportions (together accounting for ca. 68% of dissolved ions); magnesium Mg^{2+}, sulphate SO_4^{2-}, sodium Na^+, chloride Cl^- (together ca. 27%), and potassium K^+ (which is relatively abundant in Balkhash, 2%).[16] In addition, silicon Si^{4+} represents an important dietary requirement for various organisms with a concentration of 2–3 mg/L, in spite of its low reactivity. Micro-components or substances occurring in very low concentrations (1 μg/L $< x <$ 1 mg/L) represent 3% of the total ionic mass in Lake Balkhash. However, they are very significant as most important nutrients including various inorganic and organic nitrogen compounds such as ammonium NH_4^+, nitrite NO_2^- and nitrate NO_3^- (0.01–0.7 mg/L), phosphorus (phosphate PO_4^{3-}), and iron Fe^{3+} (0.1–0.5 mg/L).

Salinity of Balkhash's waters mainly depends on the concentrations of Ca^{2+}, Mg^{2+} and HCO_3^- and thus, on water hardness. Salinity increases from 0.2 to 5.0 g/L from

[15] 1 mg/L is equivalent to 1 ppm (parts per million).

[16] The ionic composition of Lake Balkhash's water is relatively distinct if waters of different basins are compared: "The proportion of chloride (9–21 equiv. percent) is 2–3 times lower than the proportion of chloride in the sea. However, the proportions of potassium, calcium, magnesium, sulphate and carbonate/bicarbonate ions are significantly higher. In Eastern Balkhash, the proportion of potassium ions (2.9 equiv. percent) is very high in comparison to other waters (e.g., 0.6 equiv. percent in the ocean and the Aral Sea). The lower proportion of calcium ions, especially in comparison with the Aral and Caspian seas, is also notable" (Aladin and Plotnikov 1993, 5).

West to East (Fig. 5.8c).[17] The Ca^{2+}/Mg^{2+} ratio is a good indicator of salinity in Lake Balkhash (Fig. 5.8c).[18]

The water of Lake Balkhash is alkaline, with acidity (pH, power of hydrogen) decreasing from a pH of 8.65 in the West to 9.15 in the East.

In the Balkhash water, carbonate anions dominate over chloride (14%) and sulphate (18%) anions; and their concentration increase from West to East together with changes of associated cations (Ca, Mg, Na, K).

Among carbonates, the ratio of bicarbonate (HCO_3^-) and carbonate (CO_3^{2-}) depends on acidity: bicarbonate dominates in the Western Balkhash with pH values of 7–8.5, and carbonate dominates in the Eastern Balkhash at pH values > 8. In Western Balkhash bicarbonate associates with calcium and precipitates as calcite ($CaCO_3$), resulting in decreasing concentrations of both ions from West to East in Lake Balkhash. Bicarbonate concentrations decrease from 35–57 to 8%, and calcium concentrations from 11.8 to 1.2 mg/L from West to East. In Eastern Balkhash, carbonate associates mainly with magnesium, and magnesium concentrations increase from 40 to 400 mg/L from the West in the lake to its East due to the different sedimentary loads of the eastern river inflows and the higher effects of evaporative concentration in the East. The carbonate-magnesium combination favors the formation of dolomite [$CaMg(CO_3)_2$] and aragonite ($CaCO_3$). Similarly towards the East, salinity increases from 0.7–1.1 to 3.2–5.2 g/L.

The hydrochemical conditions are reflected by the mineralogy of the respective sediments on the lake floor (Fig. 5.8b) which contain 50% of calcite in the Western Balkhash, calcite and dolomite in similar proportions in the western basin of the Eastern Balkhash (i.e., in the Middle Basin; basin III in Fig. 5.3a), and 60% of dolomite and 7% of aragonite in the central basin of the Eastern Balkhash (i.e., in the Lepsinsky Basin; basin IV in Fig. 5.3a).

5.5 Modern Balkhash: Biota

5.5.1 Biological Productivity

Lake Balkhash has favorable temperature ranges and dissolved oxygen levels, but a low productivity. The low productivity results from the high concentrations of

[17]The amount of salts in the waters of the endorheic lake would be significantly higher given the sedimentary inputs if geological processes would not partly remove salts (Sect. 5.2.1 and Footnote 9).

[18]The amount of ions is directly correlated with the electrical conductivity (EC) of the water, which is also a good indicator of salinity (TDS = EC x 0.64), moreover by the fact that it keeps in consideration temperature gradients. EC is expressed in Siemens/metre (S/m or mho/m), conventionally calibrated at 25 °C, with 1 S/m equivalent to a salinity of 5 g/L. Going from West to East in Lake Balkhash, EC values at T = 1.1 °C increase from 1 S/m to more than 15 S/m. EC values increase 3% per 1 °C temperature increase and at a maximum water temperature of 26 °C reach up to 2 S/m in the westernmost and up to 27 S/m in the easternmost part of the lake.

potassium and magnesium in the Eastern Balkhash which are unfavorable for its biota (Karpevich 1975). Further increases in salinity and the concentrations of these ions will directly cause considerable reductions of biomass (Aladin and Plotnikov 1993, 6). Also other factors such as the relative isolation of the lake and the adaptation stress induced on organisms by frequent fluctuations of lake level and salinity reduce Balkhash's productivity to a medium-low rate, decreasing from West to East with increasing turbidity and salinity.

5.5.2 Floral and Faunal Species

5.5.2.1 Aquatic Flora

The aquatic flora of Lake Balkhash consists of microphytes (microalgae) living in the water column (planktonic) or on or in sediments (benthic), and macrophytes growing in or near the water. Phytoplankton and phytobenthos are represented by 350 species and varieties, including: 200 species of diatom microalgae, mainly benthic and almost constituting the totality of the benthic algae; ca. 65 species of green algae; ca. 50 of blue-green algae; eight of dinoflagellates; four of golden algae; one of yellow-green algae; six of euglenids; 18 of zygnematophyceans and some species of charophyceans (Fig. 5.9). Most of the algae forms are freshwater types (oligohalobionts) or adapted

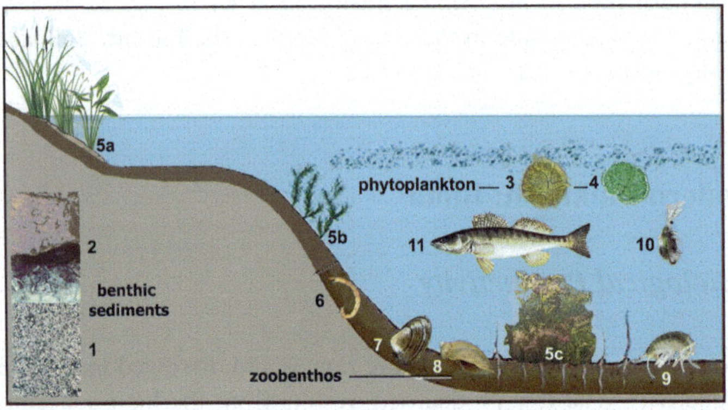

Fig. 5.9 Typical aquatic ecosystem of Lake Balkhash. Bottom sediments: 1 limy dolomitic white silt (upper strata of the lake-bottom cores of Eastern Balkhash; Sapozhnikov 1951); 2 silty sand with a peat layer, at depths between 8–60 cm in the central-western basin. Phytoplankton: 3 Dinophyte *Peridinium*, the most widespread dinoflagellate in the western basin; 4 Cyanophyte *Snowella lacustris* (blue-green algae). Macrophytes: 5a hydrophytic plants; 5b Charophyta green algae; 5c Chlorophyta green algae. Zoobenthos: 6 *Chironomus salinarius* (bloodworm midge); 7 bivalve *Pisidium henslowanum*; 8 gastropod *Radix ovata*; 9 amphipod *Gammarus lacustris*. Zooplankton: 10 *Daphnia galeata*, a common micro-crustacean in Eastern Balkhash. Ichtyofauna: 11 Balkhash perch, an endemic fish previously dominant that became rare (Red List IUCN) after the introduction of the bream

to a wide range of salinities (euryhaline). Benthic algae are mainly diatoms (Abrosov 1973; Karpevich 1975; Alekin 1984).

The basis of phytoplankton is made of a low number of species (Abrosov 1973). Before the 1950s, they were the blue-green algae *Microcystis flosaquae, Snowella lacustris, Planktolyngbya contorta, P. limnetica, Nodularia spumigena*; the dinoflagellates *Ceratium hirundinella, Peridiniopsis borgei*; the green algae *Pediastrum duplex, Pseudopediastrum boryanum*; and the diatoms *Aulacoseira granulata, Campylodiscus clypeus, Coscinodiscus lacustris, Cymatopleura elliptica* and *Entomoneis paludosa*. In the Eastern Balkhash, diatoms such as *Chaetoceros* spp. and the dinoflagellates *Chrysosporum bergii* and *Glenodinium berghii* were common.

Later, changes occurred in the phytoplankton assemblage. In 1973–1985, the common species were the blue-green algae *Merismopedia minima, M. tenuissima, Snowella lacustris, Microcystis pulverea, Coelosphaerium dubium*, the diatoms *Cyclotella comta* and *C. meneghiniana*, and the green alga *Oocystis submarina*. In Eastern Balkhash, species of genera *Cyclotella, Snowella* and *Chaetoceros* dominated (Abrosov 1973).

In the 2000s, the most common phytoplankton were the dinoflagellate *Peridinium* sp., the diatoms *Cyclotella meneghiniana* and *Navicula* sp., the euglenophyte *Trachelomonas* sp., the green alga *Franceia* sp., and the blue-green algae *Snowella lacustris, Gomphosphaeria aponina* and *Gloeocapsa* sp. (Barinova et al. 2017).

Diatoms dominate in spring and autumn, green algae during periods of increasing temperature, blue-green algae in summer. Freshwater and euryhaline forms dominate in the western part of the lake whilst freshwater forms disappear in the eastern part and become replaced by halophiles and mixed assemblages (holobionts) which spread to the West in phases of increasing salinity.

By the end of the 1970s, the average biomass of the phytoplankton in Western Balkhash fell by more than 50% as a result of increased salinity. Freshwater species became rare and substituted by more salt-resistant forms (Alekin 1984).

The variety of macrophyte flora in Lake Balkhash is relatively low. If the aquatic plants of the Ili River Delta are included, 35 species of higher plants and seven species of charophytic macroalgae occur. The causes for this low species number are the relatively high and variable salinity with macrophyte numbers decreasing with increasing salinity (Abrosov 1973), the relatively high turbidity of the water, the strong impact of coastal waves and the geographic isolation from other basins. High turbidity of water prevents the development of charophytic macroalgae. Charophytes are only found in those parts of the lake where the water transparency is high (Abrosov 1973). These parts are mainly in the Western Balkhash.

The hydrophytic plants restricted to low salinities in Lake Balkhash are the white water lily *Nymphaea candida*, the arrowhead *Sagittaria sagittifolia* and the duckweeds *Lemna minor* and *L. trisulca*. The higher plants most resistant to high salinity are the reed *Phragmites australis*, the sago pondweed *Stuckenia pectinata*, the horned pondweed *Zannichellia palustris*, the watermilfoil *Myriophyllum spicatum*, the bulrush *Scirpus kasachstanicus* and the beaked tasselweed *Ruppia maritima*. The charophytic macroalgae *Chara tomentosa* and *Nitellopsis obtusa* also tolerate relatively high salinities (Abrosov 1973; Barinova et al. 2017; Krupa et al. 2017).

Thickets of hydrophytes diminished in Lake Balkhash due to the increasing shallowing and salinization. Reed covered a 250-km long and 10–20-km wide stretch in the Western Balkhash in the past which was reduced to a 1–2 km wide reed belt in the 1970s (Abrosov 1973).

5.5.2.2 Zooplankton

The natural fauna of Lake Balkhash is considered as qualitatively and quantitatively poor due to its geographical isolation. Several allochthonous species have been artificially introduced during the last century.

The basic original zooplankton consisted of ciliates (*Codonella cratera*), rotifers (*Synchaeta* spp., *Filinia longiseta, F. longiseta* var. *limnetica, Polyarthra platypiera, Pompholyx sulcata, Keratella quadrata, K. quadrata* var. *valga, K. cochlearis, K. cochlearis* var. *tecta, Chromogaster ovalis, Hexarthra oxyuris*), copepods (*Arctodiaptomus salinus, Thermocyclops crassus, Mesocyclops leuckarti*), and cladocerans (*Daphnia galeata, Diaphanosoma lacustris, Chydorus sphaericus* and *Leptodora kindtii*; Rylov 1933; Fig. 5.9). Leading were rotifers, and *K. quadrata* predominated. Among copepods, the leading form was *A. salinus* which was more abundant in the Eastern Balkhash. Widespread, though not very abundant, was *M. leuckarti* (Abrosov 1973).

By 1978–1980, significant changes in the zooplankton had occurred due to the falling lake level and the increasing salinity, accompanied by a sharp decrease in nutrient inputs from river runoff. The number of freshwater and brackish-water rotifer species decreased, and to a lesser degree also the numbers of cladocerans and copepods. Of the latter two, *A. salinus* and *D. lacustris* became the dominating species.

By 1983–1985, with the continuously falling lake level and rising salinity, the total number of zooplankton species declined by more than a half compared to the end of the 1960s. Small forms or "fine" detritophages filter-feeders disappeared completely, reducing the number of species most valuable as food for juvenile commercial fish. Instead, the "rough" filter-feeders *D. lacustris* and *A. salinus*, characterized by relatively low nutritional values for juvenile fish, began to dominate. The cladoceran *Sida cristallina* became abundant and the typical inhabitant of the brackish waters.

By the end of the 1990s, the zooplankton composition of Lake Balkhash stabilized. The main species were *Brachionus calyciflorus, Euchlanis dilatata, Keratella cochleans tecta, K. quadrata quadrata, Hexarthra oxyuris, Daphnia galeata, D. cucullata* Sars, *Diaphanosoma lacustris, Eucyclops vicinus, Mesocyclops leuckarti* and *Arctodiaptomus salinus.*

The current zooplankton of the lake includes 123 species and subspecies of invertebrates: 82 rotifers, 22 cladocerans and 19 copepods (without harpacticoids). Only a few are found in high numbers: *Polyarthra dolichoptera dolichoptera, Euchlanis dilatata dilatata, Keratella cochlearis cochlearis* and *K. quadrata quadrata* among the rotifers; *Diaphanosoma lacustris* and *Daphnia galeata*, especially in

Eastern Balkhash, among the cladocerans; and *Arctodiaptomus salinus, Mesocyclops leuckarti* and *Thermocyclops crassus* among the copepods (Rylov 1933; Saduakasova 1972; Karpevich 1975; Krupa et al. 2013). The three cladocerans *Leptodora kindtii, Ceriodaphnia reticulata* and *Polyphemus pediculus* are rare. A few species of the cladoceran families Chydoridae and Macrothricidae and of the copepod genera *Eucyclops, Microcyclops* and *Macrocyclops* occur near the river mouths.

The specific ionic composition of Balkhash's water, especially the higher concentration of potassium and magnesium compared to other large saline lakes is unfavorable for hydrobionts. Potassium and magnesium are toxic to organisms but their toxicity is weakened by calcium and sodium. The abundance of planktonic crustaceans is largely controlled by the concentrations of these ions in the Eastern Balkhash. It decreases with increasing K^+/Na^+ and K^+/Ca^{2+} ratios. *Daphnia galeata* is the most resistant to higher potassium concentrations among the dominant species (Krupa et al. 2008).

5.5.2.3 Zoobenthos

The original zoobenthos of Lake Balkhash initially had a very poor species composition, mainly represented by larvae of terrestrial insects, especially chironomids (more than 30 forms), dragonflies, mayflies, stoneflies, caddisflies and oligochaete worms (Abrosov 1973; Karpevich 1975). Gastropods, bivalves and malacostraca (shrimps and amphipods) were also relatively abundant members of the native benthic fauna.

Oligochaetes are widespread in the lake. Nine species were recorded: *Potamothrix hammoniensis, P. bavaricus, Limnodrilus profundicola, L. hoffmeisteri, Tubifex tubifex, Uncinais uncinata, Nais pardalis, Spirosperma ferox* and *Stylaria lacustris* (Abrosov 1973). Leeches are represented by the three species *Piscicola geometra, Protoclepsis meyeri* and *Glossiphonia complanata* (Abrosov 1973). The free-living nematodes of Lake Balkhash did not receive proper studies and also only few information was gathered with respect to aquatic mites (Abrosov 1973).

The native mollusks included only freshwater species. The most common species were the bivalve *Pisidium henslowanum* and the gastropods *Valvata piscinalis* and *Planorbis planorbis* in the past, but their biomass is low today and their presence limited to the shallow fresher areas of the lake and the proximity of the river mouths (Samonov 1966; Karpevich 1975). *Pisidium henslowanum* is the only native bivalve in Lake Balkhash, but the native gastropods include the pond snail *Limnaea stagnalis, Radix auricularia, R. ovata* and *Galba truncatula*, the ramshorn snail *Gyraulus albus*, and the lake limpet *Acroloxus lacustris* in addition to the predominating *V. piscinalis* and *P. planorbis* (Fig. 5.9).[19]

Today, crustaceans are mostly represented by the shrimp *Palaemon superbus* and two species of amphipods—*Rivulogammarus lacustris* and, in Western Balkhash, *Dikerogammarus haemobaphes*.

[19]A number of researchers reported the presence of the gastropod *Bithynia caerulans*, but this endemic gastropod became apparently extinct (Zhadin 1952; Tyutenkov 1959).

Ostracods are represented by the six species *Ilyocypris* sp., *Neglecandona neglecta, Candona* sp., *Darwinula stevensoni, Cyprideis torosa* and *Limnocythere dubiosa* (Abrosov 1973). However, this list might be incomplete as crustaceans have been insufficiently investigated.

In 1953–1966, several species of invertebrates were introduced from the Caspian and the Azov seas as valuable food for fishes. The anellid polychaetes *Hypania invalida* and *Hypaniola kowalevski*, the bivalve mollusk *Monodacna colorata*, and the crustacean amphipod *Corophium curvispinum* and mysids (shrimp-like crustaceans) *Paramysis lacustris, P. intermedia, P. ullskyi* and *P. baeri* were naturalized. In addition to these intentional introductions, the freshwater bivalves *Anodonta cygnea* and *A. cellensis* were accidentally introduced together with the zander fish from the Ural River in 1957–1958 because their glochidia larvae were on the gills of fish (Karpevich 1975).

Since 1996, the zoobenthos biomass began to increase sharply as a result of the explosive development of the bivalve *Monodacna colorata* which is now underutilized by benthophagous fishes.

A total of 93 native and introduced invertebrate species are recorded for today's macro-zoobenthos of Lake Balkhash. During 2009–2013, the main taxonomic group represented native larval and adult insects including the large native larval chironomids *Chironomus salinarius* and *C. plumosus*, followed by introduced representatives of the Ponto-Caspian fauna.

The composition of the benthic fauna differs significantly in the Western and Eastern Balkhash due to the different salinity levels. Species diversity is lower in the more saline Eastern Balkhash, and the amphipod *Dikerogammarus haemobaphes* and some native and introduced freshwater mollusks such as *Monodacna colorata* do not occur anymore. With the exception of chironomids and other Diptera, insects are also less diverse in the Eastern Balkhash. Moving from West to East, mysids which are sensitive to water pollution and sometimes used as bioindicators to monitor water quality, progressively disappear. At first, *Paramysis baeri* disappears, then *P. lacustris*, further East in the lake follows *P. ullsky,* and finally *P. intermedia* which is not found in the easternmost part of the lake anymore (Alekin 1984).

5.5.2.4 Ichthyofauna

The ichthyofauna of Lake Balkhash has been intensively manipulated during the twentieth century with the introduction of species from other hydrological basins of the Soviet Union. Introductions caused major changes in the dominating species, replacing native ones by imported taxa. In general, the fish-species diversity decreased, and benefits of the substitution of native species by alien species on the potential total catch are doubtful (Mitrofanov and Petr 1999).

The modern ichthyofauna of Lake Balkhash includes 26 species. The Balkhash marinka *Schizothorax argentatus*, the Ili marinka *Sch. pseudaksaiensis pseudaksaiensis*, the Balkhash perch *Perca schrenkii*, the spotted stone loach *Triplophysa strauchi*, the plain thicklip loach *Barbatula labiata* and the Balkhash minnow

Lagowskiella poljakowi are native.[20] All other fishes are recent introductions or invaders.

Commercially valuable among native fishes are the Balkhash marinka, Ili marinka and Balkhash perch (Abrosov 1973; Karpevich 1975). The Balkhash marinka maintained commercial value and was found throughout the lake until the mid-1960s; but it is now, together with spotted stone loach, absent in the lake and only remaining in some rivers. The Balkhash perch is now found in small numbers, occurring only in the deltaic lakes of the Ili River, in some bays of Lake Balkhash, and in the Ayaguz River (Mamilov et al. 2013).

The ichthyofauna of the Balkhash and of the rivers of the basin was enhanced by 22 fish species during the twentieth century as a result of human activities, intentionally introduced or incidentally brought in together with intentionally introduced fish (Karpevich 1975).

- The first invader was the carp *Cyprinus carpio*. Originally (1905), it entered accidentally into the Ili River from a fish pond and then appeared in the lake. By the end of the 1920s, it became the main commercial species of the Balkhash.
- The Siberian dace *Leuciscus leuciscus* is also an auto-acclimatizant. It was initially brought into the Ayaguz River where it was discovered in 1928, and it appeared in the lake in 1950.
- In 1931, the Aral barbel *Barbus brachycephalus* was brought from the Syr Darya into the Ili from where it soon penetrated and settled into the lake (Karpevich 1975). The Aral barbel is currently not found in the lake anymore (Mamilov et al. 2013).
- In 1933–1934, the ship sturgeon *Acipenser nudiventris* was introduced from the Aral Sea. It became one of the widespread commercial species prior to the construction of the Kapchagai Reservoir (Karpevich 1975), but it is relatively rare today (Mamilov et al. 2013).
- In 1948, the tench *Tinca tinca*, although not recommended for introduction, was brought from the Zaisan Lake into the Ili River Basin.
- In 1949, the eastern bream *Abramis brama orientalis* (also not recommended) was introduced from the Syr Darya. It became a mass and commercial species. It is a food competitor of carp and marinka.
- In 1954, the Prussian carp *Carassius auratus gibelio* (also not recommended) was released into Karatal River and it then entered and settled in the lake, becoming a commercial species (Karpevich 1975).
- In 1957–1958, the zander *Sander lucioperca* was introduced as the first fish species recommended for acclimatization.
- Together with the zander, the predators wels *Silurus glanis*, asp *Aspius aspius* and the Volga zander *Sander volgensis* were accidentally also brought. They acclimatized successfully and became commercial fish species.
- In 1958 and in 1962, the recommended herbivorous grass carp *Cthenopharhyngodon idella* was introduced. This fish settled across the lake and penetrated into rivers, mostly in their deltas overgrown with aquatic vegetation (Karpevich 1975).

[20]The Ili marinka, Balkhash perch and spotted stone loach are endemics of Balkhash-Alakol Basin.

- Together with the grass carp, a coarse fish, the Chinese freshwater sleeper *Micropercops cinctus*, was brought into Balkhash (Seleznev 1974; Karpevich 1975; Reshetnikov 2010).
- In 1965, the not recommended vobla *Rutilus caspicus* and the Talas dace *Leuciscus lindbergi* were introduced accidentally.

5.5.3 Development of Fishery in Lake Balkhash During the Last 100 Years

Fisheries in Lake Balkhash developed in the early 1930s with the introduction of allochthonous species and the organization of collective fishing brigades. Catches, negligible during the 1920s (68–106 tons per year), soon increased and reached 14,650 tons in 1932, with common carp (*Cyprinus carpio*, accidentally introduced in 1905) accounting for 58–86% of the catch and the native perch and marinka for 4–30%, varying over the years.

During the 1940s and 1950s, following a natural reduction of the lake level and the lake's productivity, the average annual capture decreased by 40%. It recovered and reached former high numbers by the early 1960s. Then, the introduced common carp represented 76% of the total catch, native species such as perch and marinka 12 and 11%, respectively, and the freshwater bream the remaining 1%.

The building of the Kapchagai Reservoir (1970–1986) did not significantly affect the total catch but it changed the natural cycle of re-occurring spring floods. The hydrological regime of the Ili River was especially altered during the filling of the reservoir, and spawning and feeding areas of the commercially valuable *Cyprinus carpio* were reduced. Its populations consequently shrank by more than 90% and its yearly catch decreased from 12,000 to less than 2000 tons without recovery to high levels afterwards. The loss in catch of *Cyprinus carpio* was compensated by a higher catch of the freshwater bream which is a very adaptable and competitive but less valuable cyprinid. Its harvest grew from 100 to 8000 tons per year over the same period (Pueppke et al. 2018).

A clear drop of the annual catch accompanied the economical crisis that followed the end of the Soviet regime, attributed to several factors: (1) the decrease of lake productivity and reduction of spawning areas provoked by non-compliance with discharge schedules from the Kapchagai Reservoir; (2) the uncontrolled over-exploitation of the fish stocks; and possibly (3) unregistered and unofficial catches by recreational trophy fishing. The re-establishment of controls and regulations temporarily improved the situation and the total catch recovered to 12,000 tons per year between 2002 and 2006. Afterwards, it fell again and the decline continued in most recent years. Between 2010 and 2017, the fish harvest decreased by 17%, mainly as a result of lower catches of the most valuable species (Pueppke et al. 2018).

5.6 Outlook

Lake Balkhash is facing serious risks in the near future totally depending on the collective will and decisions of the responsible agencies in Kazakhstan and China.

The most disastrous scenario would be an increase of the total water withdrawal from tributary rivers by 9 km^3, which would cause a lake-level drop to below 336 m asl and the division of Lake Balkhash in three to six different and partly dry basins and unbearable levels of salinization in the remaining water bodies.

Intermediate levels of water subtraction would also induce serious stress to the lake's ecosystem because the related increase of water salinity would restructure the ecological communities and deteriorate the reproduction conditions of aquatic organisms. The total gross production of phytoplankton, zooplankton and zoobenthos would be reduced (Krupa et al. 2013), and the overall decline in trophic state would have a devastating impact on the fish productivity of the lake.

Additional water withdrawal from the Ili River and non-compliance with discharge schedule from the Kapchagai Reservoir would inevitably affect the fish spawning areas. In the fresher western part of the lake, spawning areas would move following the retreat of the water, but in the eastern part, they would be reduced to the estuarine freshwater areas of the Karatal and Lepsy rivers with increasing lake-water mineralization.

The degradation of Lake Balkhash's ecosystem would favor the development of non-commercial and low-value fish species, while the number of fast-growing forms of carp, bream, asp and other commercial species would rapidly decline. As a result, fish resources of the lake would rapidly decline.

The water deficit will also lead to the degradation of the Ili River Delta (Dostay et al. 2012), which will lose most of its value for fishery. Shoaling and swamping of deltaic water bodies would be accompanied by overgrowth of tough vegetation, salinization and increasing concentration of humic acids with detrimental effects on the development of fish and other aquatic organisms.

Acknowledgements The study of Lake Balkhash in the frame of the "Ili Project" (2007–2012) was funded by the Research Institute for Humanity and Nature (Kyoto), Japan. The work of NVA and ISP was supported by the theme of the State assignment for 2019–2021 "AAAA-A19-119020690091-0: Studies of biological diversity and the mechanisms of the impact of anthropogenic and natural factors on the structural and functional organization of ecosystems of continental water bodies. Systematization of the biodiversity of salt lakes and brackish-water inland seas in the zone of critical salinity, study of the role of brackish-water species in ecosystems".

References

Abdrasilov SA (1996) Formirovanie i dinamika vnutrikontinental'nykh delt (na primere reki Ili i ozera Balkhash) [Formation and dynamics of intracontinental deltas (on the example of the River Ili and Balkhash Lake)]. Doctorate dissertation, Almaty: Institute of Geography, 39p. (in

Russian) Available online: http://earthpapers.net/formirovanie-i-dinamika-vnutrikontinentalnyh-delt-na-primere-reki-ili-i-ozera-balhash

Abrosov VN (1973) Ozero Balkhash [Lake Balkhash]. Leningrad, Nauka (in Russian)

Akhmedsafin UM, Dzhabasov MK, Oshlakov GG (1973) Southern-Pribalkhashski artesian basin. In: Formation and hydrodynamics of artesian waters of Southern Kazakhstan. Alma-Ata, Nauka, 18–22 (in Russian) Translated in English and re-published online at the Econet website (UNESCO Central Asia and Institute of Hydrogeology and Hydrophysics of the Ministry of Science and Higher Education of the Republic of Kazakhstan) at http://www.water.unesco.kz/bal_ch_5_e.htm

Aladin NV, Plotnikov IS (1993) Large saline lakes of former USSR: a summary review. Hydrobiologia 267:1–12

Alekin OA (ed) (1984) Prirodnye resursy bol'shikh ozer SSSR i veroyatnye ikh izmeneniya [Natural resources of large lakes of the USSR and their probable changes]. Leningrad, Nauka (in Russian)

Atlas Kazakhskoi SSR (1982) T.1 Prirodnye usloviya i resursy [Atlas of Kazakh SSR. T.1 Natural conditions and resources]. Moscow, p 59 (in Russian)

Aubekerov B, Koshkin V, Sala R, Nigmatova S, Deom JM (2009a) Prehistorical and historical stages of development of lake Balkhash. In: Watanabe M, Kubota J (eds) Reconceptualizing cultural and environmental change in Central Asia. RIHN, Kyoto, pp 49–76. Available online: http://www.chikyu.ac.jp/ilipro/page/18-publication/workshop-book/workshop-book_individual%20files/2-1_Aubekerov.pdf

Aubekerov BZ, Nigmatova SA, Sala R, Deom JM, Endo K, Haraguchi T (2009b) Complex analysis of the development of the Balkhash lake during the last 2000 years. In: Watanabe M, Kubota J (eds) Reconceptualizing cultural and environmental change in Central Asia. RIHN, Kyoto, pp 77–92

Aubekerov BZ, Yerofeeva IV, Sala R, Nigmatova SA, Deom JM (2010) Geoarkheologicheskoe izuchenie ozera Balkhash kak geograficheskogo tsentra kochevykh kultur Kazahstana [Geoarchaeological studies of the Balkhash Lake as geographical centre of nomadic culture of Kazakhstan]. In: Rol' nomadov v formirovanii kul'turnogo naslediya Kazakhstana. Nauchnyye chteniya pamyati N.E. Masanova. Sbornik materialov Mezhdunarodnaya nauchhnaya konferentsiya [Role of the nomads in shaping the cultural heritage of Kazakhstan. Scientific readings in memory of NE Masanov. International scientific conference material collection]. Almaty: Print-S, pp 37–45 (in Russian)

Barinova S, Krupa E, Kadyrova U (2017) Spatial dynamics of species richness of phytoplankton of Lake Balkhash (Kazakhstan) in the gradient of abiotic factors. Transylvanian Rev Systematical Ecol Res 19(2):1–18

Berg LS (1904) Predvaritel'nyy otchet ob issledovanii ozera Balkhash letom 1903 g. [Preliminary report on research of the Balkhash lake during summer 1903]. Izvestiya Russkogo geograficheskogo obshchestva 40(4) (in Russian)

Bolch T (2015) Glacier area and mass changes since 1964 in Ala Archa Valley, Kyrgyz Ala-Too, northern Tien Shan. Led i Sneg [Snow and Ice] 1(129):28–39

Chen F, Yu Z, Yang M, Ito E, Wang S, Madsen DB, Huang X, Zhao Y, Sato T, Birks HJB, Boomer I (2008) Holocene moisture evolution in arid central Asia and its out-of-phase relationship with Asian monsoon history. Quat Sci Rev 27(3–4):351–364

Chiba T, Endo K, Sugai T, Haraguchi T, Kondo R, Kubota J (2016) Reconstruction of Lake Balkhash levels and precipitation/evaporation changes during the last 2000 years from fossil diatom assemblages. Quat Int 197:330–341

Deom JM, Sala R, Laudisoit A (2019) The Ili River Delta: Holocene Hydrogeological Evolution and Human colonization. In: Yang LE, Bork H-R, Fang X, Mischke S (eds) Socio-Environmental Dynamics along the Historical Silk Road. Springer, pp 69–97

Dostay ZD (2009) Upravlenie ghidroekosistemoy basseyna ozera Balkhash [Management of the hydro-ecosystem of the lake Balkhash basin]. Almaty (in Russian)

Dostay Z, Alimkulov A, Tursunova A, Myrzakhmetov A (2012) Modern hydrological status of the estuary of Ili river. Appl Water Sci 2:227–233

Dzhurkashev TN (1964) Proliv Uzunaral i nekotorye noveishei istorii ozera Balkhash [The Uzunaral strait and some questions about the recent history of the Balkhash lake]. Izvestiya, AN Kaz SSR, Geological Series, vol 4 (in Russian)

Dzhurkashev TN (1972) Antropogenovaya istoriya Balhash-Alakolskoi vpadiny [The anthropogenic history of the Balkhash-Alakol depression]. Alma-Ata, Nauka (in Russian)

Endo K, Chiba T, Sugai T, Haraguchi T, Yamazaki H, Nakayama Y, Yoshinaga Y, Miyata K, Ogino S, Arakawa K, Nakao Y, Komori J, Kondo R, Matsuoka H, Aubekerov BZ, Sala R, Deom JM, Sohma H, Kubota J (2010) Reconstruction of lake level and paleoenvironmental changes from a core from Balkhash lake, Kazakhstan. In: Watanabe M, Kubota J (eds) Reconceptualizing cultural and environmental change in Central Asia, Kyoto, RIHN, pp 93–104

Endo K, Sugai T, Haraguchi T, Chiba T, Kondo R, Nakao Y, Nakayama Y, Suzuki S, Shimizu I, Sato A, Montani I, Yamasaki I, Matsuoka H, Yoshinaga Y, Miyata K, Minami Y, Komori J, Hara Y, Nakamura A, Kubo N, Sohma I, Deom J-M, Sala R, Nigmatova SA, Aubekerov BZ (2012) Lake level change and environmental evolution during the last 8000 years mainly based on Balkhash Lake cores in Kazakhstan, Central Eurasia. In: Kubota J, Watanabe M (eds) Toward a sustainable society in Central Asia: An historical perspective on the future. RIHN, Kyoto, pp 35–48

Feng ZD, Wu HN, Zhang CJ, Ran M, Sun AZ (2013) Bioclimatic change of the past 2500 years within the Balkhash Basin, eastern Kazakhstan, Central Asia. Quat Int 311:63–70

Hill DJ (2018) Climate Change and the Rise of the Central Asian Silk Roads. In: Yang LE, Bork H-R, Fang X, Mischke S (eds) Socio-Environmental Dynamics along the Historical Silk Road. Springer, pp 69–97

Karpevich AF (1975) Teoriya i praktika akklimatizatsii vodnykh organizmov [Theory and practice of aquatic organisms acclimatization]. Moscow: Pischevaya promyshlennost (in Russian)

Kezer K, Matsuyama H (2006) Decrease of river runoff in the Lake Balkhash basin in Central Asia. Hydrol Proc Int J 20(6):1407–1423

Khrustalev YP, Chernousov YG (1992) K golotsenovoi istorii razvitiya ozera Balkhash [On the Holocene history of the Balkhash lake]. Izvestia Russ Geogr Soc 124(2):164–171 (in Russian)

Kostenko NN (1946) K istorii Balkhasha [About the Balkhash history]. Izvestiya Kaz fil AN SSSR, ser. Geologiya [News of the Kazakh Filial of Academy of science SSSR] 8(26):96–104 (in Russian)

Krupa EG, Stuge TS, Lopareva TY, Shaukharbaeva DS (2008) Distribution of planktonic crustaceans in Lake Balkhash in relation to environmental factors. J Inland Water Biol 1(2):150–157

Krupa EG, Tsoy VN, Lopareva TY, Ponomareva LP, Anureva AN, Sadyrbaeva NN, Assylbekova SZ, Isbekov KB (2013) Mnogoletnyaya dinamika gidrobiontov ozera Balkhash i ee svyaz s faktorami sredy [Long-term dynamics of hydrobionts in Lake Balkhash and its connection with environmental factors]. Vestnik AGTU Ser Rybnoe Khoz 2:85–96 (in Russian)

Krupa EG, Barinova SS, Tsoy VN, Sadyrbaeva NN (2017) Formirovaniye fitoplanktona ozera Balkhash (Kazakhstan) pod vliyaniyem osnovnykh regional'noklimaticheskikh faktorov [Formation of phytoplankton of lake Balkhash (Kazakhstan) under the influence of major regional-climatic factors]. Adv Biol Earth Sci 2(2):204–213 (in Russian)

Krylov S, Nurgaliyev DK, Yasonov PG (2014) Ob istorii razvitiya ozera Balkhash (Kazakhstan) po seysmoakusticheskim dannym [History of the Balkhash lake development based on seismo-acoustic data. Uchenye zapiski Kazanskogo universiteta [Scientific reports of the Kazan University], Nat Sci 156(1):128–136 (in Russian)

Kurdin RD (1976) O vekovykh kolebaniyakh urovnya vody krupnykh yestestvennykh vodoyemov Kazakhstana i Sredney Azii [About the secular water level fluctuations of large natural basins of Kazakhstan and Central Asia]. In: Trudy IV Vsesoyuz. gidrol. s"yezda [Works of the IV all Union Hydrological summit], T.5. Gidrometeoizdat, Leningrad, pp 98–107 (in Russian)

Kurdyukov KV (1958) K geologicheskomu razvitiyu Pribalkashya v pozdnem kainozoe [About the geological development of the Pre-Balkhash during late Cenozoic]. BMOIP, New Series, series Geology 33(3) (in Russian)

Maksimov EV (1961) Stadialnyi kharakter otstupaniya vyurmskikh lednikov v Dzhungarskom Alatau i v nekotorykh drugikh gornykh sistemakh Azii. [Stadial character of melting of Wurm

glaciers in the Jungarian Alatau and some other Asian mountain systems]. Doklady AN SSSR [Reports of the Academy of Sciences SSSR], 136(1) (in Russian)
Mamilov NS, Balabieva GK, Mitrofanov IV (2013) Problemy sokhraneniya aborigennoy ikhtiofauny Ili-Balkhashskogo basseina [Problems of preservation of aboriginal ichthyofauna in Ili-Balkhash basin] (in Russian). Published online by Green Salvation: http://esgrs.org/wp-content/uploads/2015/01/2011fishMamilov.pdf. Accessed 17 May 2018
Mischke S, Zhang CJ, Plessen B (2020) Lake Balkhash (Kazakhstan): Recent human impact and natural variability in the last 2900 years. J Great Lakes Res 46:267–276
Mitrofanov VP, Petr T (1999) Fish and fisheries in the Altai, Northern Tien Shan and Lake Balkhash (Kazakhstan). Fish and fisheries at higher altitudes: Asia. Rome: FAO Fisheries Technical Paper, vol 385, pp 149–167
Myrzakhmetov A, Dostay Z, Alimkulov S, Madibekov A (2017) Level regime of Balkhash Lake as the indicator of the state of the environmental ecosystems of the region. Int J Adv Res Sci Eng Technol 4(9):4554–4563
Propastin P (2012) Patterns of Lake Balkhash water level changes and their climatic correlates during 1992–2010 period. Lakes Reservoirs Res Manag 17(3):161–169
Propastin P (2013) Assessment of climate and human induced disaster risk over shared water resources in the Balkhash Lake drainage basin. In: Climate change and disaster risk management. Springer, Berlin, Heidelberg, pp 41–54
Pueppke SG, Iklasov MK, Beckmann V, Nurtazin ST, Thevs N, Sharakhmetov S, Hoshino B (2018) Challenges for sustainable use of the fish resources from Lake Balkhash, a fragile lake in an arid ecosystem. Sustainability 10(4):1234
Reshetnikov AN (2010) The current range of Amur sleeper *Perccottus glenii* Dybowski, 1877 (Odontobutidae, Pisces) in Eurasia. Russ J Biol Invasions 1(2):119–126. https://doi.org/10.1134/S2075111710020116
Rusakov MP (1933) Geologicheskii ocherk Pribalkhashiya i ozero Balkhash, poleznye iskopaemye raiona [Geological treatise concerning the Pre-Balkhash region and Balkhash Lake, mineral resources district]. Moscow, Leningrad: Tsvetmetizdat (in Russian)
Rylov VM (1933) K svedeniyam o planktone ozera Balkhash [Information on the Lake Balkhash zooplankton]. Issledovaniya ozer SSSR 4:57–70 (in Russian)
Saduakasova RE (1972) Zooplankton ozera Balkhash [Zooplankton of Lake Balkhash]. Rybnye resursy vodoemov Kazakhstana i ih ispolzovaniye 7:97–100 (in Russian)
Sala R, Deom JM, Nigmatova S, Endo K, Kubota J (2016) Soviet, recent and planned studies of the behavior of the Balkhash lake. News of the Academy of Sciences of the Republic of Kazakhstan. Series of Geology and Technical Sciences 2(416):76–86
Samonov AM (1966) Molluski ozera Balkhash [Mollusks of Lake Balkhash]. In: Biologicheskie osnovy rybnogo khozyaistva na vodoemah Sredney Azii i Kazakhstana, pp 131–144 (in Russian)
Sapozhnikov DG (1951) Sovremennye osadki i geologiya ozera Balkhash [Balkhash Modern sediments and geology of Lake Balkhash]. Trudy Inst Geol Nauk [Works of the Institute of Geological Sciences] (132) Moscow, AN SSSR (in Russian)
Seleznev VV (1974) Malotsennyye i sornyye vidy ryb kitayskogo kompleksa v Kapchagayskom vodokhranilishche [Low-value and coarse fish species of Chinese complex in the Kapchagai Reservoir]. Rybnye resursy vodoemov Kazakhstana i ikh ispolzovanie. 8:143–148 (in Russian)
Semenov VA, Kurdin RD (eds) (1970) Resursy poverkhnostnykh vod SSSR. Tsentral'nyi i yuzhnyi Kazakhstan. Vypusk. 2. Bassein oz. Balkhash [Surface water resources SSSR, vol 13. Central and South Kazakhstan. Issue 2. Balkhash lake basin]. Leningrad: Gydrometeoizdat (in Russian)
Sevastyanov DV (ed) (1991) Istoriya ozer Sevan, Issyk-Kul', Balkhash, Zaysan i Aral [History of the lakes Sevan, Issyk-Kul, Balkhash, Zaisan and Aral]. Leningrad, Nauka (in Russian)
Severskiy I, Vilesov E, Armstrong R, Kokarev A, Kogutenko L, Usmanova Z, Morozova V, Raup B (2016) Changes in glaciation of the Balkhash-Alakol basin, Central Asia, over recent decades. Ann Glaciol 57(71):382–394
Shaporenko SI (1995) Balkhash Lake. In: Mandych AF (ed) Enclosed Seas and Large Lakes of Eastern Europe and Middle Asia. SPB Academic Publisher, Amsterdam, pp 155–197

Shnitnikov AV (1957) Izmenchivost' obshchey uvlazhnennosti materikov Severnogo polushariya [Variability of total moisture of the Northern Hemisphere]. Zapis Vsesoyuz Geograficheskoi obschestva, new series 16 (in Russian)
Shnitnikov AV (1973) Water balance variability of lakes Aral, Balkhash, Issyk-Kul and Chany. In: Proceedings of the International Symposim on the Hydrology of Lakes, Helsinki. IAHS Publication 109:130–140. https://iahs.info/uploads/dms/iahs_109_0130pdf
Smolyar VA, Mustafaev S (2007) Gidrogeologiya basseyna ozera Balkhash [Hydrogeology of the basin of Balkhash lake]. Almaty (in Russian)
Solomina O, Alverson K (2004) High latitude Eurasian paleoenvironments: introduction and synthesis. Palaeogeogr Palaeoclimatol Palaeoecol 209(1–4):1–18
Sugai T, Montani H, Endo K, Haraguchi T, Shimizu H, Chiba T, Reisuke K (2011) Holocene environment changes in Lake Balkhash reconstructed by high-resolution XRF core analysis and geomorphic survey. Japan Geoscience Union Meeting. http://www2.jpgu.org/meeting/2011/yokou/HQR023-17_E.pdf
Tarasov MN (1961) Gidrochimiya ozera Balkhash [Hydrochemistry of Balkhash lake]. Moscow (in Russian)
Thevs N, Nurtazin S, Beckmann V, Salmyrzauli R, Khalil A (2017) Water consumption of agriculture and natural ecosystems along the Ili River in China and Kazakhstan. Water 9(3):207
Tursunov AA (2002) Ot Arala do Lobnora. Gidroekologiya besstochnykh basseinov Tsentralnoi Azii [From the Aral to the Lob Nor. Hydroecology of inland basins of Central Asia]. Almaty, pp 238–253 (in Russian)
Tyutenkov SK (1959) Bentos ozera Blakhash i ego znachenie v pitanii ryb [Benthos of Lake Balkhash and its importance in the diet of fish]. Sbornik rabot po ikhtiologii i gidrobiologii 2:45–79 (in Russian)
Venus BG (1985) Osobennosti razvitiya ozernykh kotlovin v gummidnoy i aridnoy zone [Particularities of the development of lacustrine depressions in humid and arid zones]. In: Martison GG (ed) Paleolimnologiya ozer v aridnykh i gumidnykh usloviyakh [Paleolimnology of lakes in arid and humid zones]. Leningrad, Nauka, pp 5–29 (in Russian)
Watanabe M, Kubota J (eds) (2010) Reconceptualizing cultural and environmental change in Central Asia. RIHN, Kyoto
Xu JL, Liu SY, Guo WQ, Zhang Z, Wei JF, Feng T (2015) Glacial area changes in the Ili River catchment (Northeastern Tian Shan) in Xinjiang, China, from the 1960s to 2009. Adv Meteorol 2015:(847257)
Zhadin VI (1952) Molluski presnykh i solonovatykh vod SSSR [Mollusks of fresh and brackish waters of the USSR]. Moscow, Leningrad, AN SSSR (in Russian)

Chapter 6
Lake Issyk-Kul: Its History and Present State

Andrei O. Podrezov, Ari J. Mäkelä and Steffen Mischke

Abstract Lake Issyk-Kul in Kyrgyzstan is one of the deepest and largest mountain lakes of the world. The closed-basin lake is slightly brackish and waters are dominated by sodium, and chlorite and sulphate. Although large tourism resorts and many camps without appropriate sanitary facilities were established on Lake Issyk-Kul's shores after the Soviet era, the lake is still oligotrophic with high concentrations of dissolved oxygen down to its great depths. The large volume of the lake in comparison to surface and groundwater inflows inhibited major reductions of the water quality in the central and deep part of the lake so far. However, decreasing lake levels since the beginning of systematic observations in the year 1927, contamination of littoral waters due to inefficient or lacking wastewater treatment, and the influx of fertilizers and pesticides from farmlands on the lakeshore plain are major concerns for sustaining the fragile ecosystem of Lake Issyk-Kul during times of ongoing global warming and economic development in the region.

Keywords Brackish water · Mountain lake · Lake depth · Lake mixing · Pollution

A. O. Podrezov
Department of Meteorology, Ecology and Environmental Protection, Kyrgyz-Russian Slavic University, 6 Chui Ave., 720000 Bishkek, Kyrgyzstan
e-mail: andrey_podrezov@mail.ru

A. J. Mäkelä (✉)
Finnish Environment Institute, Freshwater Centre, Latokartanonkaari 11, 00790 Helsinki, Finland
e-mail: Ari.Makela@environment.fi

S. Mischke
University of Iceland, Institute of Earth Sciences, Sturlugata 7, 101 Reykjavík, Iceland
e-mail: smi@hi.is

S. Mischke (ed.), *Large Asian Lakes in a Changing World*, Springer Water,
https://doi.org/10.1007/978-3-030-42254-7_6

6.1 General Information About Lake Issyk-Kul and Its Basin

6.1.1 Geographical Characteristics of the Issyk-Kul Basin

Lake Issyk-Kul is a closed-basin lake in Kyrgyzstan, Central Asia (Fig. 6.1). It is the seventh deepest lake in the world and the tenth largest lake in the world by volume (Table 6.1). The lake fills a fault-controlled intramontane basin which resulted from far-field effects of the Cenozoic India-Eurasia collision on the reactivated tectonics of the northern Tien Shan Mountain range. Lake sediments were probably accumulated in the basin during most of the Miocene and Pliocene, although the lake in its modern form possibly exists since the middle Pleistocene (Trofimov 1990). The catchment area of Lake Issyk-Kul covers 22,080 km^2 at moderate altitude in the northern Tien Shan. The main mountain chains surrounding the basin are the Terskei Ala-Too (also Terskey Alatau) in the south and the Kungei Ala-Too (also Küngöy Ala-Too or Kungey Alatau) in the north which have highest peaks at 5216 m and 4770 m, respectively (Fig. 6.2). Mean elevations are 4300 m and 4200 m, respectively. At an altitude of ca. 2 km, the Issyk-Kul Basin has relatively low mountain bulkheads at its western (Boom Gorge) and eastern (Santash Passage) ends. The lowest parts of bulkheads have the form of narrow wind gorges. The rivers Chu (also Chui or Chuy) and Tyup flow through these gorges that prevent free air exchange with surrounding regions.

The current water level of Lake Issyk-Kul lies at ca. 1606 m above sea level. The lake's surface-water area is 6236 km^2 (28% of drainage basin) and the length of its shoreline is 688 km (Table 6.1). The average sounding depth is 278 m and the deepest point reaches 668 m below the lake surface. In total, the lake-water volume is 1738 km^3 with comparatively little annual variation.

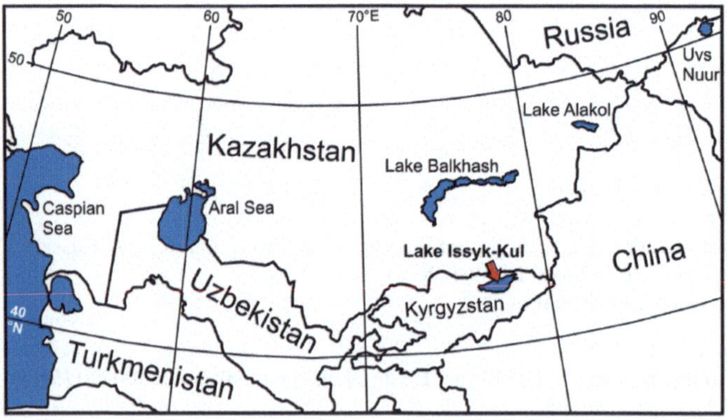

Fig. 6.1 Location of Lake Issyk-Kul (red arrow) in Central Asia

Table 6.1 Main characteristics of Lake Issyk-Kul

Parameter	Data	Source
Catchment area	22,080 km^2	Lyons et al. 2001
Lake area	6236 km^2	Lyons et al. 2001
Volume	1738 km^3	Lyons et al. 2001
Altitude	1606 m	Vollmer et al. (2002)
Mean depth	278 m	Lyons et al. 2001
Maximum depth	668 m	Lyons et al. 2001
pH	8.00–8.75	Giralt et al. (2004)
Surface water temperature range	2 °C (January) to 20 °C (July)	Giralt et al. (2004)
Secchi depth in open lake	15–20 m (up to 53 m)	Romanovsky and Shabunin (2002)
Salinity	~6 g/kg	Vollmer et al. (2002)
Major anions	$Cl^- \geq SO_4^{2-} \gg HCO_3^-$	Lyons et al. 2001; Vollmer et al. (2002)
Major cations	$Na^+ > Mg^{2+} > Ca^{2+} > K^+$	Lyons et al. 2001; Vollmer et al. (2002)

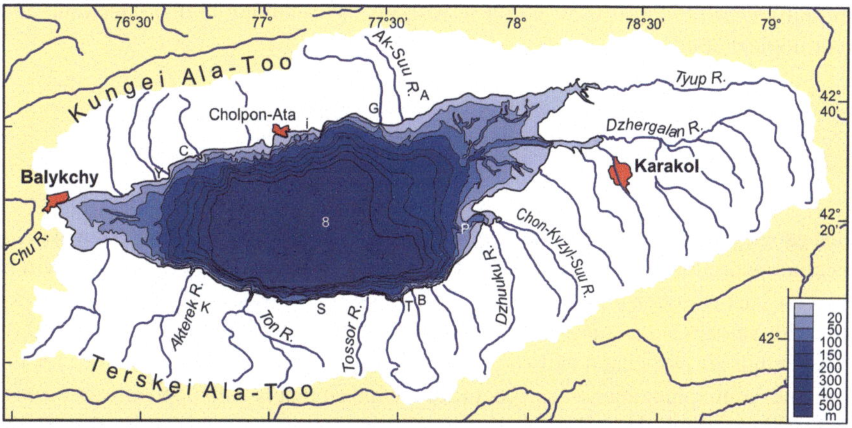

Fig. 6.2 Lake Issyk-Kul and its catchment area (white) in the Tien Shan Mountains. Letters mark locations mentioned in text: A Ananievo, B Barskoon, C Choktal, G Grigorievka, i Bosteri, K Kyzyl-Choku Mountains, P Pokrov Gulf, S Kadji Sai, T Tamga, Y Chyrpykty, 8 Station 8

Five areas are differentiated in Lake Issyk-Kul due to specific geomorphological characteristics:

(1) The western terminus, the Balykchy Gulf, is relatively shallow with 0.5–30 m depth and weakly indented coastline. The water is permanently affected by

winds that cause full mixing and high contents of suspended particles and high turbidity.

(2) The northern coastline has relatively small bays which are not deeper than 40–50 m. The most indented coastline here is the Cholpon-Ata Gulf (Fig. 6.2).

(3) The southern coastline includes large gulfs representing flooded river valleys which formed during lake-level lowstand periods in the past. The bays are more than 50 m deep and morphologically separated from the lake basin.

(4) Less brackish estuary-like gulfs in the eastern part of the lake form a deeply indented coastline with narrow backwater areas. The depth of Tyup Gulf is 35 m and those of Dzhergalan Gulf is 45 m. The submerged channels are 0.7–1.2 km and ca. 2 km wide, and 52 km and 43 km long, respectively (Savvaitova and Petr 1992).

(5) The central part of the lake has a depth of 50–668 m and represents 95% of its volume.

The surface of Lake Issyk-Kul lies currently 14 m below the 'Kutemaldinsky Threshold' (1620 m) which separates the lake from a potential outlet to the Chu River in the west. Ancient shorelines at 1620, 1640 and 1660 m suggest that Lake Issyk-Kul experienced significant lake-level rises, and that the outlet threshold must have been higher in the past (Rosenwinkel et al. 2017). The higher shorelines and submerged river deltas and channels indicate that the lake level fluctuated above and below its present level in the past most likely in response to changing climate, tectonic and hydrological conditions.

6.1.2 Climate

The geographic situation of the Issyk-Kul Basin is unique in Kyrgyzstan with a specific nature of climate conditions. The western Ulan and eastern Santash winds enter the basin through topographically low regions in the west and east of Lake Issyk-Kul with velocities reaching 40 m/s (Podrezov 1985; Podrezov and Pavlova 2003). These winds are associated with the intrusion of Arctic air from the north. As a result, a unique nature phenomenon can occur: two wind channels with opposite directions. The western wind channel Ulan is typically more powerful; it covers the whole western basin plain and western part of the lake from Boom Gorge to Chyrpykty village and sometimes as far east as to Cholpon-Ata at the central-northern shore (Fig. 6.2). The eastern wind channel, Santash, covers the eastern part of the basin from Santash Passage extending up to 10–20 km over the lake. The winds control the local climate by creating arid conditions in the western part of the basin where mean annual precipitation is only slightly higher than 100 mm. Air moving from west to east over the 182 km long axis of the lake collects water due to evaporation from its surface, and then rises convectively or near the basin slopes to often form cumulonimbus (thunder) clouds. As a result, mean annual precipitation increases significantly up to 500–600 mm towards the east.

The deep and permanently unfrozen Lake Issyk-Kul ("Issyk-Kul" means "warm lake" in the Kyrgyz language) acts as a large local heat exchanger. It significantly mitigates temperature conditions in the basin in comparison to other regions far from world oceans and at similar altitude. Summer temperatures are relatively low and winter temperatures relatively high resembling properties of marine climate.

The main feature of Issyk-Kul's local climate specifics is the presence of different climate types within a relatively small region: (1) desert or semi-desert conditions which are common in Central Asia, (2) alpine conditions supporting mountain forests, and (3) marine climate in a narrow lakeside zone. Such diversity of climate conditions results from the orographic isolation, the large altitudinal gradient and the presence of a large and permanently unfrozen lake. There are four areas with clearly different temperature conditions, precipitation amounts and other meteorological parameters in the ca. 3100-km^2 large lakeside-plain region.

Climate area 1: The western part of the basin is the climatically most unfavourable region. The precipitation is lowest in this area with mean annual amounts ranging from ca. 100–150 mm. Only a minor fraction, ca. 10 mm, arrives during the cold period of the year. In some years, there is no precipitation at all during the cold season.

Climate area 2: The northern shore region forms a distinct climate zone between the villages Chyrpykty and Ananievo, almost equivalent to the entire northern shoreline of the lake (Fig. 6.2). It receives more precipitation and wind speed is typically lower, favouring agricultural development and land cultivation. Mean annual precipitation is ca. 300–350 mm, with up to 50% occurring in June to August. A stable thin (<3 cm) snow cover usually forms by the end of November and remains for 30–40 days. The average length of the frost-free season is about 180 days per year. The climate here is mild, without hot summers and with moderately cold winters. The most intense cloud coverage occurs during spring months, and the least occurs in September.

Climate area 3: The eastern shoreline section differs climatically from the other areas notably by the larger amount of precipitation. Mean annual precipitation is 420–450 mm near the shoreline and up to 700–800 mm or even higher in the eastern mountains. Winters typically bring more snow than in other parts of the basin. A stable snow cover usually forms by the end of October and prevails for about 110 days; its average thickness achieves 18 cm by the beginning of February. The average duration of the frost-free period is about 145 days per year. The predominant wind comes from the east funnelled through the gap of the latitudinal intermountain valley between the Kungei and Terskei Ala-Too mountain ranges. Western winds are rarely observed with a frequency of less than 10%. The mean July temperature is ca. 16 °C. The absolute maximum temperature can be as high as 35 °C, and minimum temperatures in winter decrease to −22 °C. The most dense cloud coverage occurs in spring from March to June, and the least cloud cover exists in August and September.

Climate area 4: The fourth climate area is the southern shoreline region between the Akterek River in the west and the Chon-Kyzyl-Suu River in the east. The climatic conditions are characterized by moderately cold winters, comparable to those in the northern shoreline area, and relatively warm summers with typically higher precipitation in a range from 260 to 300 mm. In the area of Tamga resort, the annual range of mean monthly temperatures is ca. 20 °C which is typical for a transitional climate between continental and marine conditions. The lake breeze and mountain-and-valley winds are predominant in this area. Winds often blow along the gorges toward the lake basin or in the reversed direction. However, the average wind speed is low with ca. 2 m/s.

6.1.3 Glaciation in the Lake Catchment and River Runoff

The Issyk-Kul Basin includes 834 glaciers with total area of 650.4 km^2 or 3% of the total catchment area. The glaciated area in the Issyk-Kul Basin represents 6.4% of total glaciated area of the Tien Shan. The glaciers are located at elevations of 3000–5000 m.

Repeated aerosurveys showed that the area of 179 glaciers from 210 glaciers of the northern Tien Shan decreased during the period from 1943 to 1977 (Bakov 1983; Podrezov et al. 2001). During the same period, 28 glaciers were stable and three were advancing. The glaciation reduction intensified in recent years. The investigation of 22 glaciers in the region between the rivers Tossor and Chon-Kyzyl-Suu revealed that seven glaciers retreated 90–110 m, six glaciers 60–89 m and nine glaciers 25–59 m.

The number of rivers and temporary streams feeding Issyk-Kul is estimated between 84 and 118. Of these, 30 rivers are permanent. Most rivers of the Issyk-Kul Basin are fed by snow and ice with highest runoff occurring in July. Meltwater from snow and glaciers contributes ca. 30% to total annual runoff in the basin, and the meltwater contribution can be as high as 61% during summer months. The north-facing slopes of the Terskei Ala-Too include ca. 80% of all glaciers in the catchment of Lake Issyk-Kul. Thus, they generate the majority of glacial meltwater in the basin. Issyk-Kul's largest tributary is the Dzhergalan (also Dzhirgalan, Jergalan, Jyrgalan, Djyrgalan) which drains its southeastern catchment area (Fig. 6.2). The second largest river, the Tyup River, drains its northeastern catchment. A first flood period occurs in spring due to snow melt, and a second period in summer as a result of meltwater discharge from glaciers. The flood season is dominated by snow melt, and highest discharge is recorded in spring. Mean annual water discharge of the Dzhergalan and Tyup rivers are 22.6 and 11.5 m^3/s, corresponding to 0.71 and 0.36 km^3, respectively. The two rivers contribute more than half of the total surface discharge to the lake. Runoff of the Dzhergalan and Tyup rivers together can be as high as 67.4 m^3/s in summer, and as low as 9.4 m^3/s in winter. The large runoff seasonality results from accelerated glacier melting during summer which was generally recorded for the rivers of the Issyk-Kul Basin after 1972. For example, July discharges of Dzhuuku (also Zauka) and Ak-Suu rivers increased by 3.8 and 3.3 m^3/s during 1973–1991 in comparison to preceding years. As a result, mean annual discharge of the two rivers grew from 6.2 up to 7.0 m^3/s for the Dzhuuku River and from 2.9 up to 4.1 m^3/s for the Ak-Suu River. Mean annual discharge of all rivers feeding Lake Issyk-Kul increased from 117 m^3/s (3.69 km^3/a) between 1935 and 1972 to 129 m^3/s

(4.07 km^3/a) between 1973 and 1991. This discharge increase is equivalent to a rise of direct precipitation onto the lake surface from an annual amount of 590 mm to as much as 650 mm if all incoming water was provided as direct precipitation. The discharge increase resulted from the melting of ca. 4 km^3 of glacier ice which represents a decrease of the catchment's total ice volume by 8.3%. The timing of runoff maxima changed accordingly. The snow-melt flood of the Tyup River shifted from May to April after 1972. With diminishing glacier ice resources, continuing climate warming will lead to decreasing summer-water discharges of rivers. This threat is more relevant for the southern slope of the Kungei Ala-Too mountain range where glaciated regions are relatively small and where summer-water discharge to Lake Issyk-Kul will first decrease and will first impact the agricultural sector based on irrigated crop cultivation.

6.1.4 Landscapes in the Issyk-Kul Catchment, and Terrestrial Flora and Fauna

6.1.4.1 Landscapes

According to the geomorphologist E. K. Azykova, the Issyk-Kul catchment includes steppe landscape (1800–2500 m), forest-meadow-steppe landscape (1800–2900 m), subalpine-meadow and meadow-steppe landscapes (2900–3100 m), alpine-meadow and meadow-steppe landscapes (3100–3400 m), and high mountain tundra landscape (3400–4300 m; Azykova 1991). Due to the more pronounced dryness in the west of the basin, desert landscape exists on the lakeside plains, followed by semi-desert and forestless meadow-steppe higher up. In contrast, forest-meadow-steppe is the lowest unit in the wetter eastern part of the basin. Thus, a transition from deserts to semi-deserts and steppe exists along the west-east oriented shorelines. Desert landscape extents as far east as Choktal village on the northern shore of Lake Issyk-Kul and as far as the Kyzyl-Choku Mountains in the south of the lake.

6.1.4.2 Flora

The low precipitation (130–150 mm per year) and high evaporation (>1000 mm per year) at the arid northwestern shoreline of the lake lead to the accumulation of carbonates and other salts in soils which are not frequently leached by percolating water. As a result, halophilous (salsuginous) plants such as niterbush (*Nitraria sibirica*) and glasswort (*Salicornia europaea*) dominate. The vegetation is poor on the hammadas in the western part of the Issyk-Kul Basin and represented by *Reaumuria songarica, Sympegma regelii, Kalidium caspicum, Artemisia tianschanica, Stipa caucasica, Halimodendron* and *Achnatherum splendens*. Sea-buckthorn (*Hippophae rhamnoides*), clematis (*Clematis orientalis*), rosehip (*Rosa albertii*), false tamarix (*Myricaria alopecuroides*) and reed (*Phragmites communis*) grow at the semiarid northern shores of the lake, on sandy and pebbly beaches. Farther away from the shoreline, sheep's fescue (*Festuca sulcata*), tarragon (*Artemisia dracunculus*), ziziphora (*Ziziphora clinopodioides*), sedge (*Carex stenophylloides*) and gum-succory (*Chondrilla*) grow on the sandy and pebbly Holocene lake terraces. Wild ruttishness (*Elymus*), liquorice (*Glycyrrhiza*), thermopsis (*Thermopsis*) and bromegrass (*Bromus*) grow on sandy substrates of the higher parts of Holocene terraces.

The grassy meadows of the steppe regions on the lakeside plain are dominated by alpine bistort (*Bistorta vivipara*), astragalus (*Astragalus tibetanus*), bluegrass (*Poa*), tarragon (*Artemisia dracunculus*), hicory (*Cichorium*) and other weeds.

Spruce forests occupy one third of the forest-meadow-steppe belt with planted spruce forests mostly growing in the lower part of this belt. The larger part of this belt is mostly covered by rosehip, barberry, spirea and representatives of gramineous meadows such as heath false brome (*Brachypodium pinnatum*), cattail grass (*Phleum pratense*) and airy oat grass (*Avena pubescens*). The Russian blue flag (*Iris*), rock geranium (*Geranium*), bedstraw (*Galium*) and meadow rue (*Thalictrum vulgaris* and *T. foetidum*) grow on the meadows.

The middle part of the forest-meadow-steppe belt is dominated by spruce forests with a moss cover on the soils and with relatively few shrubs in the understorey. Two types of mosses, *Hylocomium splendens* and *Rhytidiadelphus triquetrum*,and various lichens are characteristic. Open space between spruce trees is covered by honeysuckle (*Lonicera*), rosehip (*Rosa canina*), whitebeam (*Sorbus*), currant (*Ribes*) and spirea (*Spiraea*).

Spruce forests in the upper part of the belt form a parkland vegetation. Shrubbery on the southern slopes consists of Turkestan juniper (*Juniperus turcomanica*) and on northern slopes of willow (*Salix*) and pea shrubs (*Caragana*).

Meadow-grass vegetation communities of the subalpine belt are dominated by short (<20 cm) grasses and geranium (*Geranium*) and lady's-mantle (*Alchemilla*). *Ziziphora, Kobresia*, sheep's fescue (*Festuca sulcata*), edelweiss (*Leontopodium alpinum*), thermopsis (*Thermopsis*) and starwort (*Aster*) occur in summer.

The cryophilic cushion plant formation of the high mountain tundra consists of gentian (*Gentiana*), oxytrope (*Oxytropis*) and spearwort (*Ranunculus*), growing on moraines in the glacial-nival belt. Alpine motley grasses represented by globeflower (*Trollius*), Altaic violet (*Viola*) and marsh gilled (*Coronaria flos-cuculi*) grow on the warmer slopes with southern exposition.

6.1.4.3 Fauna

The different landscapes and large climatic gradients in the catchment of the lake cause a high faunal diversity. The permanently unfrozen Lake Issyk-Kul is an important waterfowl habitat (Vereschagin 1991). Birds such as pheasants (Phasianidae), grebe (*Colymbus* spp.), dabbling ducks, ruddy shelduck (*Tadorna ferruginea*), merganser (*Mergus*), baldicoot (*Fulica atra*), red-crested pochard (*Netta rufina*), bittern (*Ixobrychus*), moorhen (*Gallinula*), rail (*Rallus*), grass-drake (*Crex crex*), demoiselle

(*Anthropoides virgo*), peewit (*Vanellus*), redshank (*Tringa totanus*), black-winged stilt (*Himantopus*), fantail snipe (*Gallinago gallinago*), black-headed gull (*Larus ridibundus*), common tern (*Sterna hirundo*), kingfishers (Alcedinidae), bank swallow (*Riparia riparia*) and wagtail (*Motacilla*) build nests on the lake shore (DDWSSD 2018). Eurasian whitethroat (*Sylvia communis*), scarlet grosbeak (*Carpodacus erythrinus*), great reed warbler (*Acrocephalus arundinaceus*), eastern turtle dove (*Streptopelia orientalis*), turtle dove (*Streptopelia turtur*), great tit (*Parus major*), bearded tit (*Panurus*), Eurasian long-tailed titmouse (*Aegithalos caudatus*), azure tit (*Parus cyanus*), common blackbird (*Turdus merula*), kestrel (*Falco tinnunculus*), long-eared owl (*Asio otus*), cuckoo (*Cuculus*), common nightingale (*Luscinia megarhynchos*), goldfinch (*Carduelis carduelis*), starling (*Sturnus vulgaris*), magpie (*Pica*), sparrow hawk (*Accipiter nisus*) and marsh harrier (*Circus aeruginosus*) inhabit sea-buckthorn brushwoods and wetlands (DDWSSD 2018).

Mammals of these habitats are represented by common field mouse (*Apodemus sylvalicus*), narrow-headed vole (*Microtus gregalis*), common red fox (*Vulpes vulpes*), common weasel (*Mustela erminea*), least weasel (*Mustela nivalis*), shrewmice (Soricidae), badger (*Meles meles*), tolai hare (*Lepus tolai*) and muskrat (*Ondatra zibethicus;* Shukurov 1991).

The tesselated snake (*Natrix tessellata*), Dione's ratsnake (*Elaphe dione*), and sand lizard (*Lacerta agilis*) are found on pebbly or rocky shores.

The mammal fauna of desert and semi-desert regions in the basin is represented by the eared hedgehog (*Hemiechinus*), tolai hare (*Lepus tolai*), Tien Shan souslik (*Citellus relictus*), and steppe polecat (*Mustela eversmannii*). Reptiles are represented there by copperhead snake (*Agkistrodon*), multi-ocellated racerunner (*Eremias multiocellata*) and racerunner (*Eremias arguta*); amphibians by green toad (*Bufo viridis*); and birds by Pallas's sandgrouse (*Syrrhaptes paradoxus*), black-bellied sandgrouse (*Pterocles orientalis*), little owl (*Athene noctua*), Daurian partridge (*Perdix dauurica*), dorhawk (*Caprimulgus*), trumpeter bullfinch (*Bucanetes githagineus*), rock sparrow (*Petronia petronia*), gray-hooded bunting (*Emberiza buchanani*), larks (Alaudidae), Richard's pipit (*Anthus novaeseelandiae*), wheatear (*Oenanthe*) and black redstart (*Phoenicurus ochruros*).

The mammals of the forests include shrewmice (Soricidae), tolai hare (*Lepus tolai*), squirrel (*Sciurus*), large-eared pika (*Ochotona macrotis*), grey hamster (*Cricetulus migratorius*), common field mouse (*Apodemus sylvalicus*), common wolf (*Canis lupus*), common red fox (*Vulpes vulpes*), common weasel (*Mustela erminea*), badger (*Meles meles*), wild pig (*Sus scrofa*), roe deer (*Capreolus capreolus*) and rock marten (*Martes foina*). Forests are inhabited by birds such as kite (*Milvus*), sparrow hawk (*Accipiter nisus*), goshawk (*Astur palumbarius*), buzzard (*Buteo*), hobby (*Falco subbuteo*), kestrel (*Falco tinnunculus*), grouse (*Lyrurus*), ring dove (*Columba palumbus*), rufous turtle dove (*Streptopelia orientalis*), cuckoo (*Cuculus*), hawk owl (*Surnia ulula*), three-toed woodpecker (*Picoides tridactylus*), tree pipit (*Anthus trivialis*), common magpie (*Pica pica*), nutcracker (*Nucifraga*), carrion crow (*Corvus corone*), wren (*Troglodytes*) and tit-warbler (*Leptopoecile*).

Forestless slopes, flood plains and rocks are inhabited by pale harrier (*Circus macrourus*), golden eagle (*Aquila chrysaetos*), partridge (*Alectoris*), quail

(*Coturnix*), ibis-bill (*Ibidorhyncha*) pigeon (*Columba*), eagle owl (*Bubo*), swifts (Apodidae), larks (Alaudidae), water ouzel (*Cinclus*), wheatear (*Oenanthe*), wall creeper (*Tichodroma muraria*) and rock thrush (*Monticola saxatilis*).

The muflon (*Ovis ammon*) and wild mountain goat (*Capra ibex*) are found in the alpine meadow and tundra belts. The relatively high abundance of mammals in these belts results from favourable forage resources and the rough terrain that inhibits accessibility. The true "landlord" of high mountain regions is the snow leopard (*Panthera uncia*). Its seasonal migration follows those of its main prey, the Alpine ibex. The avian fauna of the high mountain ranges is characterized by the Himalayan snowcock (*Tetraogallus himalayensis*), ruddy shelduck (*Tadorna ferruginea*), lammergeier (*Gypaetus barbatus*), black vulture (*Aegypius monachus*), griffon vulture (*Gyps fulvus*), Himalayan griffon (*Gyps himalayensis*), horned lark (*Eremophila alpestris*), rock pipit (*Anthus spinoletta petrosus*), red-billed chough (*Pyrrhocorax pyrrhocorax*), Alpine chough (*Pyrrhocorax graculus*), accentor (*Prunella*), redstart (*Phoenicurus*), finchs (Fringillidae) including the white-winged snow finch (*Montifringilla nivalis*), and the mountain linnet (*Acanthis flavirostris*).

6.2 Lake Issyk-Kul in the Past and Its Current Conditions

6.2.1 History of Lake-Level Changes Prior to the Twentieth Century

The investigated lake-level history of Issyk-Kul can be divided into five phases according to the work of Sevastianov (1991) and subsequent studies.

1. Highest lake-levels existed at Lake Issyk-Kul in the late Pleistocene. Paleolithic landmarks were recorded on a lake terrace 30–35 m above the present lake (absolute elevation 1640 m; Sevastianov 1991). Radiocarbon dating of lacustrine sediments at this position above the lake yielded an age of ca. 26 thousand years (ka; Romanovsky 2002a). Correspondingly, geomorphological and dating analyses by Rosenwinkel et al. (2017) suggest that higher lake levels existed in the late Pleistocene ca. 45–22 ka ago, and that a catastrophic dam break (or breaks) and outburst floods occurred ca. 20.5–18.5 ka.
2. A significant regression of Lake Issyk-Kul occurred at the end of the late Pleistocene and during the early Holocene as a result of dry and cold climate conditions accompanied by the accumulation of snow and ice in the Tien Shan Mountains and the reduction of meltwater discharge (Sevastianov 1991). Submerged terraces and channels which reach down to 110 m below Issyk-Kul's current level are possibly associated with this regression (De Batist et al. 2002). The significantly lowered lake level caused the exposure of a 3–5 km wide area along the lake's northern shore, of a 0.5–2.0 km wide area along its southern shore and of a 35–40 km wide section in the west and east of the lake in comparison to today's area. Desert and semidesert vegetation with members of the goosefoot family

(Chenopodiaceae) and sagebrush (*Artemisia*) as dominant plants spread widely in the Issyk-Kul Basin during this period. Brushwood cover in river-valleys and regions with spruce forests in the mountains were reduced (Sevastianov 1991).

3. Issyk-Kul experienced a transgression in the first half of the Holocene caused by climate warming and the decrease of mountain glaciation generating high meltwater discharge. The maximum lake level reached the terrace at ca. 30–35 m above its current level. Humidity was significantly higher during this period with a mean annual precipitation of up to 400 mm in the southern part of the basin and of up to 600 mm in its northeastern part (Sevastianov 1991). Brushwood and forests spread again due to higher humidity and temperatures in the mountains around Lake Issyk-Kul.

   Based on the study of lake sediment cores by Ricketts et al. (2001), higher lake levels and open-lake conditions with the establishment of an outlet were inferred for the period from 8.3 to 6.9 ka. Following a transitional period until 4.9 ka, closed-basin conditions similar to the situation today prevailed afterwards. This new regression was probably caused by the increasing aridification of the region in the middle-late Holocene and the Bronze Era. A low lake level existed until ca. 3 ka ago.

4. A significant increase in humidity led to a new transgression of Lake Issyk-Kul ca. 2 ka ago. The lake reached a position between its present level and a terrace ca. 10–12 m above its current position (absolute elevation 1620 m). Ancient chronicles and maps suggest that Lake Issyk-Kul was an open-basin lake with an outlet ca. 3–2 ka ago (Romanovsky 2002a). However, the increase in available moisture caused the widespread development of steppe vegetation in the lake basin. A relatively short open-basin period existed probably 1.4–1.2 ka (Ricketts et al. 2001). In contrast, levels 3–6 m below its present level in the thirteenth to fifteenth century as a result of drier climate conditions were inferred from submerged soil, tree and settlement remains (Ricketts et al. 2001; Romanovsky 2002a).

5. The most recent open-basin period of Lake Issyk-Kul existed in the eighteenth century. Historical maps show the lake in the years 1698, 1724, 1756, 1760 and sometime in one of the years before 1823; and all apart from the one for the year 1724 include an inflow of the Chu River and an outlet in its north (Romanovsky 2002a; Narama et al. 2010). The lake level apparently started to decline during the second part of the nineteenth century. This regression lasted until the beginning of the twenty-first century. A slight increase is recorded since that time.

   A seismic survey of Lake Issyk-Kul's sediments revealed river delta deposits as deep as 500 m below the present lake level and provided evidence for four different regression periods which all caused lake-level declines beneath 200 m in comparison to its present surface. However, the timing of these major lake-level falls remains elusive (Gebhardt et al. 2017).

6.2.2 *Thermal Regime, Currents and Mixing of Lake Issyk-Kul*

The mean annual surface water temperature of Lake Issyk-Kul is ca. 11 °C. Mean annual surface water temperatures are slightly lower in the western and eastern parts of the lake with ca. 10.3 °C, and slightly higher with 11.4 °C near the northern shore.

Measurements of bottom water temperatures between 1928 and 1932 were conducted during expeditions of the Academy of Science of the USSR led by L. S. Berg. Temperatures at 200 m and 600 m depth were 4.3 and 4.2 °C in 1928, 4.0 and 3.7 °C in 1930, and 4.2 and 3.7 °C in 1932, respectively. Bottom water temperatures were regularly monitored on board of the research vessel "Moltur" by the Kyrgyz Administration on Hydrometeorology during the period from 1981 to 1988. Water temperatures at 650 m depth were in a range from 3.7 to 4.1 °C. Significant temperature deviations in comparison to those measured between 1928 and 1932 were not recorded at 200 m depth. Following a gap in Issyk-Kul surveys due to the economic decline after the collapse of the Soviet Union, water temperature measurements in 2001 and 2003 revealed increased values of up to 4.4 °C at 650 m depth. Compared with data for the 1980s, temperatures rose by 0.4 °C at 650 m, by 0.6 °C at 200 m and by 0.8 °C at 100 m depth. This deep-water temperature increase likely results from global warming in recent decades (Peeters et al. 2003; Salamat et al. 2015).

The current system of Lake Issyk-Kul is mostly controlled by thermal conditions and the winds blowing in the basin. In winter, surface water cooling is typically accompanied by an anticyclonic current with clockwise water movement in the lake. In contrast, the western (Ulan) and eastern (Santash) winds cause a counterclockwise water movement due to the southward turn of the western wind and the northward turn of the eastern wind after entering the basin. As a result, a cyclonic water current forms in the shoreline area of the lake, and an anticyclonic circulation in its central part with a zone of water convergence in between. In the western and eastern parts of the lake, the convergence zone aligns with the position of a steep thermal bar with colder waters closer to the lake shore (Bulin and Romanovsky 1975; Romanovsky and Shabunin 2002). Measurements of current velocities at 10, 15 and 25 m water depth of buoy-based stations showed decreasing velocities at greater depth. Average current velocity is in a range from 20 to 38 cm/s, and maximum velocity at 10 m depth is 65 cm/s. Current velocities of 32, 13 and 10 cm/s were measured during periods of unstable conditions in the lake at 100, 150 and 300 m depth, respectively.

The vertical stability of the water column and possible mixing mechanisms and deep-water renewal processes were most recently studied by Vollmer et al. (2002) in September 2000. They identified a thin well-mixed surface water layer with temperatures between 16.5 and 18.5 °C, a sharp thermocline between 10 and 60 m water depth, and a nearly isothermal hypolimnion with temperature variations of <0.7 °C between 150 m and the maximum depth (Fig. 6.3). Lowest temperatures were 4.29 °C near the lake floor at several stations. Tracer analyses by Hofer et al. (2002) and Kipfer and Peeters (2002) revealed that deep-water renewal in Issyk-Kul is rapid and that residence time of deep water is <13 years. Measurements of vertical

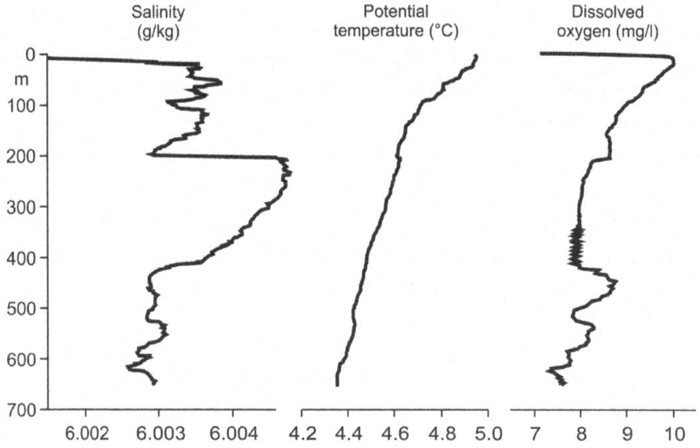

Fig. 6.3 High-resolution depth readings of salinity, temperature and dissolved oxygen concentration at central positions in Lake Issyk-Kul in March 2001 (Hofer et al. 2002; Peeters et al. 2003). Salinity changes, although minor, result from different water masses due to deep-water formation by cooling of surface waters in the shallow bays in winter

transects in the lake in March and August 2001 showed that deep-water renewal results from differential cooling of surface waters and accumulation of cold water in deeply incised channels in the shallower parts of the lake basin in winter that moves then down the channels to greater depth (Peeters et al. 2003). These results extent earlier suggestions that cold river water moves down in the deeply incised channels towards to centre of the lake and displaces warmer waters above (Stavissky 1977).

Salinity differences are generally small in lake waters of Issyk-Kul and not considered to contribute significantly to mixing processes (Peeters et al. 2003; Fig. 6.3). Salinity measurements for 19 water samples collected in September 2000 at station 8 of the study by Vollmer et al. (2002) between 655.1 and 2.7 m depth in the lake ranged from 5.89 to 6.21 g/kg with a mean and standard deviation of 6.06 ± 0.07 g/kg. Lake-water mixing as a result of surface-water cooling in the shallow parts of the lake will be possibly reduced in the future if winter air temperatures continue to increase (Salamat et al. 2015).

So far, lake waters in Issyk-Kul generally have high concentrations of dissolved oxygen. Depth profiles do not show large vertical gradients (Fig. 6.3). For example, high-resolution dissolved oxygen-depth profiles measured in March 2001 at the deepest part of the lake revealed maximum concentrations of ca. 10 mg/l in the uppermost 20 m of the water column and minimum concentrations of ca. 7.5 mg/l beneath 600 m depth (Hofer et al. 2002; Peeters et al. 2003; Fig. 6.3). Surface waters in the shallower parts of the lake (above 50 m depth) can be significantly oversaturated with respect to dissolved oxygen especially in summer due to the high phytoplankton and macrophyte productivity. Highest oxygen saturation values were measured for surface waters by V. L. Kadyrov and V. P. Matveev in the Tyup Gulf (121.5 and 185.9%, respectively; Matveev 1935). Dissolved oxygen saturation of surface (0–10 m) water is typically 112% in summer, 106–108% in autumn, 99% in winter, and 107–109% in spring, reflecting the seasonal temperature variations and the related growth of phytoplankton. Oxygen saturation is typically not lower than ca. 70% even in the deepest part of the lake.

The wave disturbance of lake sediments in Issyk-Kul ranges from negligible rip currents driven by weak winds and waves of few centimeters height to major resuspension and wave-erosion events during storms. Waves of up to 4 m height were measured during storms of >20 m/s wind speed at the wave-height pole 2.5 km from the Kara-Bulun headland located at 10 m water depth.

6.2.3 Hydro-Physics and Hydrochemistry of the Lake

Water transparency, measured by Secchi Disk, is generally relatively high in Lake Issyk-Kul. Average transparency is 12.4 m near the northern shore and 13.9 m near the southern shore. The higher transparency near the southern shore results from the steeper bathymetry in the south with a narrower shallow-water zone (Fig. 6.2). The littoral zone is widest in the eastern part of the lake basin starting at the Kara Bulan headland where transparency is typically in a range from 8 to 12 m. The two largest rivers, Dzhergalan and Tyup, transport suspended particles to the eastern part of the basin where transparencies drop after glacier meltwater pulses and heavy rains. Transparency is only 0.2 to 1.0 m in the proximal part of the Dzhergalan River mouth, it increases to 2 m 120 m farther away and up to 8 m in a distance of 17 km.

In Balykchy Bay in the western part of the lake, water transparency is typically higher than 12 m in spite of the wide shallow zone in the bay and the occurrence of storm waves as a result of western winds. Transparency is mostly in a range from 15 to 17 m. Close to the shore, in a distance of 1.5 to 2.0 km, transparency is reduced to 7.5–10.0 m. The higher transparency in the western shallow part of Issyk-Kul in comparison to its eastern analogue results from the lack of major inflows and the dense cover of charophytes down to 40 m water depth.

Water transparency in the central part of the lake is only controlled by phytoplankton concentration and typically in a range from 15 to 20 m in summer.

Water with deep blue color or colors I-III on the Forel-Ule (FU) color scale typically occurs in the winter in Issyk-Kul when transparencies are in a range from 25 to 53 m. According to V. P. Matveev's observations in 1928, blue water color (I–III) was registered in Issyk-Kul even during summer (Matveev 1935). Currently, water in summer has mostly FU colors IV–VI that also have blue tints. In the Tyup Gulf, where Tyup River delivers large quantities of suspended sediments and where the phytoplankton population reaches densities ca. 8–9 times higher than in the open lake, water predominantly has green color complying with FU colors VII–X. Some backwater regions of the Tyup Gulf are characterized by a green-brown water color (FU color XI–XVI) in summer.

According to Kadyrov (1986), the density of Issyk-Kul water in the open part of the lake is 1005.2 kg/m^3. The density decreases to 1003.2–1004.0 kg/m^3 in the east of the lake near the Tyup and Dzhergalan inflows. The water density increases with depth in the gulfs of the two main tributaries below ca. 10 m showing that the inflows form a lighter surficial water layer on the lake water prior to mixing. The temperature of maximum density of 2.6 °C is typically never reached in the open lake area due to the large heat capacity of the lake and the mild winters in the region (Kipfer and Peeters 2002; Vollmer et al. 2002). Cooling of waters to 2.6 °C and even below occurs in the littoral zone of the lake in winters (Vollmer et al. 2002). The water-ice transition temperature is −0.3 °C.

The salinity distribution in the lake is controlled by the two large inflows in the eastern part of the basin. Runoff of Dzhergalan and Tyup rivers has a salinity of 0.6 g/kg. Salinity in the Tyup Gulf is between 2.6 and 2.8 g/kg at 3 km distance to the river mouth. At 9 km distance, surface waters have a salinity of 4.3 g/kg, water at 10 m depth has a salinity of 5.6 g/kg and water at 25 m depth has a salinity of 5.8 g/kg. Salinity in the Dzhergalan Gulf is mostly in a range from 3.1 to 5.9 g/kg. In contrast to the eastern part of the lake, the Balykchy Gulf in its western part is not affected by major inflows, and salinity is 5.8–6.0 g/kg, similar to the open central part of the lake. Salinity in the Cholpon-Ata Gulf at the northern shore and in the gulfs along the southern shore (e.g., Tamga, Ton, Akterek) is between 5.5 and 5.8 g/kg. Salinity in the central part of the lake is ≥5.8 g/kg. A homogenous salinity distribution pattern is also recorded on a vertical scale (Table 6.3). Salinity in the central part of the lake at Cholpon-Ata's longitude (77.1° E) is 5.8 g/kg at 50 m water depth and 5.9 g/kg at 650 m depth, or 6.0 and 5.9 g/kg at 50 and 290 m depth, respectively, ca. 30 km further to the east at 77.5° E. The horizontal and vertical homogeneity of the salinity values in Lake Issyk-Kul reflects the efficient exchange of water between different parts of the lake. However, the ratio of the relatively small annual total runoff volume to the total lake volume is 1/530 resulting in relatively minor effects of inflows on the salinity distribution in Lake Issyk-Kul.

The salinity of Lake Issyk-Kul increased since the measurements of V. P. Matveev in 1928 who determined an average ($n = 34$) of 5.823 g/kg (Matveev 1935; Fig. 6.4). Average salinity was increased to 5.968 g/kg in the early 1960s (Kadyrov and Karmanchuk 1964). Measurements of the Kyrgyz Meteorology Agency in the middle of the 1980s yielded an average salinity of 6.01 g/kg. In 2000, the average salinity of the water column in the center of the lake was 6.06 g/kg (Vollmer et al. 2002). Kadyrov's (1986) measurement of a high salinity of 6.11 g/kg for near-bottom waters in 1947 is based on a different methodological approach (either the dry residue or the sum-of ions method) and is possibly not comparable with the other data (Fig. 6.4). However, the observed steady salinity increase of Lake Issyk-Kul reflects diminishing runoff entering the lake and continuous climate warming in the region since the 1930s, resulting in higher evaporation effects on a closed-basin lake (Salamat et al. 2015).

According to the salinity balance of Issyk-Kul determined by the Novocherkassk Hydrochemical Institute, the majority of dissolved salts in Lake Issyk-Kul reaches the lake as groundwater inflow (959 thousand tons of dissolved salts per year equal to 72% of its total influx; Romanovsky et al. 2004). The remaining portion arrives with runoff (296 thousand tons per year, 22%) and direct precipitation (75 thousand tons per year, 6%). About half (50.2%) of the total influx of dissolved salts of 1330

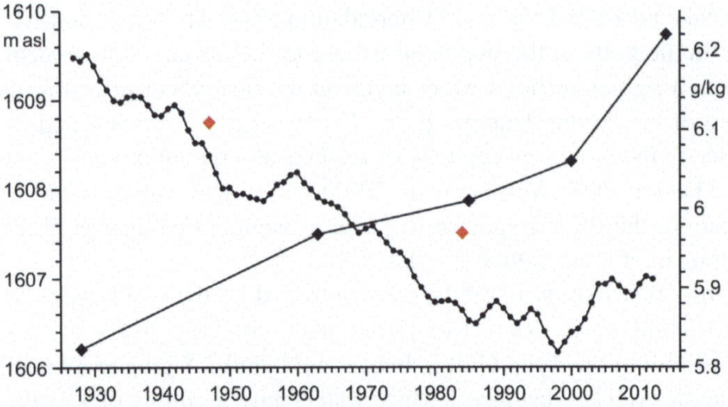

Fig. 6.4 Observed level changes of Lake Issyk-Kul for the period of 1927–2012, and salinities measured in the lake (Abuduwaili et al. 2019). Salinity data marked red are based on either the dry residues or sum-of ions whilst others are mostly calculated from measured temperature, electric conductivity and pressure

thousand tons per year is removed from the water column by precipitation which is dominated by the formation of monohydrocalcite and calcite (Giralt et al. 2004). The other half of the annual influx of dissolved salts to the lake results in an annual salinity increase by 0.4 mg/kg. The relatively low salinity of Lake Issyk-Kul indicates that the present lake without an outlet is geologically young. Geomorphological and archaeological data confirm that the lake in its present closed-basin state is a generally young phenomenon (Bondarev 1978; Sevastianov 1991; Rosenwinkel et al. 2017). According to maps of Jesuit missionaries, the present closed-basin conditions were not established before the 1820s (Romanovsky 1990; Narama et al. 2010).

First reliable hydrochemical data for the water of Lake Issyk-Kul were provided for Matveev's survey of the lake in 1928 (Matveev 1935). More systematic hydrochemical analysis started in 1966. Dominating anions in Lake Issyk-Kul are Cl^- and SO_4^{2-}, dominating cations are Na^+ and Mg^{2+} (Table 6.2). Thus, Issyk-Kul water is characterized as sodium chloride and sulfate water (Kutzeva 1980). The water column of Lake Issyk-Kul shows only insignificant variations of the concentrations of major ions with depth providing additional evidence for the well-mixed water masses (Table 6.3). With respect to the dominating cations Na^+ and Mg^{2+}, Issyk-Kul water is similar to ocean water. In contrast, the dominance and similar concentrations of Cl^- and SO_4^{2-} are typical for an inland saline lake.

The pH value of Issyk-Kul water is 8.00–8.75 (Giralt et al. 2004). Alkalinity of lake water (content of bicarbonate and carbonate ions) is high in comparison to other large water bodies with ca. 5.8 meq/l (Vollmer et al. 2002).

Table 6.2 Average chemical composition of Issyk-Kul water (Kyrgyzhydromet: Agency of Hydrometeorology under the Ministry of Emergency Situations of the Kyrgyz Republic)

Ions	g/kg	mg/equivalent	% equivalent
Cl^-	1.585	44.70	24.1
SO_4^{2-}	2.115	44.03	23.8
HCO_3^-	0.240	3.93	2.1
Common anions	3.940	92.66	50.0
Ca^{2+}	0.114	5.69	3.1
Mg^{2+}	0.294	24.18	13.0
K^+	0.068	1.74	0.9
Na^+	1.407	61.04	33.0
Common cations	1.883	92.65	50.0
Total	5.823	185.31	100.0

Table 6.3 Vertical distribution of major cations and anions (in mg/l) in Lake Issyk-Kul between 1966 and 1992

Depth (m)	Ca^{2+}	Mg^{2+}	$Na^+ + K^+$	HCO_3^-	SO_4^{2-}	Cl^-	Sum
5	115.1	279.9	1465.1	247.8	2138.4	1556.4	5802.7
10	115.5	286.4	1471.1	247.9	2145.0	1579.3	5845.2
25	116.5	290.3	1446.2	248.0	2118.1	1573.4	5792.5
50	116.0	283.3	1461.9	248.8	2121.3	1573.9	5805.2
100	116.1	287.7	1475.5	249.2	2148.9	1587.4	5864.8
200	116.1	284.4	1463.0	249.7	2127.7	1574.0	5814.9
300	115.1	248.7	1462.6	249.3	2121.2	1571.7	5768.6
400	116.1	288.6	1458.0	249.1	2129.5	1577.5	5818.8
500	116.4	291.4	1448.1	248.2	2121.3	1577.5	5802.9
650	115.6	285.3	1463.4	249.6	2121.0	1581.0	5815.9

6.2.4 Water Balance of Lake Issyk-Kul and Lake-Level Changes Since the Twentieth Century

The water balance of Lake Issyk-Kul was described by Romanovsky (2002a) as

$$x + y + y_1 - Q - z \pm \Delta h = 0$$

where x is precipitation on the lake surface, y is surface water inflow, y_1 is groundwater inflow, Q is water loss due to withdrawal for irrigation, z is evaporation from the lake surface and Δh is lake-level change. About 20 versions of Lake Issyk-Kul's water balance have been developed by different scientists (Romanovsky 2002a). Assumed or calculated components of the water balances vary over large ranges.

Annual direct atmospheric precipitation on the surface of the lake varies in these calculations in a range from 230 to 350 mm. The annual surface water inflow covers a range from 184 to 567 mm, i.e. highest assumed values are as high as three times the lowest assumed discharge numbers. Annual groundwater inflow varies from 241 to 330 mm in the water balances, and annual evaporation from the surface of the lake ranges from 700 to 910 mm.

Gronskaya (1983) and Krivoshei and Gronskaya (1986) suggested a water balance of Lake Issyk-Kul for the period from 1935 to 1978 which is considered as most reliable. Based on many observational data, they used 274 mm as annual precipitation on the lake surface, 295 mm as surface water inflow, 299 mm as groundwater inflow, 836 mm as evaporation from the lake, 78 mm as water loss for irrigation and 50 mm for the lake-level change. The resulting residual is 4 mm.

A water balance for the period from 1979 to 2000 uses 306 mm as annual precipitation on the lake surface, 627 mm as surface and groundwater inflow, 910 mm as evaporation from the lake, 46 mm as water loss for irrigation and 18 mm for the lake-level decline. The resulting residual is −5 mm. All climate-controlled components of the water balance increased in comparison to the balance for the period from 1935 to 1978, and the increase is higher for the income components precipitation and inflows, resulting in a significantly lower lake-level decline during 1979–2000 (Fig. 6.4).

The level of Lake Issyk-Kul fluctuates ca. 20 cm annually due to higher precipitation and runoff including meltwater during spring and summer, and lower precipitation and runoff in autumn and winter. During the period of instrumental observations from 1927 to 2012, the lake level decreased by 249 cm (Fig. 6.4). The average annual decline is 2.9 cm. The maximum annual decline was recorded with 20 cm in 1948, and the highest rise occurred with 18 cm in 2003.

The highest average annual level of 1609.52 m was recorded in 1929, and the lowest level of 1606.17 m occurred in 1998. The general decline was mostly caused by regional climate warming and increasing evaporation from the lake, and the increasing withdrawal of water from tributaries for irrigation farming (Salamat et al. 2015). However, the decline was punctuated by wetter periods and related lake-level rises in the years 1929, 1935–1936, 1941–1942, 1956–1960, 1970–1971, 1981–1982, 1987–1989, 1993–1994, and 1999–2003 (Fig. 6.4). The longest periods of lake-level increase were observed in 1956–1960 and 1999–2003. Higher precipitation and meltwater discharge caused lake-level rises of 32 and 55 cm over a period of five years, respectively. The first longer period of increasing lake levels was preceded by a long continues lake-level decline from 1943 to 1955 when the lake's level fell 110 cm due to lower precipitation and higher air temperatures in the Issyk-Kul Basin.

The lowest water level of Lake Issyk-Kul at 1606.08 m was registered in the end of 1997 and in the beginning of 1998. The level reached 1606.80 m in the end of 2003 after a rise by 72 cm since the beginning of 1998. This last significant lake-level rise caused intensive shore abrasion. Sand beaches near the Sukhoi Bereg Peninsula which were 100–250 m wide in the 1980s were washed away in September 2003, and sea-buckthorn shrubs destroyed by waves. Future global warming and stronger evaporation, together with diminishing meltwater runoff due to decreasing glacier-ice resources in the upper reaches of the lake and continuing withdrawal of water from tributaries for irrigation farming represent a major threat for the stability of Lake Issyk-Kul's ecosystem in the future.

6.2.5 Plankton, Benthos, and Fish of Lake Issyk-Kul

Phytoplankton in a lake is crucial for its ecosystem due to its role as primary producers in the food web. The first comprehensive study of Issyk-Kul's phytoplankton was conducted by I. A. Kiselev. The results of Kiselev's (1932) studies showed that the algae composition of Lake Issyk-Kul is more diverse than those of many other water bodies. A total of 309 algae taxa were identified in the phytoplankton of the lake. Most algae taxa were recorded in the more shallower parts near the shores of the lake. The less diverse phytoplankton in the open part of the lake is mainly represented by *Botryococcus braunii, Peredinium borgei* and *Amphiprora paludosa* var. *issykkulentis*. *Botryococcus* is associated with upper waters in the water column whilst *Amphiprora* dominates at greater water depth. T. G. Matukova studied benthic algae of the Tyup Gulf in the end of 1950s. She identified 240 taxa and forms of diatoms of which only 13 taxa are planktonic (Matukova 1958). In the beginning of the 1960s, S. M. Mambetalieva conducted investigations of the benthic algal flora of the northern shore of Lake Issyk-Kul. She listed 346 taxa and forms of algae in Issyk-Kul's waters. The phytoplankton in the open lake was studied in more detail by A. A. Kulumbaeva in the years 1972–1975. According to her work, 111 species and forms of phytoplankton were determined: 44 taxa are blue-green algae, two taxa are yellow-green algae, 19 taxa are diatomic algae, 40 taxa are green algae, and six taxa are dinoflagellates (Kulumbaeva 1982). Phytoplankton concentrations in the open lake are 166–488 thousand cells/l, corresponding to a biomass in a range from 73 to 153 mg/l. Concentrations and corresponding biomass are higher in the Tyup Gulf: 1513–4126 thousand cells/l and 207–398 mg/l, respectively (Kulumbaeva 1982). Blue-green algae such as *Merismopedia punctata* are most abundant among the phytoplankton (44% of total cell numbers), and concentrations up to 12,356 thousand cells/l were determined in the open lake during the summer season.

The maximum phytoplankton concentration and highest biomass in Lake Issyk-Kul occurs at 15–50 m water depth. However, algae concentrations are also high during most of the year in a depth range of 100–150 m. In contrast to many other water bodies with a significant seasonal change of the algae composition, seasonal changes are minor in Lake Issyk-Kul. In addition, the majority of the phytoplankton in Lake Issyk-Kul are small or nannoplanktonic forms of blue-green and green algae that represent the major part of lake productivity. The cell diameter of these algae is typically not exceeding 2–3 μm resulting in a relatively low biomass in spite of relatively high phytoplankton concentrations in the lake. Very likely, the dominance of nannoplanktonic algae is an adaptive response to low concentrations of nutrients in Lake Issyk-Kul.

The list of Issyk-Kul's zooplankton prepared by L. A. Folian in 1973 consists of 154 species and forms, 76 of which are protozoans, 59 are rotifers, 11 are cladocerans, and eight are copepods. The zooplankton predominates in the shallow coastal zone, and only three rotifer taxa and two copepods (*Arctodiaptomus salinus* and *A. viridis*) are abundant in the open lake (Folian 1981). Most abundant in Lake Issyk-Kul are nine rotifers, two cladocerans and four copepods, and it is the copepod *A. salinus* which produces mass occurrences and dominates the zooplankton in all areas of the lake (Savvaitova and Petr 1992). During the year *A. salinus* amounts to 75–95% of the total zooplankton population and 95–99% of the total zooplankton biomass. The maximum density of *A. salinus* populations is typically reached with 6–8 thousand individuals/m^3 (equal to a biomass of 0.17–0.22 g/m^3) in the upper water layer (0–100 m) in the end of August and beginning of September. The minimum density of the *A. salinus* population (ca. 2.0–2.5 thousand individuals/m^3, biomass ca. 0.12–0.15 g/m^3) typically occurs in the end of February and beginning of March.

According to Pavlova (1964), the zoobenthos of Lake Issyk-Kul comprises 224 species and forms. Included are three taxa of shellfish, 35 taxa of chironomid larvae, two other dipterous larvae taxa, two water beetles, five water bugs (Gerromorpha and Nepomorpha), seven caddisfly larvae, one stone fly larva, four dragonfly larvae, five acarians, five amphipods, 12 ostracods, five leeches, five oligochaetes, and 35 protozoans. In addition, turbellarian and nematode worms occur. Zoobenthos is most abundant in the spring season. It is reduced in autumn due to the emergence of the chironomids and predation by fish. Zoobenthos biomass in Pokrov Gulf was 22.14 g/m^2 in spring, 8.8 g/m^2 in summer, 5.5 g/m^2 in autumn and 11.9 g/m^2 in winter (Pavlova 1964). Near the shoreline, representatives of the Chironominae family are abundant, dominated by *Strictochironomus pictulus* and *Micropsectra* sp. Lake floor covered by stoneworts (macro-algae) is mostly inhabited by Orthocladiini larvae. The amphipod *Gammarus ocellatus* M. is typically also very abundant among stoneworts, whilst *G. bergi* M. is abundant over gravelly and rocky substrates. Among shellfishes, the most important representative with respect to biomass is *Radix auricularia* (Pavlova 1983). The average biomass of zoobenthos in the "stonewort belt" of the littoral area is 8–10 g/m^2. The zoobenthos diversity and biomass is decreasing to a half or a third from the "stonewort belt" to a depth of 60–70 m. In this depth range, four taxa of the Chironominae family predominate: *Chironomus anthracinus, Ch. plumosus, Strictochironomus pictulus* and *Micropsectra* sp. In addition, *Radix auricularia* is abundant at this depth. The average annual biomass in this depth belt is 2.5–3.5 g/m^2. At water depth exceeding 70 m, the zoobenthos diversity and biomass are low. Zoobenthos mainly consists here of the two endemic oligochaetes *Enchytraeus przewalskii* Hrabe and *E. issykulensis* Hrabe, and the amphipod *Issykkogammarus hamatus* Chevreux (Romanovsky 2002b). The average annual biomass of zoobenthos in this zone is $\leq$0.2–0.3 g/m^2.

Radix auricularia, which is widely spread in the coastal zone down to ca. 60 m water depth, is the main food source of fish in Lake Issyk-Kul. In contrast, the main food source in the Tyup Gulf are chironomid larvae that account for ca. 60% of the zoobenthos' biomass there. The high caloric value of these invertebrate larvae enables high growth rates of gold fish and carp in the Tyup Gulf.

In addition to her research work on the algae of Lake Issyk-Kul, T. G. Matukova studied also its macrophyte flora in detail. Plant associations including emerged and submerged macrophytes show especially large differences in the gulf regions such as the Tyup Gulf due to different hydrochemical conditions and substrate types. The emergent flora of Lake Issyk-Kul consists of common reed grass (*Phragmites communis* Trin.), blackamoor (*Typha latifolia* L.), reed (*Scirpus tabernaemontani* Gmel.), and clubroot (*Bolboschoenus affinis* Drob. and *B. compactus* Drob.).

Common submerged macrophytes are fennel-leaved pondweed (*Potamogeton pectinatus* L.), amblyophyllous (*P. otusifolius* M.K.), amplexicauline (*P. perfoliatus* L.), sea naiad (*Najas marina* L.), meakin (*Myriophyllum spicatum*), hornweed (*Ceratophyllum demestrum*), common bladderwort (*Urticularja vulgaris* L.), buttercup (*Ranunculus natans* C.A.M.), poolmate (*Zannichellia palustris* L.), and sea-grass (*Ruppia maritima* L.).

Stoneworts (charophytes) grow in the littoral belt of the lake between 6–8 and 35–40 m or sometimes even 50 m depth. They occupy large areas and produce the largest phytomass, and form the main type of aquatic vegetation in Lake Issyk-Kul. Charophytes are included in all associations of water vegetation and most often dominate the plant and macro-algae associations (Savvaitova and Petr 1992). Thus, Issyk-Kul is a charophyte lake. The "stonewort belt" along the shore of the lake is wider in the north in comparison to the south as a result of the steeper slope in the south. The belt is as wide as 15 km in the flat littoral zone of the Balykchy Gulf, near Prishyb Mountain and Tsor village, or as narrow as 200 m in regions with steep slopes. Bays protected from waves and far from inflows as sources of suspended sediments are most densely covered by charophytes. In contrast, the sandbanks of the peninsulas Dolinsk, Bosteri, Kara-Bulun and the Sukhoi Mountain range are not covered by charophytes above a water depth of 5–10 m due to their exposure to large storm waves. Due to the lower water transparency in the Tyup Gulf and the steep estuary-like bathymetry, charophytes grow here at the depth of 6–8 m and form a 20–30 m narrow belt. Charophytes are not found near inflows with water transparencies of $\leq$0.5 m.

Charophytes in Lake Issyk-Kul are represented by 17 species: *Nitellopsis obtusa* (Desv.) Groves, *Tolypella nidifica* (Müll.) Leonh., *Chara altaica* A. Br., *Ch. aspera* (Dethard.) Willd., *Ch. aspera* (Dethard.) Willd. var. *brevispina* Mig., *Ch. tomentosa* L., *Ch. connivens* Salzm., *Ch. contraria* A. Br., *Ch. contraria* A. Br. *hispidula* A. Br., *Ch. crinita* Wallr., *Ch. delicatula* Ag. var. *bulbillifera* A. Br., *Ch. fischeri* Mig., *Ch. intermedia* A. Br., *Ch. schaffneri* (A. Br.) Robinson, *Ch. crinitoides* Hollerb., *Ch. vulgaris* L., and *Ch. fragifera* Durun (Romanovsky 2002b). The tallest (up to 2 m height) macro-alga in Lake Issyk-Kul which is also most widely distributed is *Chara tomentosa*. It grows both in less brackish shallow zones of gulfs and in the open lake. In general, *Ch. tomentosa* forms dense vegetation at a depth range from 6–8 to 25–30 m with high cover (90–100%). Its phytomass reaches 35 kg/m^2. At larger depth down to 50 m, *Ch. tomentosa* grows in isolated spots.

The fish fauna of Lake Issyk-Kul includes 12 or 13 native species and more than ten introduced taxa. The most diverse native fish family in the lake are the carps (Cyprinidae; Pivnev 1990). The native members are Schmidt's dace (*Leuciscus schmidti* (Herzenstein)), the Issyk-Kul dace (*L. bergi* Kaschkarov), the Issyk-Kul minnow (*Phoxinus issykkulensis* Berg), the Issyk-Kul gudgeon (*Gobio gobio latus* Anikin), the Issyk-Kul marinka (*Schizothorax issykkulensis* Berg), the scaly osman (*Diptychus maculatus* Steindachner), and the naked osman (*Gymnodiptychus dybowskii* (Kessler)). The origin of the common carp (*Ciprinus carpio* L.) in Lake Issyk-Kul is not known (Mikkola 2012). Other native taxa are four stone loaches (Nemacheilidae: the Tibetan stone loach *Triplophysa stolickai* (Steindachner), the gray loach *T. dorsalis* (Kessler), the spotted thick-lip loach *T. strauchii* (Kessler) and the Issyk-Kul naked loach *T. ulacholica* (Anikin), and one catfish (Sisoridae), the Turkestan catfish *Glyptosternon reticulatum* McClelland (Mikkola 2012).

The Sevan trout *Salmo ischchan* Kessler was introduced as early as in the 1930s to establish commercial trout fishery in Lake Issyk-Kul (Alamanov and Mikkola 2011). The zander *Sander lucioperca* (L.) and the oriental bream *Abramis brama orientalis* Berg were introduced in the lake in 1954–1956. Grass carp (*Ctenopharyngodon idella* (Valenciennes)) and tench (*Tinca tinca* (L.)) followed soon afterwards. The common whitefish (*Coregonus lavaretus* (L.)) and the Arctic cisco (*C. autumnalis* (Pallas)) were introduced in Lake Issyk-Kul in the 1970s (Alamanov and Mikkola 2011). Additional introduced carps are the striped bystranka (*Alburnoides taeniatus* (Kessler)), the stone moroko (*Pseudorasbora parva* Temm. & Schleg.), the silver carp (*Hypophthalmichthys molitrix* (Valenciennes)), and the golden carp (*Carassius auratus gibelio* (Bloch)). The rainbow trout (*Oncorhynchus mykiss* (Walbaum)), the Valaam whitefish (*Coregonus widegreni* Malmgren), the plain thicklip loach (*Triplophysa labiata* (Kessler)), and the freshwater sleeper (*Micropercops cinctus* (Dabry de Thiersant)) were also successfully introduced in the lake. Not fully confirmed yet is the successful establishment of the introduced Caucasian scraper (*Capoeta capoeta* (Güldenstädt)), the asp (*Leuciscus aspius* (L.)), the vendace (*Coregonus albula* (L.)), and the peled (*C. peled* (Gmelin)) in the lake (Mikkola 2012). As a result of the successful establishment of alien predatory fish in Lake Issyk-Kul such as the rainbow trout *Oncorhynchus mykiss*, the endemic fish fauna is largely endangered today (Alamanov and Mikkola 2011). The alien fish taxa in Lake Issyk-Kul represent 73% of the total fish fauna of the lake today.

6.3 Population, Natural Resources and Environmental Issues of the Issyk-Kul Basin

6.3.1 Population in Lake Issyk-Kul's Catchment

Archaeological evidence suggests that man became active in the Issyk-Kul region sometime in the Middle Paleolithic (ca. 300–40 ka; Vishnyatsky 1999). Petroglyphs from the Bronze Age and later periods are especially abundant along the northern lake shore and show that relatively dense human occupation of the region dates back to more than 3000 years (Hermann 2017). Medieval (tenth to thirteenth century CE) remains of settlements in the Issyk-Kul region including caravanserai and fortification buildings indicate the importance of trading along one of the ancient Silk Road routes (Vinnik et al. 1978; Mokrynin and Ploskikh 1988). However, today's settlements in the basin were established in the second half of the nineteenth century and afterwards. The foundation of modern Karakol is dated to the summer of 1869. The first permanent settlement of people in the area of modern Balykchy was established in 1871 due to the creation of a postal station. In 1913, the population of the Issyk-Kul Basin was not exceeding 250 thousand people. Kyrgyz people predominated, and minorities accounted for ca. 25% of the population.

In the year 2014, the Issyk-Kul Basin was inhabited by 458.5 thousand people (Abuduwaili et al. 2019). The majority of the population is distributed on the lakeside plain where the average population density is 130 people per km^2 (average population density in Kyrgyzstan is ca. 30 people per km^2). Most important urban centres in the basin are the three cities Karakol (population of 74.1 thousand), Balykchy (46.0 thousand) and Cholpon-Ata (12.4 thousand) where 33% of the basin's population lives.

6.3.2 Natural Resources of the Issyk-Kul Basin

The Issyk-Kul Basin has rich natural resources with respect to recreational conditions, fisheries, fertile arable land, and mineral resources (gold, uranium). Recreational resources clearly have priority in the Issyk-Kul region. The combination of a diverse and fascinating landscape including the sandy beaches of the lake and steep mountains in its catchment, and favourable weather conditions near the lake (long sunshine duration, mild air temperatures without large fluctuations, low variations of atmospheric pressure) provides a unique setting for recreational purposes. In addition, clean mountain air and lake water, large mineral water reserves and therapeutic muds make Lake Issyk-Kul a unique sanatorium in the Central Asian region and Siberia. For example, natural thermal springs in the eastern part of the Terskei Ala-Too range have great potentials for the growing tourism sector in the Issyk-Kul Basin.

In 1990, 135 recreation facilities for rest, tourism and sanatorium treatment with a total capacity of ca. 40 thousand people existed in the Issyk-Kul region. Among them were nine sanatoria, 12 rest houses and 30 pioneer camps (summer camps for adolescents in Soviet times). The annual number of holidaymakers exceeded 350 thousand people which is comparable with the total number of permanent residents in the region. Today, ca. 400 thousand travellers visit the lake per year, mostly using tourism resorts in and near Cholpon-Ata on the northern shore of the lake (Fig. 6.5).

The Issyk-Kul State Reserve was established as the first nature reserve in Kyrgyzstan in 1948. In 1975, the lake became Kyrgyzstan's first internationally recognized and protected habitat for wintering of waterfowl with its declaration as a Ramsar site.

Fig. 6.5 The seaside resort Bosteri east of Cholpon-Ata, the peaks of the Kungei Ala-Too in the back

The Biosphere Reserve Issyk-Kul with an area of 43 km^2 was established as administrative entity of the Issyk-Kul Oblast (province) in 1998. It joined the UNESCO World Network of Biosphere Reserves in 2001, and is currently considered as Protected Nature Area at national level. The reserve is divided into core, buffer, transition and rehabilitation zones.

6.3.3 Current Environmental Issues of the Issyk-Kul Basin

Lake Issyk-Kul received increased attention in recent years due to two reasons: the declining lake level and pollution issues. The lake level decreases steadily as a result of global warming and enhanced evaporation from the lake, and the agricultural development in the region and related increasing diversion of water from the tributaries of the lake (Salamat et al. 2015). In 1966, 9% of the total river discharge in the Issyk-Kul Basin was diverted for agricultural purposes. This number increased to 34% in 1982 and 43% in 1998. Ca. 65 km^2 of farmland were irrigated in the catchment of Lake Issyk-Kul in the beginning of the twentieth century, whilst 500 km^2 were irrigated in the 1930s and ca. 1600 km^2 in the beginning of the twenty-first century (Lyons et al. 2001; Mikkola 2012). Water losses from irrigation farming are large due to the prevailing furrow irrigation and the lack of more sophisticated techniques (Romanovsky 1990).

The agricultural development was accompanied by pollution of water resources from irrigated fields (pesticides and nutrients in return flow) and as a result of intensified cattle breading, and by pollution due to the increasing construction and subsequent maintenance of recreational facilities. Lake Issyk-Kul's northern shore is a particularly problematic region now because it became popular as unregulated camping site with temporary canteens and food stands lacking appropriate sanitary facilities. Illegal poaching and cutting of bushes and trees became serious threats for the local fauna and flora, and are causes of soil erosion in other regions of the catchment area too.

Organic compounds such as hydrocarbons, detergents and phenols, and heavy metals accumulate in the closed-basin Lake Issyk-Kul as a result of improperly treated wastewaters, diffuse and subtle influxes from farmlands and accidental spillages.

Issyk-Kul is an oligotrophic lake with low productivity. Nutrients which define the biological productivity of the lake arrive mostly with river waters. Analyses of lake water samples in the years 1998–2000 revealed nitrite-nitrogen concentrations between 0.0 (Cholpon-Ata Gulf) and 0.082 mg/l (Balykchy Gulf), nitrate-nitrogen concentrations between 0.09 (Tyup Gulf) and 0.26 mg/l (Balykchy Gulf), and phosphorus concentrations between 0.0 (Balykchy Gulf) and 0.046 mg/l (Ton Gulf). Concentrations of nutrients in the open part of the lake are low. Nitrite-nitrogen concentrations were 0.0 mg/l, nitrate-nitrogen concentrations between 0.09 and 0.16 mg/l, and phosphorus concentrations between 0.0 and 0.009 mg/l.

Ammonia-nitrogen concentrations in water samples collected from the coastline of Lake Issyk-Kul were between 0.0 and 0.12 mg/l (Balykchy Gulf) in 1998–2000. Between 2015 and 2018, increased ammonia-nitrogen concentrations between 0.35 and 0.84 mg/l were recorded near the ship repair facility in the Balykchy Gulf. Highest ammonia-nitrogen concentrations were up to 2.2 times its maximum allowable concentration (MAC). Nitrite-nitrogen concentrations were 1.8–7.5 times its MAC of 0.024 mg/l near the ship repair facility and 5.0 times the MAC near the petroleum storage depot in 2015. Pollution near the ship repair facility in the Balykchy Gulf probably originated from the old fertilizer storage place "Selkhozhimiya" nearby where fertilizer reserves for the Issyk-Kul und Naryn Oblasts are stored. Heavy rains caused the contamination of runoff which entered the lake. However, ammonia-nitrogen concentrations remained below detection limit in the deep open part of the lake during 2015–2018, and MACs for nitrite nitrogen, nitrate nitrogen and phosphorus were not exceeded. Thus, the local contamination did not affect the main part of the lake due to its large volume in comparison to relatively minor inflow volumes.

However, the main pollutants of water in Lake Issyk-Kul are hydrocarbons and heavy metals. Contamination by oil products mostly results from refuelling of heating systems near the lake, and refuelling of vehicles and other equipment near and on the lake. Boating on Lake Issyk-Kul does not require permissions and is not efficiently controlled. According to data for 1975–1982, accidental spillage of hydrocarbons caused pollution in the shoreline zone of the lake where average concentrations of oil products ranged from 0.0 to 0.69 mg/l. Seasonal fluctuations varied from 0.0 to 1.87 mg/l (Karmanchuk and Kadyrov 1986). The concentrations of oil products in the deep-water region of the lake were lower with averages in a range from 0.05 to

0.20 mg/l, and seasonal maxima not exceeding 0.50 mg/l. Highest concentrations of oil products were recorded in the Balykchy Gulf (1.4–37 times higher than the MAC of 0.05 mg/l), in the Tyup Gulf (1.2–9.6 times MAC), the eastern Cholpon-Ata Bay (1.4–15 times MAC) and in the Kadji Sai River mouth (1.4–6.4 times MAC).

During the period from 1985 to 1992, the contamination of lake waters by oil products had decreased due to the wide introduction of wastewater treatment facilities in resorts and other recreational housing complexes, the establishment of sanitation zones near the lake, the ban of hydrocarbon shipment by water-borne transportation means, and the movement of oil storage facilities away from the shoreline.

During 1998–2000, concentrations of oil products had decreased to 0.001–0.005 mg/l in the Balykchy Gulf, to 0.0–0.003 mg/l at the northern shore of Cholpon-Ata, to 0.0–0.18 mg/l at the southern shoreline, and to 0.0–0.02 mg/l in the Tyup and Dzhergalan gulfs. The decreasing contamination by oil products was mainly substantiated by economic difficulties of the Kyrgyz Republic following the independence from the Soviet Union in 1991. The termination of operational activities of industrial enterprises and the reduction in tourist numbers caused a decrease of discharged wastewaters and fostered the lake rehabilitation.

Monitoring during 2015–2018 showed that oil-product concentrations were below the MAC across the whole lake area apart from the Balykchy Gulf. There, concentrations of oil products ranged from 0.07 to 0.14 mg/l (or 1.4–2.8 times MAC). In the Cholpon-Ata Bay near the Choktal and Grigorievka villages, the oil-product concentrations increased, even during the period prior to the tourism season. Concentrations of oil products exceeded the MAC after the tourism season when values of 1.3–2.4 times the MAC were recorded in coastal waters. The increase of oil-product concentrations during and after the tourism season mostly results from the more intensive operation of water-born transportation means using leaded petrol and diesel.

Phenols in near-shore waters have concentrations between 0.0 and 0.014 mg/l. Concentrations of phenols were below detection limit in the open part of the lake.

According to monitoring data during the years 2015–2018, the biological oxygen demand (BOD) as proxy of the degree of organic pollution of water is within norms (MAC 3.0 milligrams of oxygen consumed per litre of sample during five days of incubation at 20 °C (mg/l)) in Lake Issyk-Kul. The BOD is in a range from 0.19 to 2.88 mg/l. Values are 0.30–1.29 mg/l along the southern shore, 0.12–1.71 mg/l near the northern shore, 0.55–1.86 mg/l in the Balykchy Gulf, and 2.10–2.54 mg/l in the Tyup Gulf and near Karakol city.

Concentrations of heavy metals such as copper, zinc, lead, mercury and cadmium were monitored in surface waters (upper 0.5 m) in the year 2000. Analysis of surface waters in the shoreline zone of Kadji Sai village and in the eastern bay of Cholpon-Ata revealed maximum copper concentrations of 0.009 and 0.008 mg/l, respectively, clearly exceeding the MAC of 0.001 mg/l. Higher concentrations of copper were observed in the deeper part of the lake where concentrations of 0.004 mg/l were recorded on average and 0.026 mg/l at maximum. Monitoring results from the period 2015–2018 have shown that concentrations of heavy metals such as zinc, cobalt, cadmium, lead, chromium and arsenic in the water of Lake Issyk-Kul are below MACs.

One of the country's largest dump sites for radioactive waste originating from uranium mining during the Soviet era is situated at the Kadji Sai village on the southern shore of Lake Issyk-Kul. Remediation of the uranium mining tailing dump continues until today to reduce the radiation load on the local population. According to Mikkola (2012), monitoring of the uranium contamination of water is not done in a consistent way and the potential contamination of fish stocks was not assessed yet.

A major accident related to the gold mining at the Kumtor gold mine ca. 80 km south of Issyk-Kul occurred on May 20, 1998. A truck overturned on the road to the mine and caused a major spill of some 1700–1800 kg of highly poisonous sodium cyanide, used for gold extraction, in the Barskoon River. Approximately 4000 residents of the Barskoon village some 8 km downstream of the spillage and near the southern shore of Lake Issyk-Kul had to be evacuated of which 2000 sought medical care afterwards. Following clean-ups with hypochlorite, analyses by a laboratory of the national sanitary-epidemiological station under the Agency for Geology and Natural Resources revealed that cyanide concentrations in the river, in irrigation channels and Lake Issyk-Kul were below the MAC on May 27, 1998, and later on. However, the accident highlights the vulnerability of Lake Issyk-Kul and its surroundings to sources of potential severe pollution.

Long-term monitoring of water pollution in the periods from 1975–1982 to 1985–1992 showed that anthropogenic activities in the Issyk-Kul Basin affected the chemical composition of lake water in the coastline area. Negative effects of insufficient wastewater treatment and accidental spillages caused increased concentrations of oil products, detergents and phenols, and heavy metals. Most affected regions of the lake are the Balykchy Gulf, the eastern bay of Cholpon-Ata, and the waters near the Kadji Sai village and in the Ton Gulf on the southern shore. Higher concentrations of pollutants occur also in the eastern part of the lake in the Tyup and Dzhergalan gulfs.

So far, Issyk-Kul waters are mostly affected by anthropogenic activities in the coastal region of the lake with depth not exceeding 50 m. Lake water in its central and deep parts is not significantly impacted yet, as a result of the slow dynamics in a large and deep lake. However, better wastewater treatment, better controls of potential contamination with respect to households and industrial and agricultural sources, improved monitoring and the better implementation of existing environmental regulations are required to inhibit the slow and continuous degradation of Lake Issyk-Kul's environmental status.

References

Abuduwaili J, Issanova G, Saparov G (2019) Hydrographical and physical-geographical characteristics of the Issyk-Kul Lake Basin and use of water resources of the basin, and impact of climate change on it. In: Abuduwaili J, Issanova G, Saparov G (eds) Hydrology and limnology of central Asia, Springer, pp 297–357

Alamanov A, Mikkola H (2011) Is biodiversity friendly fisheries management possible on Issyk-Kul Lake in the Kyrgyz Republic? Ambio 40:479–495

Azykova EK (1991) Landscapes. Issyk-Kul, Naryn. Encyclopedia. GRKSE, Frunze, pp 56–62 (in Russian)
Bakov EK (1983) Trends of movement and dynamics of glaciers of the Central Tien Shan. Ilim, Frunze 157pp (in Russian)
Bondarev LG (1978) Fluctuations of level based on archaeological and historical data. In: Koralev VG (ed) Lake Issyk-Kul. Frunze, Ilim, pp 112–117 (in Russian)
Bulin VM, Romanovsky VV (1975) Thermal bar of Lake Issyk-Kul. The problems of Kyrgyzstan's geography. Llim, Frunze, pp 122–123 (in Russian)
DDWSSD (Department of Drinking Water Supply and Sewerage Development under the State Agency for Architecture, Construction and Public Utilities under the Government of Kyrgyz Republic), 2018. Issyk-Kul Wastewater Management Project (RRP KGZ 50176). Report for the the Asian Development Bank, 194pp
De Batist M, Imbo Y, Vermeesch P, Klerkx J, Giralt S, Delvaux D, Lignier V, Beck C, Kalugin I, Abdrakhmatov KE (2002) Bathymetry and sedimentary environments of Lake Issyk-Kul, Kyrgyz Republic (Central Asia): a large, high-altitude, tectonic lake. In: Klerkx J, Imanackunov B (eds) Lake Issyk-Kul: its natural environment. Springer NATO Science Series IV, Earth and Environmental Sciences, vol 13, pp 101–123
Folian LA (1981) Zooplankton of Lake Issyk-Kul. Biological basis of fishery of Central Asia and Kazakhstan water bodies. Thesis of XVII scientific conference. Llim, Frunze, pp 389–390
Gebhardt AC, Naudts L, De Mol L, Klerkx J, Abdrakhmatov K, Sobel ER, De Batist M (2017) High-amplitude lake-level changes in tectonically active Lake Issyk-Kul (Kyrgyzstan) revealed by high-resolution seismic reflection data. Clim Past 13:73–92
Giralt S, Julià R, Klerkx J, Riera S, Leroy S, Buchaca T, Catalan J, De Batist M, Beck C, Bobrov V, Gavshin V, Kalugin I, Sukhorukov F, Brennwald M, Kipfer R, Peeters F, Lombardi S, Matychenkov V, Romanovsky V, Podsetchine V, Voltattorni N (2004) 1000-year environmental history of lake Issyk-Kul. In: Nihoul JCJ, Zavialov PO, Micklin PP (eds) Dying and dead seas climatic versus anthropic causes. Springer NATO Science Series IV, Earth and Environmental Sciences, vol 36, pp 253–285
Gronskaya TP (1983) Water balance and anticipated water levels in Lake Issyk Kul. Abstract, Dissertation thesis, Leningrad, 16p (in Russian)
Hermann L (2017) Rock art sites on the northern bank of Lake Issyk-Kul, in Kirghizstan. INORA 78:1–10
Hofer M, Peeters F, Aeschbach-Hertig W, Brennwald M, Holocher J, Livingstone DM, Romanovski V, Kipfer R (2002) Rapid deep-water renewal in Lake Issyk-Kul (Kyrgyzstan) indicated by transient tracers. Limnol Oceanogr 47:1210–1216
Kadyrov VK (1986) Hydrochemistry of Lake Issyk-Kul and its basin. Ilim, Frunze 212p (in Russian)
Kadyrov VK, Karmanchuk AR (1964) About content of some elements in Lake Issyk-Kul water. Microelement in water of Lake Issyk-Kul. Frunze (2):101–106 (in Russian)
Karmanchuk AR, Kadyrov VK (1986) Influence of anthropogenic activities on water quality in the shoreline zone of Lake Issyk-Kul. Study of natural water's chemical composition formation process under conditions of anthropogenic impacts. Hydrometisdat, Leningrad, 71pp (in Russian)
Kipfer R, Peeters F (2002) Using transient conservative and environmental tracers to study water exchange in Lake Issyk-Kul. In: Klerkx J, Imanackunov B (eds) Lake Issyk-Kul: its natural environment. Springer NATO Science Series IV, Earth and Environmental Sciences vol 13, pp 89–100
Kiselev IA (1932) In the phytoplankton of Lake Issyk-Kul. Mémoires de l'Institut Hydrologique 7:65–94 (in Russian)
Krivoshei ML, Gronskaya TP (1986) Water balance of Lake Issyk Kul. In: Shnitnikov AV (ed) Studies of Large Lakes in the USSR. Nauka, Leningrad, pp 276–280 (in Russian)
Kulumbaeva AA (1982) Phytoplankton of Lake Issyk-Kul. Llim, Frunze, 106pp (in Russian)
Kutzeva PP (1980) Balance of chemical elements in Lake Issyk-Kul. Studies of water balance, thermal and hydrochemical regime of Lake Issyk-Kul. Hydrometisdat, Leningrad, pp 71–77 (in Russian)

Lyons WB, Welch KA, Bonzongo J-C, Graham EY, Shabunin G, Gaudette HE, Poreda RJ (2001) A preliminary assessment of the geochemical dynamics of Issyk-Kul Lake, Kirghizstan. Limnol Oceanogr 46:713–718

Matukova TG (1958) Stonewort of Lake Issyk-Kul. Transactions of Biology and Soil Department of the Kyrgyz State University. Issue 7, Frunze, pp 43–49 (in Russian)

Matveev VP (1935) Hydrochemical studies of Lake Issyk-Kul in 1932. Lake Issyk-Kul. Materials on hydrology, ichthyology and fishery. Published by AS USSR, T. 111(2):7–56 (in Russian)

Mikkola H (2012) Implication of alien species introduction to loss of fish biodiversity and livelihoods on Issyk-Kul Lake in Kyrgyzstan. In: Lameed GSA (ed) Biodiversity enrichment in a diverse world, IntechOpen. https://doi.org/10.5772/48460

Mokrynin VP, Ploskikh VM (1988) Issyk-Kul: Sunken towns. Ilim, Frunze, 190pp (in Russian)

Narama C, Kubota J, Shatravin VI, Duishonakunov M, Moholdt G, Abdrakhmatov K (2010) The lake-level changes in central Asia during the last 1000 years based on historical map. In: Watanabe M, Kubota J (eds) Reconceptualizing cultural and environmental change in central Asia: an historical perspective on the future. Ili Project, Research Institute for Humanity and Nature, Kyoto, pp 11–27

Pavlova MV (1964) Zoobenthos of Lake Issyk-Kul gulfs and its consumption by fishes. Ilim, Frunze, 84pp (in Russian)

Pavlova MV (1983) Biology and production of *Limnaea* (*Radix auricularia* (L.)) var. *obliquata* Martens in Lake Issyk-Kul. Limnological investigations in Kyrgyzstan. Ilim, Frunze, pp 34–47 (in Russian)

Peeters F, Finger D, Hofer M, Brennwald M, Livingstone DM, Kipfer R (2003) Deep-water renewal in Lake Issyk-Kul driven by differential cooling. Limnol Ocean 48:1419–1431

Pivnev IA (1990) Fishes of Kyrgyzstan. Ilim, Frunze, 138pp (in Russian)

Podrezov OA (1985) The strong wind regime in Issyk-Kul, Kochkor and Karakufjur basins. Trans SANIGMI 23:48–53

Podrezov AO, Pavlova IA (2003) The dangerous meteorological phenomena on the territory of Kyrgyzstan. Book 1. The strong wind regime. Published by KRSU, Bishkek, 114pp (in Russian)

Podrezov OA, Dikikh AN, Bakirov KB (2001) Instability of climatic conditions and glaciation of Tien Shan for the last 100 years. Newsletter of KRSU, Chapter 1. Bishkek, pp 33–40

Ricketts RD, Johnson TJ, Brown ET, Rasmussen KA, Romanovsky VV (2001) The Holocene paleolimnology of Lake Issyk-Kul, Kyrgyzstan: Trace element and stable isotope composition of ostracodes. Palaeogeogr Palaeoclimatol Palaeoecol 176:207–227

Romanovsky VV (1990) The Natural Complex of Lake Issyk-Kul. Ilim, Frunze 168pp (in Russian)

Romanovsky VV (2002a) Water level variations and water balance of Lake Issyk-Kul. In: Klerkx J, Imanackunov B (eds) Lake Issyk-Kul: its natural environment. NATO Science Series (Series IV: Earth And Environmental Sciences), vol 13, Springer, Dordrecht, pp 45–59

Romanovsky VV (2002b) Hydrobiology of Lake Issyk-Kul. In: Klerkx J, Imanackunov B (eds) Lake Issyk-Kul: its natural environment. NATO Science Series (Series IV: Earth And Environmental Sciences), vol 13, Springer, Dordrecht, pp 27–45

Romanovsky VV, Shabunin G (2002) Currents and vertical water exchange in Lake Issyk-Kul. In: Klerkx J, Imanackunov B (eds) Lake Issyk-Kul: its natural environment. Springer NATO Science Series IV, Earth and Environmental Sciences vol 13, pp 77–89

Romanovsky VV, Kuzmichonok VA, Mamatkanov DM, Podrezov SA (2004) All about Lake Issyk-Kul: Questions and answers. Kyrgyz-Russian Slavic University Bishkek, 407pp (in Russian)

Rosenwinkel S, Landgraf A, Schwanghart W, Volkmer F, Dzhumabaeva A, Merchel S, Rugel G, Preusser F, Korup O (2017) Late Pleistocene outburst floods from Issyk Kul, Kyrgyzstan? Earth Surf Proc Land 42:1535–1548

Salamat A, Abuduwaili J, Shaidyldaeva N (2015) Impact of climate change on water level fluctuation of Issyk-Kul Lake. Arab J Geosci 8:5361–5371

Savvaitova K, Petr T (1992) Lake Issyk-Kul, Kirgizia. Int J Salt Lake Res 1:21–46

Sevastianov DV (1991) Historical and archaeological data about alteration of shores and lake level fluctuations. In: Treshnikov AF (ed) The History of Lakes Sevan, Issyk-Kul, Balkhash, Zaisan, and Aral. Nauka, Leningrad, pp 90–96 (in Russian)

Shukurov ED (1991) Fauna of Issyk-Kul basin in connection with its environmental status. Fauna and ecology of terrestrial vertebrates of Kyrgyzstan. Ilim, Bishkek, pp 37–48 (in Russian)

Stavissky YS (1977) Water temperature dynamics of Lake Issyk-Kul. Papers of the Central-Asian Scientific Research Institute of Hydrometeorology, vol 50, pp 75–80. The dangerous meteorological phenomena on the territory of Kyrgyzstan

Trofimov AK (1990) Quaternary deposits of the Issyk-Kul depression related to its tectonics. Proc Acad Sci, Kirghiz SSR 1:87–95 (in Russian)

Vereschagin AP (1991) Wintering birds of the eastern part of Issyk-Kul basin. Fauna and ecology of terrestrial vertebrate species of Kyrgyzstan. Ilim, Bishkek pp 69–85 (in Russian)

Vinnik DF, Lesnichenko NS, Sanarova AV (1978) Works at Issyk-Kul: archaeological findings in 1977. Nauka, Moscow (in Russian)

Vishnyatsky LB (1999) The Paleolithic of Central Asia. J World Prehistory 13:69–122

Vollmer MK, Weiss RF, Williams RT, Falkner KK, Qiu X, Ralph EA, Romanovsky VV (2002) Physical and chemical properties of the waters of saline lakes and their importance for deep-water renewal: Lake Issyk-Kul, Kyrgyzstan. Geochim Cosmochim Acta 66:4235–4246

Chapter 7
Lop Nur in NW China: Its Natural State, and a Long History of Human Impact

Steffen Mischke, Chenglin Liu and Jiafu Zhang

Abstract Lop Nur is a large salt-crust covered playa in the eastern part of the Tarim Basin in northwestern China. Its centre was filled by a lake with a surface area of more than 2000 km^2 in the early 1930s for the last time. Geological evidence and historical documents indicate that a large hyperhaline lake existed in the Lop Nur Basin until ca. 2000 years ago. Most regions of the former lake basin desiccated when the Chinese Empire was extended towards the arid and semi-arid regions in the west during the Han Dynasty as a result of water withdrawal from the tributaries of Lop Nur for irrigation farming. Surface water inflow in Lop Nur is probably mostly controlled by the intensity of farming activities in the catchment of the lake since that time. Current trends in population increase, claiming of arable land and industrialization upstream of Lop Nur imply that the ancient lake region will remain dry in the future.

Keywords Xinjiang · Tarim Basin · Lake history · Irrigation farming · Desiccation

7.1 Introduction

Lop Nur is a large groundwater-discharge playa and the base level of China's largest endorheic river system. The Tarim River is the largest river of the Tarim Basin which covers a vast area of 530,000 km^2 (Ma et al. 2010; Fig. 7.1). The river flows

S. Mischke (✉)
Institute of Earth Sciences, University of Iceland, Sturlugata 7, 101 Reykjavík, Iceland
e-mail: smi@hi.is

C. Liu
Institute of Mineral Resources, Chinese Academy of Geological Sciences, 26 Baiwanzhuang Road, Beijing 100037, China
e-mail: liuchengl@263.net

J. Zhang
MOE Laboratory for Earth Surface Processes, Department of Geography, College of Urban and Environmental Sciences, Peking University, Shaw Building No. 2, Haidian District, Beijing 100871, China
e-mail: jfzhang@pku.edu.cn

S. Mischke (ed.), *Large Asian Lakes in a Changing World*, Springer Water,
https://doi.org/10.1007/978-3-030-42254-7_7

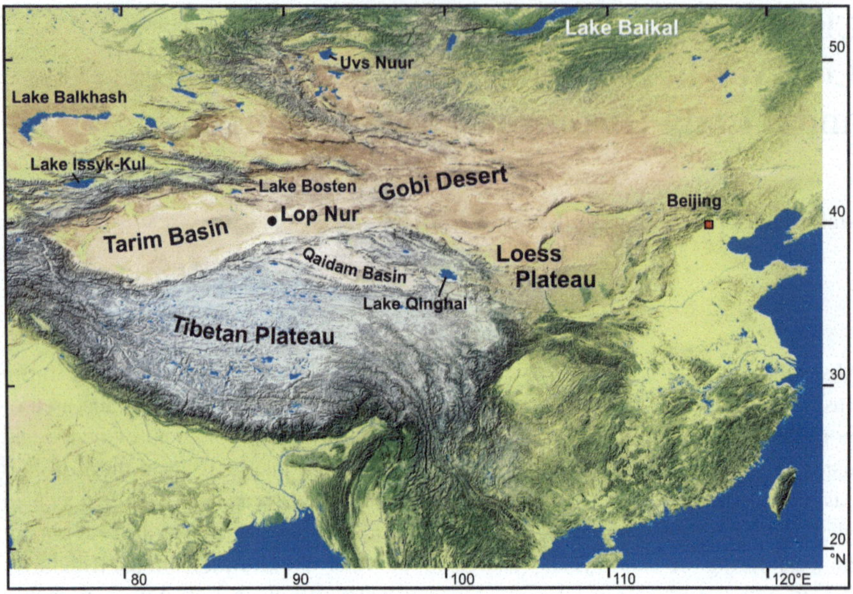

Fig. 7.1 Physiography of Central and Eastern Asia and location of Lop Nur (black dot) in the eastern Tarim Basin

eastwards along the northern margin of the Tarim Basin supported by numerous tributaries which drain the Tianshan (*shan* = mountains) in the north (Figs. 7.1 and 7.2). Mountain ranges in the west (the Pamirs Plateau), south (the Kunlunshan) and southeast (the Altun Mountains) are partly higher than 4000 m above sea level (asl). Together with lower ranges in the east (Beishan) and northeast (Kuruk Tagh, part of the eastern Tianshan), they shield the basin from moist air masses of oceanic origin. A lake existed in the eastern part of the basin until the late 1930s or early 1940s when diminishing inflows resulted in the formation of a mostly salt-covered playa in an area of 20,000 km^2 (Li et al. 2008; Fig. 7.3). The lowest region in the centre of the remarkable earlike structure when seen from space lies at ca. 780 m asl. The central and western parts of the Tarim Basin are occupied by the second largest sand sea on Earth, the Taklamakan Desert (Fig. 7.2). Dunes and gravel plains cover wide regions also in the southern vicinity of the playa whilst wind-eroded remnants of alluvial and lacustrine sediments (called *yardangs*) occur in its northern vicinity (Shao et al. 2012).

The Lop Nur region is located in a hyperarid part of Central Asia far from the seas and shaded by high mountain ranges. Mean annual precipitation for Ruoqiang 230 km to the southwest of the Lop Nur playa was 22.9 mm in the observational period from 1953 to 1990 (WorldClimate, www.worldclimate.com). Potential evaporation is as high as 3500 mm per year. Mean annual, January and July temperatures in Ruoqiang are 11.6, −8.0 and 27.4 °C, respectively.

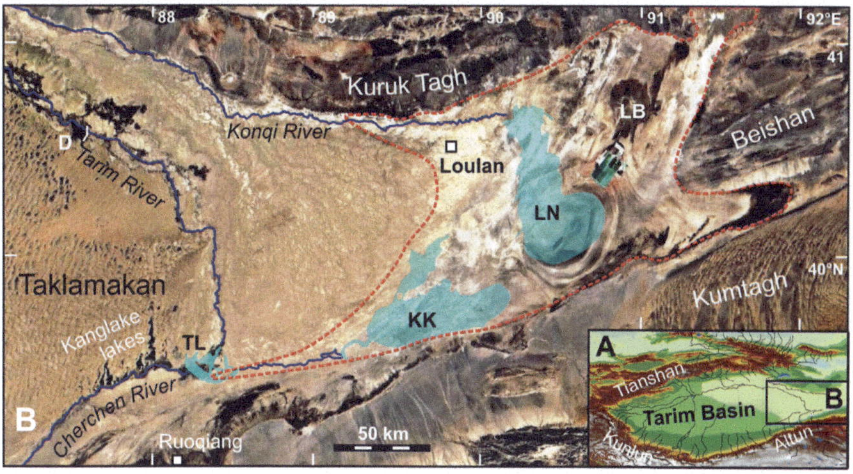

Fig. 7.2 Overview of the Tarim Basin (A) and its eastern part (B) with the Konqi, Tarim and Cherchen rivers, Lop Nur (LN, i.e. 'Great Ear' and 'Last Lake'), the Luobei sub-basin (LB) and the lake and swamp areas of Kara Koshun (KK), Taitema Lake (TL) and the Kanglake lakes. The Daxihaizi Reservoir (D) was established in 1972. The broken red line outlines the assumed area of the ancient Great Lop Nur (Ma et al. 2010). Lake and swamp areas of KK and TL are drawn according to a map of Hörner and Chen (1935), but KK was placed ca. 8 km to the east and 2 km to the north of its originally shown geographical location and TL was placed ca. 6 km to the east to fit to the shown topography here. The areal extent of LN is shown according to a survey of the lake in 1934 (Hedin 1942), but its location is placed ca. 9 km further to the east in comparison to the original map illustration. The rectangular blue area between LN and LB are potash-producing salt ponds established since 2002

7.2 The Origin of the Lop Nur Depression

The Tarim Basin is underlain by the rigid Tarim Block or the Tarim Craton which became segmented in the course of the Cenozoic collision of the Indian and Eurasian plates. The basin's western and northern parts were occupied by the epicontinental proto-Paratethys Sea until its retreat in the Eocene (Bosboom et al. 2014). Afterwards, sediment accumulation occurred in a lake setting in the western part of the Tarim Basin, whilst erosion dominated in the Lop Nur depression (i.e., the region of the ancient lake including the Luobei sub-basin; Liu et al. 2015). The northward movement of the Pamirs and its connection with the Tianshan caused uplift in the western and southern parts of the basin and altered its east-west inclination. As a consequence, the depositional centre of the Tarim Basin shifted eastwards. The Lop Nur depression became the depocentre of the Tarim Basin in the Pliocene (Liu et al. 2015). The present surface of the basin declines from elevations of ca. 1300 m asl in the west to ca. 800 m asl in the east. The modern architecture of the Lop Nur depression is mostly defined by active normal faults trending NNE-SSW and NE-SW (Woldai 1983).

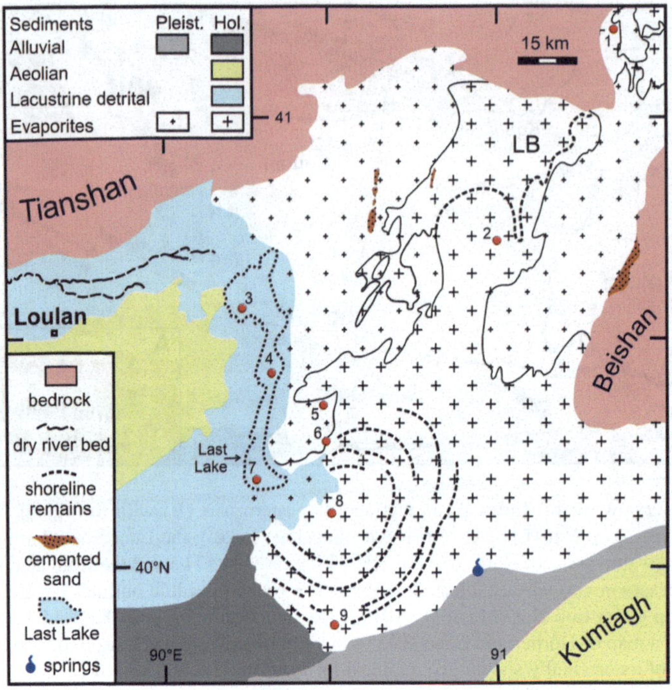

Fig. 7.3 Geomorphological sketch map of the Lop Nur region simplified following Wang et al. (2005), and locations of dug pits and drilled sediment cores (1 ZKD0014, 2 CK-2, 3 ZK95-6, 4 section Ma, 5 K1, 6 F4, 7 YKD0301, 8 L07-10, 9 Luo4). Cemented sands occur in discharge areas of springs. Location of springs in south according to Ma et al. (2010). The 'Last Lake' represents an area where a topographic map from 1942 indicated an open water surface (Li et al. 2008), however, its underlying evidence is unclear

7.3 The Ancient Pleistocene and Holocene Lake

Drilled sediment cores from the Lop Nur region provide evidence that a large lake existed from the early Pleistocene and prevailed until the earliest stage of the middle Pleistocene (Liu and Wang 1999). A 1050-m core from the western margin of the Taitema Lake area shows that predominantly fluviolacustrine silts, clays and argillaceous limestones were deposited at a relatively constant sediment accumulation rate since at least ca. 7 million years (Chang et al. 2012). Clay and silt dominate the sediments but the lack of preserved pollen since ca. 6 million years possibly indicates prevailing playa or fluvial depositional settings in the Taitema Lake area (Hao et al. 2012). Clayey and silty sediments in the 100-m long core K1 in the northern part of the 'Great Ear' (also called East Lake) suggest that a lake existed in the Lop Nur area since at least 1.2 million years (Wang et al. 2000). Drilled cores from the Luobei region indicate that shallow lacustrine, fluvial and aeolian sediments accumulated further to the northeast since the early Pleistocene (Lü et al. 2015). Research of yardangs mainly composed of fine-grained fluvial and lacustrine sediments in the Kumtagh Desert in the east of Lop Nur also suggests that a large lake existed in the region in the early and middle Pleistocene (Qu et al. 2004; Lin et al. 2005).

Geomorphological evidence for a large ancient lake in the Lop Nur region in the late Pleistocene is present as wave-cut platforms and gravelly shoreline bars. Erosional platforms and shoreline bars are well preserved in the NE section of the ancient lake (Ma et al. 2010). Wang et al. (2008) inferred that a lake of >55,000 km^2 size occupied the region ca. 130,000–90,000 years ago. A size of ca. 20,000 km^2 was suggested for a period ca. 30,000 years ago and of 9250 km^2 for the middle Holocene ca. 7500–7000 years ago (Wang et al. 2008). Analyses of sediment core CK-2 (50.14 m length) from the Luobei sub-basin indicate that deposits were apparently continuously formed in a relatively stable lake in the sub-basin between 32 and 9 ka (Luo et al. 2009). Holocene sediments recovered from dug sections and drilled cores in the 'Great Ear' region are generally dominated by fine-grained deposits that contain gypsum and other soluble salts (Sect. 7.5; Fig. 7.4). There is no evidence for desiccation periods earlier than ca. 1800 years ago, supporting the reconstruction of a large perennial and mostly hyperhaline lake in the Lop Nur region in the early and middle Holocene (Liu et al. 2016; Mischke et al. 2019).

Fig. 7.4 Early Holocene clastic sediments from core ZKD0014 (lower part) in the northernmost part of the Luobei sub-basin overlain by middle and late Holocene salt deposits (upper part; Liu et al. 2008). Sediment core location shown in Fig. 7.3

7.4 Early Human Impact on the Hydrology of Lop Nur

Lop Nur developed into a salt-crust covered playa since its final desiccation in the late 1930s or early 1940s (Li et al. 2008). Direct evidence for stagnant surface water in the Lop Nur region for the last time before its desiccation originates from detailed reports of Sven Hedin who explored the lake in a canoe in the summer of 1934 (Hedin 1942). He reported that the lake was large but mostly very shallow with a depth of only 20 cm (Fig. 7.2). Freshwater filled its northern part near the inflow of the Konqi River whilst the main water body was saline (Hörner 1932; Hörner and Chen 1935; Hedin 1942). The lake floor was dry in the years 1900 and 1901 when Hedin explored the region. Corona and Landsat imagery from 1961 and starting in 1972, respectively, captured the astonishing 'Great Ear' structure in the lowermost part of the basin but also showed that the region was dry (Li et al. 2008). The existence of surface waters in the western part of the Lop Nur Basin and a 'Last Lake' stage was suggested in a topographic map published in 1942 in the Soviet Union which was later adopted by many authors (Li et al. 2008; Fig. 7.3). However, the underlying evidence for the inference of the 'Last Lake' stage of Lop Nur is unclear.

In contrast to short-lived and very shallow (<1 m) water bodies in the Lop Nur Basin in the twentieth century, *Han Shu*, the historical book of the Han Dynasty (Han Dynasty: 206 BCE–220 CE) described a large water body in the region which was called 'Puchang Sea' (Ban, 92; i.e., originally published in 92 CE). This 'sea' covered an estimated area of 18,000–31,000 km^2 and, as the name implies, was a large brackish or saline water body (Yang et al. 2006). The modern topography of the 'Great Ear' region suggests that the 'Puchang Sea' had a maximum depth of at least 5.2 m (Li et al. 2008). The ancient Silk Road oasis Loulan flourished during these days, and branches of the Tarim River ran to the north and south of the oasis and entered the lake farther in the east (Figs. 7.2 and 7.3). The assessment of historical documents and archaeological data led to the conclusion that the Tarim River had changed its previous flow from west to east between 330 and 400 CE towards a southern direction (Fig. 7.2; Hedin 1942; Zhao and Xia 1984; Wang 1996). As a consequence, the Tarim River branches near Loulan were left dry. Near-desiccation of the 'Puchang Sea' ca. 1800 years ago was reconstructed from the first occurrence of a sand layer in the YKD0301 section following the relatively uniform accumulation of silty sediments since the onset of sediment formation at the section base ca. 9400 years ago (Liu et al. 2016; Mischke et al. 2017). The Loulan oasis became abandoned between ca. 230 $\pm$ 66 CE (youngest dated organic material from the ruined city; Lü et al. 2009) and 645 CE when the monk Xuanzang travelled the region (Xuanzang, 645; Qin et al. 2012; Mischke et al. 2017). However, the river-course change between 330 and 400 CE likely resulted in the abandonment of Loulan. Increasing aridity due to regional climate change was discussed as the most likely reason of the fall of the Loulan Kingdom and of other desert oases in the region, and of the expansion of aeolian sands (Zhang et al. 2003, 2011; Yang et al. 2006; Lü et al. 2009; Dong et al. 2012; Qin et al. 2012). In contrast, a recent assessment of palaeoclimate studies from northwestern China, southern Mongolia and adjoining regions further to the west by Mischke et al. (2017, 2018) showed that regional climate was relatively wetter during the time of Loulan's fall. As a consequence, Mischke et al. (2017, 2018) argued that water withdrawal from tributaries of Lop Nur caused diminishing run-off farther downstream and eventually led to the collapse of the Loulan Kingdom and near-desiccation of Lop Nur. Water withdrawal from these tributaries was used for intensifying irrigation farming practices which probably reached maximum activity during the westward expansion of the Chinese Empire of the Han Dynasty. Remains of intensive irrigation farming such as channels, dams and reservoirs were reported from the region and river water was probably distributed on large cultivated fields during the Han Dynasty (Yang et al. 2006; Lü et al. 2009; Zhang et al. 2011; Qin et al. 2012). Pollen grains and phytoliths of cereal grasses and Vitaceae were recorded in gypsum-encrusted B horizons of soils in the region suggesting that common millet, foxtail millet and possibly naked barley, and grapes were intensively cultivated ca. 2000 years ago (Qin et al. 2012; Zhang et al. 2012a). Thus, the 'natural' state of Lop Nur cannot be assessed based on descriptions from the last century or even the last millennium if the history of significant human impacts on Lop Nur's hydrology started already during the Han Dynasty 2000 years ago. Geological proxy records from the Lop Nur region have to be assessed to reconstruct the more or less natural conditions of the lake during times without significant man-made alterations of its hydrology.

7.5 The Natural State of Lop Nur Before the Han Dynasty

Sediments of dug sections or drilled cores from the 'Great Ear' region pre-date the Han Dynasty a few decimetres or a metre beneath the surface and probably represent lake conditions without significant human impact on Lop Nur (Ma et al. 2008; Liu et al. 2016). Unfortunately, the chronological control of shallow sections and cores from the Lop Nur region is only poorly constrained by typically one to three radiocarbon dating results per location. The only exception is the YKD0301 pit in the western part of the 'Great Ear' where four radiocarbon samples and ten optically stimulated luminescence (OSL) samples yielded Holocene ages (Fig. 7.3; Zhang et al. 2012b). The YKD0301 section and also other sections in the 'Great Ear' and 'Last Lake' region of Lop Nur (sections or cores F4, K1, L07-10, Luo4, Ma, ZK95-6; Fig. 7.3) are composed of fine-grained sediments described as silt and clay or mud (Yan et al. 1983; Zheng et al. 1991; Wang et al. 2000; Liu et al. 2003, 2016; Ma et al. 2008; Hua et al. 2009). Detailed grain-size data available for the Holocene parts of the sections or cores YKD0301, ZK95-6, Ma and K1 show that the detrital sediments have a relatively uniform mean or median grain size in the fine to medium silt range (8–31 μm; Wang et al. 2000; Liu et al. 2003, 2016; Ma et al. 2008). Carbonate, gypsum or relatively high concentrations of soluble salts are reported from these sediments (Liu et al. 2003, 2016; Hua et al. 2009). Mischke et al. (2019) inferred hyperhaline conditions with salinities exceeding 100‰ during the middle and late Holocene based on the abundance of ooids and gypsum minerals. Thus, a permanent saline lake and significant distances between section or core locations and contemporaneous delta and shore regions can be inferred for the Holocene prior to the Han Dynasty. Detrital sediment accumulation was apparently dominated by fine-grained suspension load from inflowing rivers and fine dust particles. The presence of gypsum and more soluble salts indicates a high salinity of ca. five times of those of sea-water, or even higher (Fig. 7.4). Modern brines of gypsum-bearing sediments in the region indicate the beginning of gypsum formation at a salinity of ca. 188‰ (Sun et al. 2016). Accordingly, most sediments of these sections do not contain fossils that are typically found in the deposits of freshwater to slightly brackish lakes such as gastropod or bivalve shells, ostracod valves or fish remains (Mischke and Wünnemann 2006; Mischke and Zhang 2011; Mischke et al. 2019).

In contrast, valves of the highly salt-tolerant brackish water ostracods *Eucypris mareotica* (reported as *Eucypris inflata* in Wang et al. (1999) or as '*Eucypsis inflata*' in Zheng et al. (1991), i.e., *Eucypris inflata*, a younger synonym of *E. mareotica*) and *Cyprideis torosa* (reported as *Cyprideis littoralis* in Zheng et al. (1991) and Wang et al. (1999) which is the unnoded form of *C. torosa*, i.e., *C. torosa* forma *littoralis*) were reported from early Holocene sediments of the sections YKD0301, F4 and core K1 (Liu et al. 2016). In addition, tests of the foraminifer *Ammonia tepida* morphotype were recorded in the sediments of the YKD0301 section and the K1 core (as *A. beccarii* in the latter). A brackish to hyperhaline lake with a salinity in a range from ca. 20–100‰ in the early Holocene was inferred from the wide salinity tolerance ranges of the two ostracod species, the lack of typical freshwater (sensu lato - s.l.) ostracods and the observation of living *Ammonia tepida* in a hyperhaline pool with seasonal salinity fluctuations between 40 and 55‰ in Israel (Almogi-Labin et al. 1992; Meisch 2000; Mischke et al. 2019). This inference is supported by the occurrence of elongate in addition to subspherical ooids which were likely formed around fecal pellets of brine shrimps (Mischke et al. 2019). In addition to these early Holocene remains of a brackish to hyperhaline lake, valves of freshwater (s.l.) ostracods were reported from a few early Holocene samples of section YKD0301 and a single sample of presumably late Holocene age of the section Ma (Ma et al. 2008; Liu et al. 2016). However, the absence of similar fossils from most parts of middle and late Holocene sediments in the Lop Nur region, the presence of gypsum and/or more soluble salts in these sediments and those of subspherical and elongate ooids in the YKD0301 section imply that salinities were mostly higher than levels supporting *A. tepida* in Lop Nur after the early Holocene. Salinities were probably continuously or at least from time to time higher than ca. 188‰ to enable gypsum formation in the lake. Accordingly, sediment cores from the Luobei region and its vicinity indicate that Lop Nur was a brackish lake in the early Holocene and a highly saline lake with the massive formation of halite starting in the middle Holocene (Fig. 7.4; Liu et al. 2008).

7.6 Reported Fauna and Flora of the Lake Before the Onset of Significant Human Impact

Organism remains from water bodies in the Lop Nur region before the Han Dynasty were reported for sections YKD0301 and F4 and for core K1. These locations are within the 'Great Ear' region and regarded to represent the ancient Great Lop Nur (Figs. 7.2 and 7.3).

7.6.1 Gastropoda

A single shell of *Gyraulus* sp. was recorded in sediments of section YKD0301 which were formed ca. 8500 years ago (Table 7.1; Mischke et al. 2019). *Lymnaea* shells ('*Limnæa*, at least three species'; Hörner 1932) were reported for Quaternary exposed sediments in the Lop Nur region.

7.6.2 Ostracoda

Valves of *Eucypris mareotica* and *Cyprideis torosa* were recorded in early Holocene sediments of sections YKD0301, F4 and core K1 (Table 7.1; Zheng et al. 1991; Wang et al. 1999; Liu et al. 2016). Valves of *Cyprinotus* sp. (likely *Heterocypris*, possibly *Heterocypris salina*) were recorded in addition to *E. mareotica* and *C. torosa* at the F4 site, indicating meso- or polyhaline conditions. Holocene sediments from YKD0301 which were formed ca. 8000 years ago also contain abundant valves of the species *Limnocythere inopinata*, *Cypridopsis vidua* and *Ilyocypris* sp. whilst valves of *Darwinula stevensoni*, *Pseudocandona* sp., *C. torosa*, *E. mareotica*, *Neglecandona neglecta* and *Herpetocypris* sp. were only occasionally recorded (Table 7.1). The assemblage probably represents the Holocene period with most diluted, only slightly brackish waters in Lop Nur. A few valves and fragments of *L. inopinata* and *Ilyocypris bradyi* were also recorded in early and middle Holocene sediments of core ZK95-6 (Wang et al. 2001). In addition, some valves of *N. neglecta* and *Cypricercus* were also found in a late Holocene sediment bed of 6 cm thickness at the ZK95-6 location (Table 7.1; Wang et al. 2001).

7.6.3 Foraminifera

Ammonia tepida (reported as *Ammonia beccarii*) was recorded in late Quaternary sediments of the K1 core by Wang et al. (1999; Table 7.1). The foraminifera in the sediments of the K1 core were discussed to originate from the relict fauna of

Table 7.1 Organism remains from Holocene pre-human impact sediments of Lop Nur (n.d. no data)

Taxon	Location	Reported as	Source
Gastropoda			
Gyraulus sp.	YKD0301		Mischke et al. (2019)
Lymnaea sp. 1	n.d.	*Limnæa*, three species	Hörner (1932)
Lymnaea sp. 2	n.d.	*Limnæa*, three species	Hörner (1932)
Lymnaea sp. 3	n.d.	*Limnæa*, three species	Hörner (1932)
Ostracoda			
Neglecandona neglecta	YKD0301		Mischke et al. (2019)
	ZK95-6		Wang et al. (2001)
Cypricercus sp.	ZK95-6		Wang et al. (2001)
Cyprideis torosa	YKD0301		Liu et al. (2016)
	F4	*Cyprideis littoralis*	Zheng et al. (1991)
	K1	*Cyprideis littoralis*	Wang et al. (1999)
Cypridopsis vidua	YKD0301		Liu et al. (2016)
Cyprinotus sp.	F4		Zheng et al. (1991)
Darwinula stevensoni	YKD0301		Liu et al. (2016)
Eucypris mareotica	YKD0301		Liu et al. (2016)
	F4	*Eucypsis inflata*	Zheng et al. (1991)
	K1	*Eucypris inflata*	Wang et al. (1999)
Herpetocypris sp.	YKD0301		Mischke et al. (2019)
Ilyocypris bradyi	ZK95-6		Wang et al. (2001)
Ilyocypris sp.	YKD0301		Liu et al. (2016)
Limnocythere inopinata	YKD0301		Liu et al. (2016)
	ZK95-6		Wang et al. (2001)
Pseudocandona sp.	YKD0301		Mischke et al. (2019)
Foraminifera			
Ammonia tepida	K1	*Ammonia beccarii*	Wang et al. (1999)
	YKD0301		Mischke et al. (2019)
Macrophyta			
Ruppia maritima	YKD0301		Mischke et al. (2019)
Typha	F4		Zheng et al. (1991)
	YKD0301		Liu et al. (2016)
Charophyta			
Chara sp.	K1		Wang et al. (1999)
Charophyta indet.	YKD0301		Mischke et al. (2019)
Hornichara cf. *kasakstanica*	K1		Wang et al. (1999)

(continued)

Table 7.1 (continued)

Taxon	Location	Reported as	Source
Lamprothamnium cf. *papulosum*	K1		Wang et al. (1999)

the Eocene sea in the Tarim Basin region. Abundant tests of *Ammonia tepida* were also recorded in the early Holocene sediments of the YKD0301 section. The taxon was probably also recorded in late Quaternary sediments of salt lakes of the Tibetan Plateau (from Lakes Tianshuihai and Xiao Qaidam, identified as *Elphidium*? sp. or *Elphidium* sp., respectively) and from sub-recent surface sediment samples from Gahai Lake in the Qaidam Basin (Tibetan Plateau) and the Heiyenur Pool in Inner Mongolia (as *Ammonia*) together with valves of the brackish water ostracod *Eucypris mareotica* (reported as *E. inflata*; Sun et al. 1992, 1997; Li et al. 1997). Living specimens of *Ammonia tepida* were recorded in an artificially created hyperhaline pool in Israel (salinity 39.7–54.5‰) which was colonized by means of migratory birds (Almogi-Labin et al. 1992). Thus, the monospecific occurrence of *Ammonia tepida* in Lop Nur in the early Holocene probably resulted from the colonization of the lake during a favourable stable brackish to saline water period.

7.6.4 Sub- and Emerged Macrophytes

Seeds of the aquatic macrophyte *Ruppia maritima* were recovered from early Holocene sediments of section YKD0301 in the western 'Great Ear' region (Table 7.1; Mischke et al. 2019). *Ruppia maritima* has a wide salinity range from 0.6 to 390‰ and a salinity optimum between ca. 5–22‰ (Kantrud 1991). Relatively high abundances of *Typha* pollen in early Holocene sediments of sections YKD0301 and F4 probably indicate local origins and suggest that marginal and shallow parts of the lake were covered by bulrush.

7.6.5 Charophyta

Remains of *Lamprothamnium* cf. *papulosum*, *Hornichara* cf. *kasakstanica* and *Chara* sp. were also recorded in the late Quaternary sediments of core K1 (Table 7.1; Wang et al. 1999). Charophyte oogonia and gyrogonites were also recorded in early Holocene sediments of the YKD0301 section but were not further identified yet (Mischke et al. 2019).

7.6.6 *Inference of lake conditions based on organism remains*

Holocene sediments of the YKD0301 section contain abundant valves of *E. mareotica* and *C. torosa*, *Ammonia* tests and a few *Ruppia* seeds and charophyte remains in the lower part of the sediment sequence deposited earlier than ca. 9000 years. A similar ostracod assemblage was recorded at the F4 and K1 locations with the latter similarly representing evidence for the wide-spread occurrence of *Ammonia tepida* in the brackish to saline lake. The subsequently deposited middle and late Holocene sediments which were formed before the onset of significant human impact only occasionally contain fossil ostracod valves or other remains of aquatic organisms. The lake was likely too saline to support a higher aquatic fauna and flora. Thus, hyperhaline conditions are inferred as the 'natural' state of Lop Nur for most of the middle and late Holocene.

7.7 Lop Nur Since the Onset of Significant Human Impact

7.7.1 *Lop Nur During the Period from Han to Qing Dynasties*

Historical documents suggest that a large lake existed in the Lop Nur region two millennia ago (Ban, 92; Yang et al. 2006). The first occurrence of a sand layer in deposits of the YKD0301 section ca. 1800 years ago, the fall of the Loulan Kingdom sometime between 230 and 645 CE and the contemporary and stepwise abandonment of the desert oasis along the Keriya and Niya rivers in the Tarim Basin starting in the basin's centre were used to argue that water withdrawal for irrigation farming upstream Lop Nur caused an environmental crisis similar to the modern Aral Sea disaster already almost two millennia ago (Qin et al. 2012; Mischke et al. 2017). Lop Nur apparently experienced a period of man-made near-desiccation ca. 1800 years ago. Historical books such as the *History of the Former Han (Han Shu,* 汉书*)*, the *History of the Later Han* (*Hou Han Shu,* 后汉书), *Shui Jing Zhu* (水经注) from the Northern Wei Dynasty (386–534 CE) and *Geography of Shazhou* (*Sha Zhou Tu Jing*, 沙州图经残卷) from the Tang Dynasty (618–907 CE) were examined by Yang (2004) who showed that Lop Nur covered a relatively small area in the first millennium of the common era (Fig. 7.5). The lake area was between ca. 50 and 110 km^2 which is similar to $^1/_{20}$ or $^1/_{10}$ of the size of Xinjiang's largest Lake Bosten (ca. 1000 km^2) today (Yang 2004). The *Geography of the Source Region of the Yellow River* (*He Yuan Ji Lue,* 河源纪略) and *Records and Maps of the Western Regions* (*Xi Yu Tu Zhi*, 西域图志) provide contrary information for the reign of Emperor Qianlong (1735–1796 CE) in the earlier part of the Qing Dynasty (1636–1912 CE) with reported areas of ca. 6700 and 120 km^2, respectively (Yang 2004). The discrepancy in passed on lake-area information does not necessarily result from erroneous estimations but possibly reflects a large variability over time. According to the assessment of the *Records of Water Channels in Western Territories* (*Xi Yu*

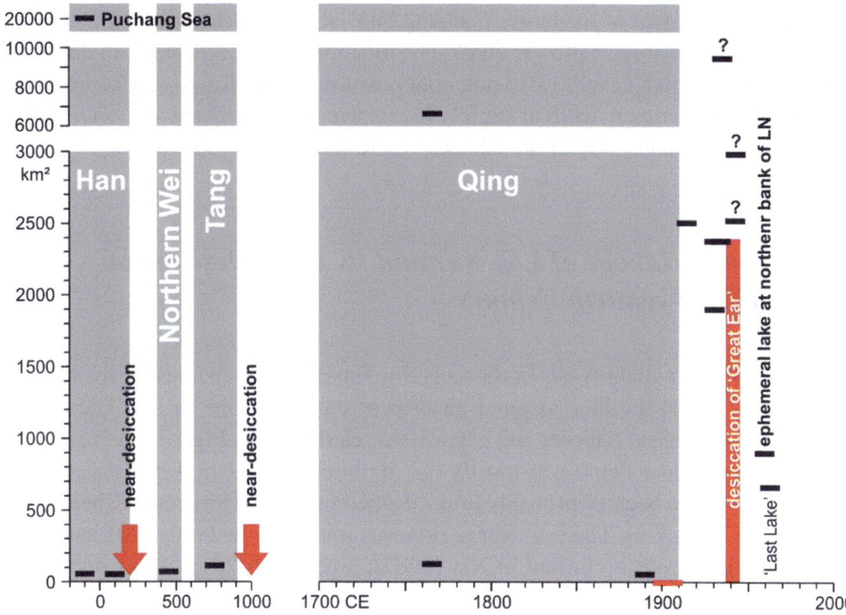

Fig. 7.5 Reported area of Lop Nur (LN) based on historical documents assessed by Yang (2004) and Yang et al. (2006) since Han Dynasty times. Near-desiccation periods were inferred from sand layers in section YKD0301 (Liu et al. 2016; Mischke et al. 2017). Desiccation of the 'Great Ear' occurred in the late 1930s or early 1940s (Li et al. 2008). The large lake areas reported for the 1930s and 1940s and tagged by question marks are probably misleading estimates (Xia 1987)

Shui Dao Ji, 西域水道记) and the *General Introduction of the Great Qing* (*Da Qing Yi Tong Zhi*, 大清一统志), and archaeological data by Yang et al. (2006), the former large 'Puchang Sea' of Han Dynasty times was replaced by the smaller Lop Nur and a number of additional lakes in the region during the Qing Dynasty. Lop Nur as the largest lake covered an area of several thousand km^2, and the total area covered by water may have still amounted ca. 11,000 km^2 which is ca. eleven times the size of Lake Bosten (Yang et al. 2006). Historic sources also imply that irrigated farmlands were significantly smaller during the Qing Dynasty in comparison with Han Dynasty times. Thus, the increase in the area of Lop Nur during Qing times probably resulted at least partly from reduced water withdrawal from its tributaries (Yang et al. 2006). A relatively small area of ca. 50 km^2 is reported in the *Illustrated Handbook of Inspection* (*Tan Cha Tu Shuo*, 探查图说) for the late Qing Dynasty between 1874 and 1911 (Fig. 7.5).

Sediments believed to have accumulated between ca. 1100 and 300 years ago at the section Ma location contain valves of freshwater (s.l.) ostracods, gastropod shells and remains of sub- and emerged macrophytes (Ma et al. 2008; Table 7.3). The ostracod and mollusc fauna and plant assemblage indicate oligohaline to slightly or moderately mesohaline conditions, and remains of wetland plants suggest near-shore or generally very shallow conditions. These inferences are consistent with the described replacement of the large ‘Puchang Sea’ of the Han Dynasty by a number of smaller and probably shallower lakes. The location of the section Ma in the ‘Last Lake’ region of Lop Nur is close to potential positions of the Konqi and Tarim River deltas implying relatively fresh to slightly brackish conditions at the section location even if the main water body of the lake was more brackish or even saline (Fig. 7.3).

7.7.2 *The Shrinkage of Lop Nur and Its Final Desiccation in the Twentieth Century*

A large body of stagnant water in the Lop Nur Basin was eyewitnessed for the last time in 1934 when the lake covered an area of ca. 2400 km^2 in the ‘Great Ear’ and ‘Last Lake’ region (Hörner and Chen 1935; Hedin 1942; Figs. 7.2, 7.3 and 7.5; Table 7.2). The water depth was mostly 0.2 m although ca. 1 m were occasionally measured. The area became probably flooded since ca. 1921 (Hörner and Chen 1935; Hedin 1942; Table 7.2). The Lop Nur region was found dry in 1900, 1901 and 1906 (Table 7.2). However, according to Xia (1987), a relatively large water body may have already existed in the basin in 1914 (Fig. 7.5).

A detailed topographic survey and assessment of aerial and satellite images of the ‘Great Ear’ region by Li et al. (2008) showed that the ‘Great Ear’ represents a former lake basin with maximum depth of 5.2 m near its centre, which dried between the late 1930s and early 1940s. The concentric rings of the ‘Great Ear’ region were identified as trails of former shorelines characterized by differences in surface roughness of the salt crust, groundwater-controlled moisture content of surficial sediments and the ratio of salt and detrital sediments (Fig. 7.6; Li et al. 2008; Dong et al. 2012). Earlier views that these rings represent morphologically distinct recessional shoreline bars (beach bars) and thus, significant changes in surface relief are not supported by field data (Zhou 1978; Dong et al. 2012). Ma et al. (2010) suggested that a permanent water body in the Lop Nur Basin eventually dried after 1952 when the Tarim River inflow was dammed and turned towards the south. Geomorphological maps of the Lop Nur region produced in the Soviet Union and China suggest that the lake still covered a large area in 1942 (Xia 1987). However, it is unclear whether the drawing of these maps was based on the descriptions of the lake from the early 1930s by Hörner and Chen (1935) and Hedin (1942) or whether they actually reflect a later state (Fig. 7.5). An ephemeral water body may have occasionally formed in the ancient terminal fan and ‘Last Lake’ regions when the Konqi River flooded these parts of the Lop Nur Basin during extraordinarily wet years. However, since Landsat imagery became available in 1972, significant flooding was recorded only once in the terminal fan region of the Konqi River in June 1973 (Ma et al. 2010).

A salt crust covers the lowest region in Lop Nur since its final desiccation which builds up a playa surrounded by dry saline mudflats (Fig. 7.6). Pore space of unconsolidated clastic sediments beneath the salt crust in the centre is filled with concentrated brines and the sediments contain primary and secondary gypsum. Discharging

Table 7.2 Reported water bodies in the Lop Nur region during the last ca. 150 years. The 'Great Ear' structure was not recognized as such before the use of airborne imagery

Region	Area (km^2)	Max. Depth (m)	Salinity	Time	References
Taitema Lake[a]	~375	1.2	Prob. Freshwater	Febr 1877	Prejevalsky (1879)
Kara Koshun[b]	~1500	4.0	freshwater	Febr 1877	Prejevalsky (1879)
Kara Koshun	n.d.	n.d.	n.d.	1876–1877	Hörner and Chen (1935)
'Great Ear'	0 (dry)	n.d.	n.d.	Mar 1900 and Mar 1901	Hedin (1942)
Kara Koshun	n.d.	0.8	n.d.	1900 and 1901	Hedin (1942)
Kara Koshun	n.d.	Reedy swamp and ponds	n.d.	Dec 1905	Huntington (1907)
'Great Ear'	0 (dry, salt covered)	n.d.	n.d.	Jan 1906	Huntington (1907)
Kara Koshun	n.d.	n.d.	Freshwater	Before 1921	Hörner and Chen, 1935
'Last Lake' and 'Great Ear'	2375	1 m	Salty but 0.3‰ in N	1930, 1931 and 1934	Hörner and Chen (1935), Li et al. (2008)
'Great Ear'	3006	n.d.	n.d.	1942	Xia et al. (2002)
'Great Ear'	Partly filled[c]	n.d.	n.d.	Aug 1958	Ma et al. (2008)
'Great Ear'	0 (dry)	n.d.	n.d.	1958	Li et al. (2008)
'Last Lake'	Partly filled	n.d.	n.d.	1958	Shao et al. (2012)

(continued)

Table 7.2 (continued)

Region	Area (km^2)	Max. Depth (m)	Salinity	Time	References
Ephemeral lake at north bank of LN basin	900	0.5	0.5‰	Fall 1959	Zhou (1978) in Ma et al. (2010)
Taitema Lake	88	0.8	n.d.	1959	Zhao and Xia (1984)
'Great Ear'	0 (dry)	n.d.	n.d.	Dec 1961	Ma et al. (2008)
'Last Lake'	Small area in S	n.d.	n.d.	Dec 1961	Ma et al. (2008)
'Last Lake'[d]	660	n.d.	n.d.	1962	Xia et al. (2002)
Mouth of Konqi River	n.d.	n.d.	n.d.	June 1973	Ma et al. (2010)

[a]called Kara Buran in original report
[b]regarded as Lop Nur in original report
[c]apparently wrong according to Li et al. (2008) and evidence of air photographs from 1958
[d]region not indicated but lake described as N-S-aligned water body

Fig. 7.6 High micro-relief of salt crust in the central part of the 'Great Ear' region (photo by Hua Zhang, Nov. 2015)

groundwater supports numerous moist salt pans in the playa where formation of evaporites resulted in a 30–120 cm thick salt crust above lacustrine sediments (Wang et al. 2001). According to the study of the radioactive hydrogen isotope tritium in subsurface brines in the Lop Nur region by Jiao et al. (2004), the Na^+ and Cl^- ions dominated shallow brine aquifer in the region mostly represents modern waters recharged in recent decades. At present, groundwater recharge occurs as a result of occasional local atmospheric precipitation and flood events in the adjacent mountain regions. Some small springs at the southern margin of the basin are fed by waters from the Altun Mountains and contribute slightly mesohaline water to the 'Great Ear' region (Fig. 7.3).

7.8 Fauna and Flora of Lop Nur Since the Han Dynasty

Information with respect to organisms in the waters of the Lop Nur region during the last two millennia is very rare. Organism remains are only reported from the uppermost 80–30 cm of section Ma in the centre of the 'Last Lake' and are assumed to represent deposition between 1100 and 300 years ago (Ma et al. 2008). However, the chronology of the 3.3-m thick section Ma is based on three radiocarbon ages without an assessment of the commonly occurring lake-reservoir effect, and the age data should be therefore regarded with caution. Additional information about living organisms in waters of the Lop Nur region was provided in reports starting in the late nineteenth century.

7.8.1 *Gastropoda*

Shells of the gastropods *Radix auricularia*, *Lymnaea stagnalis* (reported as *Lymnaca scaynalis*) and of Planorbidae were recorded in the upper 80–30 cm of the sediments of section Ma in the 'Last Lake' but more detailed information with respect to their position within the section and related age was not provided (Ma et al. 2008; Fig. 7.3; Table 7.3). The assemblage indicates oligohaline to very slightly mesohaline conditions (<7‰) consistent with a shallow lake and the river inflow near the location of the section (Glöer 2002).

7.8.2 *Ostracoda*

Sediments from section Ma which are believed to have formed 800–400 years ago revealed ostracod valves of *Eucypris mareotica* (reported as *E. inflata*; Ma et al. 2008; Table 7.3). A single sample possibly formed 500 years ago contained valves of *Cyprideis torosa*. In addition, valves of *Limnocythere inopinata*, *Darwinula stevensoni*, *Candoniella albicans* (possibly juvenile candonid valves sometimes reported as *Candoniella albicans* valves in Chinese sources), *Pseudocandona compressa* (reported as *Candona compressa* but placed into *Pseudocandona* today) and *Candona* sp. were recovered from uppermost sediments in the section which were probably accumulated ca. 300 years ago (Ma et al. 2008). Valves of *Eucypris mareotica* were the only ones found in five of the examined six samples indicating that water in the 'Last Lake' region was mostly brackish to saline during the last ca. 900 years. Only a single sample from 33 cm depth (possibly accumulated ca. 300 years ago) contained valves of the other taxa listed above, suggesting a period of relatively fresh conditions at the section location (Ma et al. 2008).

7.8.3 *Fish*

Abundant fish was reported from the shallow waters of Lop Nur in 1934 (Hedin 1942). Possibly, the fish with described lengths of up to 110 cm represent specimens of *Coregonus maraena* which were earlier described from Kara Koshun (misinterpreted

Table 7.3 Organisms and remains recorded during the human-impact period of Lop Nur

Taxon	Location	Reported as	Source
Gastropoda			
Lymnaea stagnalis	section Ma	*Lymnaca scaynalis*	Ma et al. (2008)
Planorbidae	section Ma		Ma et al. (2008)
Radix auricularia	section Ma		Ma et al. (2008)
Ostracoda			
Candona sp.[a]	section Ma	*Candoniella albicans*	Ma et al. (2008)
Cyprideis torosa	section Ma		Ma et al. (2008)
Darwinula stevensoni	section Ma		Ma et al. (2008)
Eucypris mareotica	section Ma	*Eucypris inflata*	Ma et al. (2008)
Limnocythere inopinata	section Ma		Ma et al. (2008)
Pseudocandona compressa	section Ma	*Candona compressa*	Ma et al. (2008)
Pisces			
Coregonus maraena[b]	Kara Koshun		Prejevalsky (1879)
	Lop Nur		Hedin (1942)
Cyprinidae	Kara Koshun		Prejevalsky (1879)
Triplophysa yarkandensis	Kanglake lakes		Jiang et al. (2011)
Macrophyta			
Butomus	Kara Koshun[c]		Prejevalsky (1879)
Hippuris	Kara Koshun[c]		Prejevalsky (1879)
Phragmites?[d]	Kara Koshun[c]		Prejevalsky (1879)
Potamogeton lucens	section Ma		Ma et al. (2008)
Schoenoplectus tabernaemontani	section Ma	*Scirpus tabernaemontani*	Ma et al. (2008)
Stuckenia pectinata	section Ma	*Potamogeton pectiatus*	Ma et al. (2008)
Typha	Kara Koshun[c]		Prejevalsky (1879)
Charophyta			
Chara spp.	section Ma		Ma et al. (2008)
Bacillariophyceae			
Navicula sp.	section Ma		Ma et al. (2008)

[a]Valves reported as *Candoniella albicans* are often juvenile valves of *Candona*. Ma et al. (2008) recorded also one broken valve of *Candona* sp., possibly belonging to the same species as the assumed juvenile valves. However, one specimen illustrated by Ma et al. (2008) and termed '*Condoniella* sp.' is actually a valve of *D. stevensoni*

[b]described as fish with lengths of up to 110 cm, probably *C. maraena*

[c]originally reported as Lop Nur but considered as Kara Koshun by Hörner and Chen (1935) and others

[d]reported as 'reed', taxonomic assessment not clear

as Lop Nur) by Prejevalsky (1879; Tables 7.2 and 7.3). In addition to *Coregonus maraena*, another unidentified taxon of the carp (Cyprinidae) family was reported by Prejevalsky for Kara Koshun in 1877 (Prejevalsky 1879). Jiang et al. (2011) reported that *Triplophysa (Hedinichthys) yarkandensis* occurs in the Kanglake lakes (Fig. 7.2).

7.8.4 Sub- and Emerged Macrophytes

Seeds of the aquatic submerged macrophytes *Stuckenia pectinata* (reported as *Potamogeton pectiatus* but placed into *Stuckenia* today) and *Potamogeton lucens*, and of the sedge *Schoenoplectus tabernaemontani* (reported as the younger synonym *Scirpus tabernaemontani*) were recorded in sediments of the section Ma regarded to have formed 1100–300 years ago (Ma et al. 2008; Table 7.3). The occurrence of *P. lucens* suggests a shallow water depth of 0.5-2.0 m whilst emerged wetland sedges indicate very shallow depth or patches of open-water surfaces and exposed regions (van Geest et al. 2005). The presence of *Stuckenia pectinata* suggests oligo- to moderately mesohaline waters (Kantrud 1990). Dense patches of reed accompanied by other emerged aquatic plants (*Hippuris*, *Typha* and *Butomus*) were reported for the Kara Koshun in 1877 (Prejevalsky 1879).

7.8.5 Charophyta

Charophyte oogonia or gyrogonites, assigned to *Chara* spp., were recorded in two sediment samples of section Ma believed to have formed ca. 500 and 300 years ago (Ma et al. 2008; Table 7.3).

7.8.6 Diatoms

Frustules of the diatom *Navicula* sp. were recorded in sediments of section Ma with an assumed age of ca. 400 years (Table 7.3).

7.9 Impacts on the Catchment Hydrology After Lop Nur's Desiccation

Hydrological changes in the catchment of Lop Nur during recent decades are caused by the population increase and the associated intensification of farming and reclamation of newly irrigated farmlands since the foundation of the PR of China in 1949 (Hao et al. 2008; Chen et al. 2011). The population of the Tarim Basin has increased five times between 1950 and 1995 (Xu et al. 2010). The area of cultivated farmland along the Tarim River was almost doubled between 1949 (349,820 ha) and 1994 (607,050 ha; Chen et al. 2011). Canals of more than 1000 km lengths and more than 70 reservoirs of various sizes were built (Xu et al. 2008). The total irrigated area in the Tarim Basin reached 1.63 million ha in 2008 (Chen et al. 2011). A water volume of ca. 5 billion m^3 was diverted for irrigation in the 1950s which was increased to 20.2 billion m^3 in 2008. As a consequence of such significant increase in water consumption for irrigation and other urban and industrial purposes, discharge in the lower section of the Tarim River decreased in the 1990s to ca. 25% of the runoff which was recorded in the late 1950s, although discharge in its headwaters increased during the same period of time (Hao et al. 2008). Runoff in headwaters increased as a result of global warming. A slight precipitation increase was accompanied by higher temperatures which caused enhanced meltwater discharge. Thus, the largest tributary of the Tarim River, the Aksu River experienced a runoff increase by ca. 50% between the late 1950s and the end of the twentieth century (Thevs 2011). However, this significant increase of runoff in the upper reaches of the Tarim River in recent decades was by far not sufficient to compensate for the water withdrawal in the middle and lower reaches. Following the construction of the Daxihaizi Reservoir with a capacity of 228 million m^3 and surface area of 104 km^2 in 1972, the lowermost 321 km of the Tarim River desiccated (Zhao and Xia 1984; Chen et al. 2011; Fig. 7.2). The point of zero flow moved ca. 100 km upstream until the 1990s and to the upper reaches of the river in 2009. As a result, the ca. 1200 km long section of the middle and lower reaches of the Tarim River desiccated seasonally or permanently.

The runoff decrease in the middle and lower reaches of the Tarim River was accompanied by a salinity increase due to evaporative concentration of river water in reservoirs and salinization of irrigated fields resulting from inadequate drainage systems of irrigated farmlands. Consequently, the salinity of the Tarim River increased from less than 1‰ before 1958 to 2.9‰ on average or even 6.8‰ at maximum measured in the years 1999 and 2000 downstream of the Daxihaizi Reservoir (Xu et al. 2008).

The decrease of the runoff of the Tarim River had also negative impacts on the groundwater level in the lower reaches. Following the establishment of the Daxihaizi Reservoir, groundwater levels close to the reservoir fell to ca. 5 m below the surface and to a depth of ca. 12 m farther downstream (Xu et al. 2008).

The area of many riparian forests such as the 'Green Corridor' north of the Taitema Lake region between the Taklamakan Desert in the west and the Kuluke Desert in the east decreased causing the mobilization of aeolian sands and damages to infrastructure such as roads. The total area covered by *Populus euphratica* along the lower reaches of the Tarim River decreased from 5.4×10^4 ha in the 1950s to 1.64×10^4 ha in the 1970s and to 0.67×10^4 ha in the 1990s (Xu et al. 2008). The programme "Comprehensive Governing and Managing on Tarim River" was established in June 2001 by the Central and Provincial Governments which led to the recovery of runoff in the lower reaches of the Tarim River and the re-establishment and expansion of Taitema Lake. However, highly variable annual discharge in the lower reaches of the Tarim River causes large areal changes of Taitema Lake (Ablekim et al. 2014). Discharge of the Cherchen River was also relatively high in the first years of the twenty-first century which led to the significant expansion of the Kanglake lakes and wetlands farther in the west of Taitema Lake (Ablekim et al. 2014). As a result of the expansion of these lakes and wetlands in the lower reaches of the Tarim and Cherchen rivers, sand and dust storm impacts in the Ruoqiang County in the southwest of Lop Nur were effectively reduced in recent years.

The salt-crust covered playa of Lop Nur is exploited for potash production since 2004. Potassium-rich brines are pumped to the surface and further concentrated in artificially created evaporation ponds in the northeastern part of the Lop Nur region (Figs. 7.2 and 7.7). The SDIC (State Development & Investment Corp. Ltd.) Xinjiang Luobupo Hoevellite Co. Ltd has established a system for the annual production of 1.3 million tons of potassium sulfate from the underground brines in Lop Nur (Sun and Ma 2018). However, the natural balance between groundwater recharge and

Fig. 7.7 Evaporation pond for potash production in the northeast of the 'Great Ear' (photo by Hua Zhang, Nov. 2015)

brine formation and composition is negatively affected by the mining activity now. Groundwater depression cones have formed as a result of the water extraction and the slow lateral flow of brines towards pumping stations. Thus, long-term potash production perspectives are uncertain.

The salt crust in the Lop Nur region covers a vast area and sand storms potentially cause erosion and mobilization of salt particles (Fig. 7.3). Volumes of exported salts, main dispersal paths and potential damages to farmlands were not investigated so far but lessons from other large salt-covered regions and nearby farmlands provide evidence for severe effects of wind-blown salts on crop production and health of local people (Razakov and Kosnazarov 1996).

A possible return of Lop Nur to its lake state as it was experienced in the early 1930s is not realistic. Given a continuation of present trends of population growth and associated farming activities and industrialization in the region, the Lop Nur Basin will remain a salt-crust covered playa without the re-establishment of natural perennial lacustrine conditions in the envisaged near future. Current focus of environmental protection and ecological preservation is laid on Lop Nur's former tributaries and associated riparian vegetation. The Tarim River Basin Water Resource Commission (TRBWRC) was founded in 1997 in order to implement an integrated water resource management system and to allocate quotas for river water diversion in the Tarim River catchment. Water management measures with the aim to restore riparian environments and promote sustainable development in the lower reaches of the Tarim River were established with the Regulation Scheme of the Tarim River in the year 2002 (Thevs 2011). Two canals were built in 2002 to divert water from the Konqi River to the lower Tarim River. They led to the partial restoration of vegetation along river branches in the lower reaches of the Tarim River (Xu et al. 2008). However, water withdrawal for cotton farming in the upper reaches of the Tarim River (along its main tributary, the Aksu River) is not controlled by the TRBWRC but by the Xinjiang Production and Construction Corps (XPCC). Thus, water withdrawal for the XPCC farms along the Aksu River and uncontrolled drilling of wells and groundwater extraction along the entire course of the Tarim River by local farmers inhibits the year-round flow of the river in the middle and lower reaches (Thevs 2011). Strengthening of the position of the TRBWRC and a stricter allocation of water quotas, and better control of groundwater extraction are required for a sustainable development of farming in the Tarim Basin and maintenance of a unique riparian ecosystem within a hyperarid desert. Efforts to preserve and partly restore the vegetation along the lower Tarim River will require permanent adjustments of water quotas, improvements of water-saving irrigation techniques and the partial replacement of cotton as the major crop by the utilization of naturally occurring plants due to the continuously increasing water demand, the large natural variability of precipitation and generated runoff, and also due to expected meltwater reductions of diminishing ice and snow resources in coming years of global warming.

7.10 Summary and Conclusions

Lop Nur was a perennial, large and brackish to saline lake with a few metres water depth until ca. two millennia ago. Diversion of water for irrigation-based agriculture in the upper and middle reaches of its tributaries during the expansion of the Chinese empire in the Han Dynasty caused a significant reduction of inflows and probably resulted in the near-desiccation of the lake ca. 1800 years ago. Afterwards, flooding of the lake basin or of depressions in its vicinity highly depended on irrigation farming activities on the one hand side and precipitation and flood events in wetter years, meltwater discharge in warmer summers, and river course changes due to sediment accumulation and wind erosion of exposed dry surfaces. The case of Lop Nur impressively shows that the human degradation of lakes in northwestern China is not only confined to relatively recent centuries but may have started already ca. 2000 years ago.

Acknowledgements We are indebted to Hua Zhang who kindly provided photos from the Lop Nur region and to Li Zhang for translation of Chinese publications. Research work on Lop Nur was funded by grants from China's NSF (projects 40830420, 41771004) and the Ministry of Science and Technology of China (the National 305 Project 2003BA612A-06-15).

References

Ablekim A, Kasimu A, Kurban A, Jappar T, Fan Z (2014) Monitoring the water area changes in Tetima-Kanglayka lakes region over the past four decades by remotely sensed data. J Lake Sci 26:46–54

Almogi-Labin A, Perelis-Grossovicz L, Raab M (1992) Living *Ammonia* from a hypersaline inland pool, Dead Sea area, Israel. J Foramin Res 22:257–266

Ban G (92) Han Dynasty. 92. Reprinted in 1962 by Zhonghua Book Company, Beijing

Bosboom R, Dupont-Nivet G, Grothe A, Brinkhuis H, Villa G, Mandic O, Stoica M, Huang W, Yang W, Guo Z, Krijgdman W (2014) Linking Tarim Basin sea retreat (west China) and Asian aridification in the late Eocene. Basin Res 26:621–640

Chang H, An Z, Liu W, Qiang X, Song Y, Ao H (2012) Magnetostratigraphic and paleoenvironmental records for a Late Cenozoic sedimentary sequence drilled from Lop Nor in the eastern Tarim Basin. Glob Planet Change 80–81:113–122

Chen Y, Ye Z, Shen Y (2011) Desiccation of the Tarim River, Xinjiang, China, and mitigation strategy. Quat Int 244:264–271

Dong Z, Lv P, Qian G, Xia X, Zhao Y, Mu G (2012) Research progress in China's Lop Nur. Earth Sci Rev 111:142–153

Glöer P (2002) Die Süßwassergastropoden Nord- und Mitteleuropas. Hackenheim, ConchBooks, 327pp (in German)

Hao H, Ferguson DK, Chang H, Li C-S (2012) Vegetation and climate of the Lop Nur area, China, during the past 7 million years. Clim Change 113:323–338

Hao X, Chen Y, Xu C, Li W (2008) Impacts of climate change and human activities on the surface runoff in the Tarim River Basin over the last fifty years. Water Resour Manage 22:1159–1171

Hedin S (1942) Der wandernde See. Brockhaus, Leipzig, p 295 (in German)

Hörner N, Chen PC (1935) Alternating lakes. Some river changes and lake displacements in Central Asia. Geogr Ann 17:145–166

Hörner NG (1932) Lop-nor. Topographical and geological summary. Geografiska Annaler 14:297–321

Hua YS, Jiang PA, Wu HQ, Zhong JP, Ma LC, Li BG (2009) The sedimentary characteristics of Profile L07-10 and its environmental indicator in Lop Nor "Great Ear" area. J Xinjiang Agric Univ 32:36–39 (in Chin with Engl abstract)

Huntington E (1907) Lop-Nor. A Chinese lake. Part I. The unexplored salt desert of Lop. Bull Am Geogr Soc 39:65–77

Jiang R, Lei M, Xiong J (2011) Observation on the reproductive characteristic of *Triplophysa (Hedinichthgs) yarkandensis* (Day) in kanglakes. J Aquaculture 32:30–34

Jiao P, Wang M, Liu C (2004) Characteristics and origin of tritium in the potassium-rich brine in Lop Nur, Xinjiang. Nucl Tech 27:710–715 (in Chin with Engl abstract)

Kantrud HA (1990) Sago pondweed (*Potamogeton pectinatus* L.): a literature review. U.S. Fish and Wildlife Service, Resource Publication 176, Washington DC, 89pp

Kantrud HA (1991) Wigeongrass (*Ruppia maritima* L.): a literature review. U.S. Fish and Wildlife Service, Fish and Wildlife Research 10, 58pp

Li B, Ma L, Jiang P, Duan Z, Sun D, Qiu H, Zhong J, Wu H (2008) High precision topographic data on Lop Nor basin's Lake "Great Ear"? and the timing of its becoming a dry salt lake. Chin Sci Bull 53:905–914

Li Y, Li B, Wang G, Li S, Zhu Z (1997) Quaternary foraminiferal fossil in the Tianshuihai are of the Qinghai-Xizang Plateau. Chin Sci Bull 42:1011–1014

Lin J, Zhang J, Ju Y, Wang Y, Lin F, Zhang J, Wang S, Wei M (2005) The lithostratigraphy, magnetostratigraphy, and climatostratigraphy in the Lop Nur region, Xinjiang. J Stratigr 29:317–322 (in Chin with Engl abstract)

Liu C, Jiao P, Chen Y, Wang M (2008) Late Pleistocene mirabilite deposition in the Lop Nur saline lake, Xinjiang, and its paleoclimate implications. Acta Geosci Sin 29:397–404 (in Chin with Engl abstract)

Liu C, Jiao P, Lü F, Wang Y, Sun X, Zhang H, Wang L, Yao F (2015) The impact of the linked factors of provenance, tectonics and climate on potash formation: an example from the potash deposits of Lop Nur depression in Tarim Basin, Xinjiang, Western China. Acta Geol Sin 89:2030–2047

Liu C, Zhang J, Jiao P, Mischke S (2016) The Holocene history of Lop Nur and its palaeoclimate implications. Quat Sci Rev 148:163–175

Liu CL, Wang ML (1999) Evolution of quaternary depositional environments and forming of potash deposits in Lop lake, Xinjiang, China. Acta Geosci Sin Suppl 20, 264–270 (in Chin with Engl abstract)

Liu CL, Wang ML, Jiao PC, Li S, Chen YZ (2003) Holocene yellow silt layers and the paleoclimate event of 8200 a BP in Lop Nur, Xinjiang, NW China. Acta Geol Sin (Engl Ed) 77:514–518

Lü F, Liu C, Jiao P, Yan H, Zhang H, Zhao Y, Wang L (2015) The discussion on sedimentary characteristics, phased evolution and controlling factors of saline lake in Asia interior: records from deep drill cores of LDK01 in Lop Nur, Xinjiang, northwestern China. Acta Petrol Sin 31:2770–2782 (in Chin with Engl abstract)

Lü H, Xia X, Liu J, Qin X, Wang F, Yidilisi A, Zhou L, Mu G, Jiao Y, Li J (2009) A preliminary study of chronology for a newly discovered ancient city and five archaeological sites in Lop Nor, China. Chin Sci Bull 55:63–71

Luo C, Peng Z, Yang D, Liu W, Zhang Z, He J, Chou C (2009) A lacustrine record from Lop Nur, Xinjiang, China: implications for paleoclimate change during Late Pleistocene. J Asian Earth Sci 34:38–45

Ma C, Wang F, Cao X, Li S, Li X (2008) Climate and environment reconstruction during the Medieval warm period in Lop Nur of Xinjiang, China. Chin Sci Bull 53:3016–3027

Ma L, Lowenstein TK, Li B, Jiang P, Liu C, Zhong J, Sheng J, Qiu H, Wu H (2010) Hydrochemical characteristics and brine evolution paths of Lop Nor Basin, Xinjiang Province, Western China. Appl Geochem 25:1770–1782

Meisch C (2000) Freshwater Ostracoda of Western and Central Europe. Spektrum, Heidelberg, 522pp

Mischke S, Liu C, Zhang J, Zhang C, Zhang H, Jiao P, Plessen B (2017) The world's earliest Aral-Sea type disaster: the decline of the Loulan Kingdom in the Tarim Basin. Sci Rep 7:43102
Mischke S, Wünnemann B (2006) The Holocene salinity history of Bosten Lake (Xinjiang, China) inferred from ostracod species assemblages and shell chemistry: possible palaeoclimatic implications. Quat Int 154–155:100–112
Mischke S, Zhang C (2011) Ostracod distribution in Ulungur Lake (Xinjiang, China) and a reassessed Holocene record. Ecol Res 26:133–145
Mischke S, Zhang C, Liu C, Zhang J, Jiao P, Plessen B (2019) The Holocene salinity history of Lake Lop Nur (Tarim Basin, NW China) inferred from ostracods, foraminifera, ooids and stable isotope data. Glob Planet Change 175:1–12
Mischke S, Zhang C, Liu C, Zhang J, Lai Z, Long H (2018) Landscape response to climate and human impact in western China during the Han Dynasty. In: Yang LE, Bork H-R, Fang X, Mischke S (eds) Socio-environmental dynamics along the historical Silk Road. Springer-Nature Press, Heidelberg, 45-66pp
Prejevalsky NM (1879) From Kulja, across the Tian Shan to Lob-Nor. Translated by E.D. Morgan. Including notices of the lakes of Central Asia. Sampson Low, Marston, Searle, & Rivington, London, 251pp
Qin X, Liu J, Jia H, Lu H, Xia X, Zhou L, Mu G, Xu Q, Jiao Y (2012) New evidence of agricultural activity and environmental change associated with the ancient Loulan kingdom, China, around 1500 years ago. Holocene 22:53–61
Qu J, Zheng B, Yu Q, Zhao A (2004) The yarding landform of Aqik Valley in the east of Lop-Nur and its relationship with the evolution of the Kumtagh Desert. J Desert Res 24:294–300 (in Chin with Engl abstract)
Razakov RM, Kosnazarov KA (1996) Dust and salt transfer from the exposed bed of the Aral Sea and measures to decrease its environmental impact. In: Micklin PP, Williams WD (eds) The Aral Sea Basin. NATO ASI Series (Series 2. Environment), vol 12. Springer, Berlin, 95–102pp
Shao Y, Gong H, Gao Z, Liu L, Li L, Zhang T (2012) SAR data for subsurface saline lacustrine deposits detection and primary interpretation on the evolution of the vanished Lop Nur Lake. Can J Remote Sens 38:267–280
Sun M, Ma L (2018) Potassium-rich brine deposit in Lop Nor basin, Xinjiang, China. Sci Rep 8:7676
Sun X, Liu C, Jiao P, Yan H, Chen Y, Ma L, Zhang Y, Wang C, Li W (2016) A further discussion on genesis of potassium-rich brine in Lop Nur: evaporating experiments for brine in gypsum-bearing clastic strata. Mineral Deposits 35:1190–1204 (in Chin with Engl abstract)
Sun Z, Yang F, Zhang Z, Li S, Li D, Peng L, Zeng X, Xu Y, Mao S, Wang Q (1997) Sedimentary environments and hydrocarbon generation of Cenozoic salified lakes in China. Petroleum Industry Press, Beijing, 363pp
Sun Z, Zeng X, Chen K (1992) Discovery of foraminifera in the salt lake, Chaidamu Basin and its geological significance. Acta Petrolei Sin 13:252–257
Thevs N (2011) Water scarcity and allocation in the Tarim Basin: decision structure and adaptations on the local level. J Curr Chin Aff 40:113–137
van Geest GJ, Coops H, Roijackers RMM, Buijse AD, Scheffer M (2005) Succession of aquatic vegetation driven by reduced water-level fluctuations in floodplain lakes. J Appl Ecol 42:251–260
Wang F, Ma C, Xia X, Cao Q, Zhu Q (2008) Environmental evolution in Lop Nur since late Pleistocene and its response to the global changes. Quat Sci 28:150–153
Wang M, Huang X, Liu C, Li H, Zhao Z (1999) The discovery of foraminiferal fossils in cores from hole KI in Lop Nur, Xinjiang. Geol Rev 45:158–162 (in Chin with Engl abstract)
Wang ML, Liu CL, Jiao PC, Han WT, Song SS, Chen YZ, Yang ZC, Fan WD, Li TQ (2001) Saline 3 Lop Nur. Geological Publishing House, Beijing, Xinjiang (in China)
Wang M, Liu C, Jiao P, Yang Z (2005) Minerogenic theory of the superlarge Lop Nur potash deposit, Xinjiang, China. Acta Geol Sin 79:53–65
Wang M, Pu Q, Liu C, Chen Y (2000) Quaternary climate and environment in the Lop Nur, Xinjiang. Acta Geol Sin 74:273–278

Wang S (1996) The changes of lower course of Tarim River in historical period. Arid Land Geogr 19:10–17 (in Chin with Engl abstract)
Woldai T (1983) Lop-Nur (China) studied from Landsat and SIR-A imagery. ITC J 3:253–257
Xia X (ed) (1987) The scientific expedition in Lop Nur. Science Press, Beijing, 325pp. 夏训诚主编. 罗布泊科学考察与研究. 北京: 科学出版社 (in Chinese)
Xia X, Mu G, Lei J (2002) Some new progress in scientific research on the Lop Nur Lake region, Xinjiang, China. Sci China (Ser D) 45:148–156
Xu H, Ye M, Li J (2008) The water transfer effects on agricultural development in the lower Tarim River, Xinjiang of China. Agric Water Manage 95:59–68
Xu Z, Liu Z, Fu G, Chen Y (2010) Trends of major hydroclimatic variables in the Tarim River basin during the past 50 years. J Arid Environ 74:256–267
Xuanzang (645) Datangxiyuji (Tang Dynasty). Reprinted in 1985 by Zhonghua Book Company, Beijing
Yan F, Ye Y, Mai X (1983) The sporo-pollen assemblage in the Luo-4 drilling of Lop Lake in Uygur Autonomous Region of Xinjiang and its significance. Seismol Geol 5:75–80 (in Chin with Engl abstract)
Yang Q (2004) Lop Nur is not a wandering lake: views from the evolution of Lop Nur. J Lake Sci 16:1–9 (in Chin with Engl abstract)
Yang X, Liu Z, Zhang F, White PD, Wang X (2006) Hydrological changes and land degradation in the southern and eastern Tarim Basin, Xinjiang, China. Land Degrad Dev 17:381–392
Zhang F, Wang T, Yimit H, Shi Q, Ruan Q, Sun Z, Li F (2011) Hydrological changes and settlement migrations in the Keriya River delta in central Tarim Basin ca. 2.7–1.6 ka BP: inferred from ^{14}C and OSL chronology. Sci China Earth Sci 54:1971–1980
Zhang H, Wu JW, Zheng QH, Yu YJ (2003) A preliminary study of oasis evolution in the Tarim basin, Xinjiang, China. J Arid Environ 55:545–553
Zhang J, Lu H, Wu N, Qin X, Wang L (2012a) Palaeoenvironment and agriculture of ancient Loulan and Milan on the Silk Road. Holocene 23:208–217
Zhang JF, Liu CL, Wu XH, Liu KX, Zhou LP (2012b) Optically stimulated luminescence and radiocarbon dating of sediments from Lop Nur (Lop Nor), China. Quat Geochronol 10:150–155
Zhao S, Xia X (1984) Evolution of the Lop Desert and the Lop Nor. Geogr J 150:311–321
Zheng M, Qi W, Wu Y (1991) Sedimentary environment since the Late Pleistocene and prospect of potash development in the Lop Nur Salt Lake. Chin Sci Bull 36:1810–1813 (in Chin)
Zhou T (1978) On the migration of Lop Nur. J Beijing Normal Univ (Nat Sci) 3:34–41 (in Chin)

Chapter 8
Uvs Nuur: A Sentinel for Climate Change in Eastern Central Asia

Michael Walther, Wolfgang Horn and Avirmed Dashtseren

Abstract The Uvs Nuur is located in Northwest Mongolia bordering with Russia. It is the largest lake of Mongolia (5.5 times larger than Lake Constance in Southern Germany), but not the deepest. The diameter of this round-shaped lake is ca. 80 km and its eastern shoreline is located 1300 km west-northwest of the capital city Ulaanbaatar. Studies in the Uvs Nuur Basin show that past lake-level fluctuations may have been triggered by climate change and/or by changes in the catchment hydrology such as the damming of tributaries by mobile dunes. Direct human impacts on the lake level and water quality of Uvs Nuur due to recent mining or other industrial activities were not identified till now. Increasing pasture load by livestock breeding in the lake-near areas, withdraws of river water for irrigation and increasing tourism infrastructure with insufficient wastewater management need to be under control in the future to maintain the pristine state of Uvs Nuur. The Holocene lake development shows remarkable higher and lower lake levels. There were higher lake levels at Uvs Nuur and Bayan Nuur in the middle Holocene in accordance with most Mongolian lakes. Late Holocene lower levels were identified at Uvs Nuur at -8 m and lower. The recent (2018) tendency of the lake level is slightly negative after a high level at the beginning of the new century (2000). The biological and hydrological data for Uvs Nuur confirm its natural status as a hyposaline, temperate dimictic lake with sometimes polymictic periods in the summer. The investigations show a fully adapted biological inventory of organisms and natural physical and chemical attributes for lake and river waters in accordance with the specific climate and drainage area conditions.

M. Walther (✉) · A. Dashtseren
Mongolian Academy of Sciences, Institute of Geography and Geoecology, Baruun Selbe 15, Chingeltei District, 4th Khoroo, Ulaanbaatar 15170, Mongolia
e-mail: mwaltherub@gmail.com

A. Dashtseren
e-mail: dashka.ig@gmail.com

W. Horn
Saxon Academy of Sciences at Leipzig, Research Group "Limnology of Reservoirs", Leipzig, Germany
e-mail: horn.hw@t-online.de

S. Mischke (ed.), *Large Asian Lakes in a Changing World*, Springer Water,
https://doi.org/10.1007/978-3-030-42254-7_8

Keywords Hydrology · Paleolimnology · Lake levels · Lake deposits · Late Quaternary

8.1 Introduction

The Uvs Nuur Basin is located in northwestern Mongolia (Fig. 8.1) and represents the northernmost part of the Gobi Desert (lake centre: 50.349843° N, 92.653940° E). Ulaangom, the capital city of the Uvs Aimag ("Aimag" mong. = province), is the coldest Aimag Centre of Mongolia due to its orographic location with winter inversion in the Uvs Nuur Basin (Fig. 8.2). The mean annual temperature is −3.5 °C, and the annual amplitude of 86 °C as well as the mean annual precipitation of 130 mm characterize the extreme continental climate.

The Uvs Nuur („Nuur" mong. = lake) has a water surface of 3650 km^2 (1997), an average water depth of 11.9 m (maximum depth 22 m), and a catchment area of 71,100 km^2. The lake is located in the western part of its catchment area with a lake-level altitude of 760 m above sea level (asl; 1997). The comparison of the hydrographic data with Lake Urmia (5470 km^2, 16 m max. depth, altitude 1280 m asl), Van Lake (3522 km^2, 451 m max. depth, altitude 1648 m asl) and the Aral Sea (8300 km^2 in 2015, altitude 27.5 m asl) shows the peculiarities of the Uvs Nuur. It is the largest lake in Mongolia, located in the south Siberian border zone near the Russian border. However, it is not the deepest lake in Mongolia.

Tes Gol („Gol" mong. = river) with a length of 568 km, an average discharge of 56 m^3/s and a river basin of ca. 33,400 km^2 is the main tributary river of Uvs Nuur.

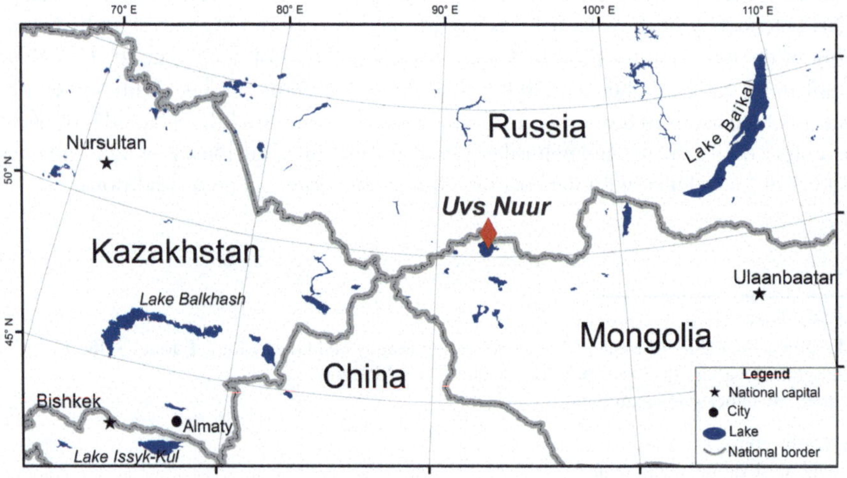

Fig. 8.1 Location of the lake Uvs Nuur in eastern Central Asia

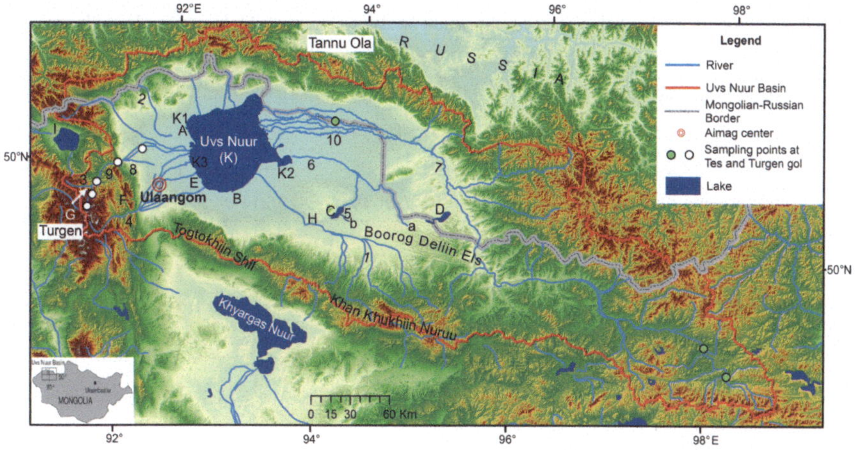

Fig. 8.2 Location, sampling sites, rivers and lakes of Uvs Nuur Basin (Locations see Table 8.1)

Smaller tributaries are the Turgen Gol and Kharkhiraa Gol in the west, rivers from the south Khan Khukhiin Mountain Range and from the north (Russian) Tannu Ola Mountain Range (Fig. 8.2). The Tes Gol delivers 85% of the total runoff to the lake (Davaa et al. 1991).

The vegetation cover is characterized by semi-desert and steppe vegetation in the basin, and forest steppe (north exposed forests) in the surrounding mountain ranges.

In order to investigate the Quaternary lake history, several sediment cores were obtained between 1995 and 1997 near the lake shore and from Bayan Nuur 30 km east of Uvs Nuur (Walther 1999; Walther et al. 2000), which is separated from Uvs Nuur by a large dune belt (a part of Boorog Deliin Els; Fig. 8.2).

Beside the lake location in the western part of the catchment, its central part is characterized by four geomorphological units: lacustrine sediments and landforms (i), eolian forms with extended dune fields (ii), pediments and river deltas (iii), and alluvial fans and estuaries (iv). Walther (1999) pointed out that these units are partly intercalated and were formed in different periods of time.

8.2 The History of Uvs Nuur Basin

According to the review of the tectonic setting and history during the Mesozoic and Cenozoic era in Central Asia by Dobretsov et al. (1996), the Altai-Sayan area and Mongolian Altai represented a zone of elevations and erosion during the Triassic. Tectonic activities increased west of the Mongolian Altai during the Lower Jurassic and intramontane and foremontane troughs developed in the Mongolian Altai and its forelands. The intramontane basins partially inherited Triassic rift depressions.

Table 8.1 Coordinates and character of important analysed rivers and lakes of the Uvs Nuur Basin

No.	Name	Coord. N	Coord. E	Alt. (m)	Type	Q (m^3/s)
1	Baruunturuun Gol	49° 39′ 06″	94° 24′ 09″	1250	2	
2	Borshoo Gol	50° 34′ 03″	91° 46′ 44″	1265	2	0.9
3	Dshibertu Gol	49° 53′ 29″	91° 21′ 08″	1867	1	1.1
4	Kharkhiraa Gol	49° 46′ 47″	91° 51′ 49″	1450	1	7.3
5	Khustay Gol	49° 56′ 05″	94° 05′ 39″	1000	3	1.1
6	Nariin Gol	50° 18′ 43″	93° 34′ 13″	775	3	4.5
7a	Tes Gol-1	49° 16′ 14″	98° 08′ 41″	1800	3	
7b	Tes Gol-4	49° 39′ 29″	95° 43′ 47″	1270	3	
7c	Tes Gol-5	50° 34′ 11″	93° 37′ 34″	800	3	48
8	Tsunkheg Gol	50° 02′ 35″	91° 37′ 20″	1650	1	
9a	Turgen Gol (spring)	49° 43′ 56″	91° 19′ 06″	2700	1	0.7
9b	Turgen Gol-0.3	49° 49′ 18″	91° 21′ 21″	2180	1	2.4
9c	Turgen Gol-1	49° 54′ 04″	91° 22′ 14″	1800	1	4.0
9d	Turgen Gol-3	50° 05′ 18″	91° 37′ 50″	1320	1	7.9
9e	Turgen Gol-4	50° 11′ 57″	91° 47′ 03″	1050	2	0.2
10	Ukhug Gol	50° 29′ 41″	93° 36′ 02″	800	3	
A	Baga Nuur	50° 23′ 14″	92° 12′ 53″	763	5	
B	Baga-South	50° 01′ 00″	92° 52′ 23″	763	6	
C	Bayan Nuur	49° 59′ 29″	93° 56′ 56″	932	4	
D	Doroo Nuur	50° 00′ 52″	94° 55′ 00″	1148	5	
E	Thermokarst Lake	50° 06′ 56″	92° 23′ 00″	760	5	
F	Khukh Nuur	49° 50′ 00″	91° 43′ 04″	1950	4	
G	Nogoon Nuur	49° 49′ 03″	91° 16′ 12″	2650	4	
H	Shavart Nuur	49° 54′ 43″	93° 43′ 37″	982	5	
I	Uureg Nuur	50° 05′ 26″	91° 03′ 00″	1425	6	
K1	Uvs Nuur (NW shore)	50° 25′ 58″	92° 14′ 16″	760	6	
K2	Uvs Nuur (E bay)	50° 16′ 07″	93° 25′ 08″	760	6	
K3	Uvs Nuur (SW shore)	50° 07′ 11″	92° 22′ 58″	760	6	
a	Spring of Nariin Gol	50° 23′ 35″	92° 10′ 27″	780	7	
b	Spring of Khustay Gol	49° 58′ 12″	91° 23′ 27″	1660	7	

Numbers and coordinates correspond to the sampling sites in Fig. 8.2; sampling sites with numbers and characters (7a–7c; 9a–9e) are only marked by dots in Fig. 8.2; some wells and springs are not shown in Fig. 8.2; type: 1 = alpine river or alpine spring stream (slope >2%), 2 = mountain river (slope 0.5–2%), 3 = plain river (slope <1%), 4 = freshwater lake, 5 = subsaline lake, 6 = salt lake, 7 = spring/well. Q = discharge as measured on sampling date. Salinity classification according to Hammer et al. (1983)

The foremontane troughs in Mongolia are bounded by the Khangai-Mongol Upland to the west and northwest, and their fault pattern progressively disappears into the structures of the Altai and Western Sayan Mountains (directly bordering north of Uvs Nuur).

Fault zones in combination with fissure eruptions had been reactivated in the Late Mesozoic (Upper Jurassic-Lower Cretaceous), and the formation of fault-related troughs occurred in both the western and the eastern zones of the Central Asian belt. According to Dobretsov et al. (1996), the Central Asian mountain belt is the largest intracontinental belt in the world. It extends for more than 5000 km from the Tien Shan (55° E) to the Stanovoy Range (125° E) with alternating mountain ranges and tectonic depressions and the northern areas of the Central Asian mountain belt (Altai, Sayan, Baikal and Stanovoy Mountain ranges). Wide and flat depressions formed along the western margin of the Khangai Mountains such us Uvs Nuur, Sangiin Dalai Nuur and Khar Nuur in the same period.

The Mesozoic faults occurred locally in a belt between the Khangai-Mongolian Block and the Mongolian Altai. They follow inherited Paleozoic faults and accompany the Pre-Altai belt of Jurassic depressions. Remnants of smooth surfaces are preserved in many places and allow the assessment of later movements.

After the Cretaceous/Early Paleocene period of tectonic stability and peneplanation, tectonic reactivation occurred in the Altai-Sayan area within the three following zones: (1) the Zaysan zone (including the Zaysan Lake in Kazakhstan) and its southeastern continuation in Dzungaria, (2) the zone between the Mongolian Altai and the Khangai-Mongolian uplift, and (3) the zone between the Northern Mongolian Khangai and Kuznetsk-Sayan uplifts. These zones also were active earlier in the Mesozoic (Dobretsov et al. 1996).

The Cenozoic north-south compression and shear movements along pre-existing northwest-trending faults led to the separation of the Altai and Kuznetsk-Sayan uplifts. The movement of these blocks northward and northwestward was impeded by structures of the southern folded margin of West Siberia. The development of the Cenozoic structure of the Altai resulted in the variable morphology of the troughs and their separating ridges (Delvaux et al. 1995).

Little is known about the extent of the lake area expansion in the 'Eopleistocene' (3–0.7 million years), which Deviatkin (1981) attributed to unfavourable outcrop conditions and a smaller expansion compared to the Pliocene. In general, however, a significantly wetter climate than today is assumed (Deviatkin 1981). Early Pleistocene sediments, which are referred to in the Russian literature as "alluvial-proluvial" deposits but are probably of lacustrine origin, were found at the foot of the Togtokhiin Shil as the southern frame of the Uvs Nuur Basin in 25 m thickness and in the south adjoining also to the depression of, The Valley of the Great Lakes' including the Khyargas Basin. There, these deposits correlate with a paleo-shoreline at 1190 m asl, from which Deviatkin and Murzaev (1989) derived an early Pleistocene transgression up to this height in the Uvs Nuur Basin. In the middle Pleistocene [ca. 360,000 years before present (BP)] drier climatic conditions led to a decline in the lake level (Deviatkin 1981). The majority of Russian and Mongolian authors assume that during the second half of an interglacial period and in the first half of a glacial period, increased precipitation must have occurred together with declining temperatures. In this context, Murzaev (1952) and Deviatkin (1981) refer to pluvial conditions in the intramontane depressions during the further course of the middle Pleistocene, whereas the westernmost range of the Mongolian and Gobi Altai, the Khuvsgul Mountains, the Khentii and the Khentii mountains were glaciated. Increased rainfall with reduced evaporation and inflow of meltwater caused a pronounced transgression of the lakes, so that Murzaev (1954) reconstructed a Middle-Pleistocene lake stage. Therefore, a closed body of water in the Great Lakes Depression with an absolute lake level at 1500 m asl and a water depth of ca. 140 m is assumed. Due to thermoluminescence dating of laminated clays, located in a lakeside terrace between Khar Nuur and Khar Us Nuur in the central part of the Great Lakes Valley south of the Uvs Nuur Basin, this transgression occurred in the period between 280,000 and 177,000 years BP (Deviatkin and Murzaev 1989). Evidence for the former establishment of a lake with ca. 140 m water depth mostly comes from morphographic studies. According to this work, both lakes—the Uvs Nuur and the southern bordering Khyargas Nuur Basin—show paleo-lake bottoms covered by proluvial debris and eolian sands (Gluchowskaja 1989). For the late middle Pleistocene, strong tectonic activities were described, which brought about significant changes in the fluvial drainage system. As a result of the uplift of the Khan Khukhiin and Togtokhiin Shil mountains, the Uvs Nuur Basin became separated from the southern Khyargas Basin, whereas there was already a regressive stage at the same time in the lakes of the Great Lakes Valley south of Uvs Nuur (Sevastianov et al. 1993, 1994).

8.3 The Late Quaternary Landscape History of Uvs Nuur Basin

Late Glacial carbonatic limnopsammites were found only in few places in the southern shoreline area of Uvs Nuur, on the pediment and in the dunes at the eastern bay. Lacustrine authigenic carbonate sands were recorded up to an elevation of 30–35 m above the present lake level; they were radiocarbon-dated to the Late Glacial of the last ice age. On this basis, a maximum high stand of 800 m asl was calculated in the entire basin. Lacustrine authigenic carbonate precipitation in the past can be compared with recent conditions at the eastern shore of Uvs Nuur on the western front side of the Boorog Deliin Els dune field. The principal genesis of carbonate is based on a strong summer biomass production (algae) with colonization by submerged macrophytes and also emersed species in past and present times. Flooded shore-near interdune valleys provide an aquatic habitat during rising groundwater tables, which finally lead to shallow water habitats with the presence of gastropods and bivalves. The Late Glacial lake carbonates of the eastern bay contain hollow stipe fragments of reed (*Phragmites)*, indicating a littoral paleoenvironment supported by the recorded molluscs (Walther 1999).

The formation and location of dunes in the Uvs Nuur Basin are strongly connected with lake development; and their distribution, genesis and chronostratigraphy had been the subject of detailed studies by Grunert et al. (1999) and Grunert (2000). Sparse or no vegetation and exposed sandy dry surfaces as source areas of eolian material are generally conditions required for dune formation. Grunert et al. (1999) distinguished three generations of dune development in different locations in the dune fields: (a) active dunes (barchans and barchanoid landforms) with little or no plant cover and no soil formation; (b) partially active dunes with a plant cover of up to 50%, which are parabolic in shape and in places incorporated remains of ancient dunes; and (c) inactive dunes with fully developed soils (Kastanozems) and a plant cover in excess of 70%. Hence, the dunes are major paleoclimatic indicators, especially in view of the fact that they are connected with lake deposits and with fossil and relict soils. The thermoluminescence dates reported by Grunert et al. (1999) support the conclusion that the youngest phase of distinct dune migration probably occurred during the Younger Dryas. More recent optically stimulated luminescence (OSL) dates by Grunert (2000) confirm this inferred phase of dune formation. However, further OSL dates from the Bayan Nuur area show that the main phase of dune formation was considerably older than the last ice age, probably middle or even Early Pleistocene. Accordingly, evidence is convincing that higher lake levels flooded an "old" dune relief at Bayan Nuur in the same way as it is happening today on the eastern shore of Uvs Nuur. This observation supports the inferences of Walther (1999), Naumann (1999), Naumann and Walther (2000) and Grunert et al. (2000) that lake sediment had been deposited in an older dune relief.

Pediments with alluvial cover beds were mostly formed during the cold semi-humid to semi-arid phases of the last ice age. Older and higher-located eroded pediments are observed in the foreland of the Turgen Mountain Range west of Uvs Nuur. Nevertheless, the pediments directly adjacent to the present-day Uvs Nuur shoreline were formed in the last glacial stage. They have been described, partly mapped, and dated by Walther and Naumann (1997), Walther (1998), Lehmkuhl (1999, 2000) and Grunert et al. (2000). Periglacial cryogenic features in the pediments show that the genesis of pediments is based on alternating active accumulation periods and episodic or rhythmic cold-dry sections with sufficient soil moisture for freezing with reduced accumulation. The extensive pediments to the south and west of Uvs Nuur received their modern shape during the Last Glacial Maximum, yet they represent a polygenetic chronosequence. Older and generally higher-lying units [P2 and P3 pediments; Walther and Naumann (1997)] were described in the foreland of west- (Turgen) and south-bordering (Togtokhiin Shil) mountain ranges covering larger continuous areas (Grunert et al. 2000).

To date, field studies indicate that the Late Glacial highstand of Uvs Nuur covered the lowermost quarter of the Uvs Nuur Basin to ca. 30 m above the present-day water level at maximum. As Walther (1998) and Walther et al. (2016) have shown for more southerly regions of Mongolia (Valley of the Gobi Lakes), wave abrasion of the lake eroded the youngest Late Glacial pediments and cut cliff edges into it.

Indicators of Holocene lake-level fluctuations at Uvs Nuur are observed and described by Walther (1999, 2010) and are represented by lacustrine deposits (carbonatic limnopsammites) and linear landforms such as recent and fossil sand bars and stretches of inactive or "fossil" cliff zones. There is now no active cliff at the present-day lake level of 760 m asl along the entire shoreline of Uvs Nuur, apart from small sand cliffs formed at the western side of the dune belt Boorog Deliin Els (eastern side of Uvs Nuur). However, if the present tendency of lake-level rise would continue and the lake level would increase by 5–9 m, pediments on the southern side would be undercut and the fossil cliff on the eastern side would be reactivated. Conditions, which existed in the Mid Holocene, would be re-established (Fig. 8.3).

For example, Holocene lake deposits overlie the alluvial fans and river deltas in the area surrounding Uvs Nuur in the north, carbonatic lake deposits occur on the pediments in the south, and lake-induced abrasion landforms are visible on the entire eastern side - the side that is strongly exposed to the westerly winds (Walther 1999, 2010). These land forms include the ingression shore on the west side of the dune field Boorog Deliin Els, continued to the north, the embayed, sand-bar shoreline of the eastern bay and a second small bay at the eastern shoreline, and the cliffs and abrasion platforms of the eastern shoreline of Uvs Nuur (Walther 1999, 2010).

Long stretches of the littoral zone represent delta regions of the Tes Gol in the east and northeast, and of tributaries from the Tannu Ola area in the north and from the Turgen Kharkhiraa catchment in the northwest. Near the shore of the northwest bay, Holocene sand bars have accumulated up to 10 m above the present lake level and fossil cliff stretches with a height of up to 10 m have formed in unconsolidated sediments (Walther 1999). From this, it is concluded that the lake level was 10 m higher after the time when the delta or alluvial fans had been formed. Therefore, the final major phase of delta deposition in the Tes Gol, Tannu Ola and Turgen-Kharkhiraa catchments probably occurred during the transitional phases between the Last Glacial Maximum and the Late Glacial or Early Holocene. However, their formation is assumed to have been polygenetic with later modifications, similar to the development of the pediments.

Both the vegetation and the sedimentological record combined with geomorphological findings clearly show that Late Glacial high lake levels in Uvs Nuur are mainly due to the rapid melting of the valley glaciers of the surrounding mountains, and to local effects such as river blocking by dunes or even tectonic movements. A dry phase apparently occurred in the lowest parts of the basin during the Younger Dryas, although there is evidence that lake levels were substantially higher than today (by as much as 30 m at Uvs Nuur and 48 m at Bayan Nuur in Zuungovi Sum; "sum" mong. = district; Fig. 8.3). Thus, lake-level fluctuations were not necessarily synchronous with the aridity/humidity development of the region. A more humid Late Glacial phase was recognized between 13,000 and 10,700 conventional (conv.) years BP (i.e., uncalibrated ^{14}C years), which, however, was interrupted by two first-order dry phases. A relatively long humid phase with moderate temperatures lasted from 9000 to 5440 conv. years BP. From ca. 5000 conv. years BP, dry and humid phases alternated rapidly until ca. 2800 conv. years BP. Between ca. 2800 and 1400 conv. years BP, conditions became more humid. Pollen analysis (Krengel 2000) and geochemical

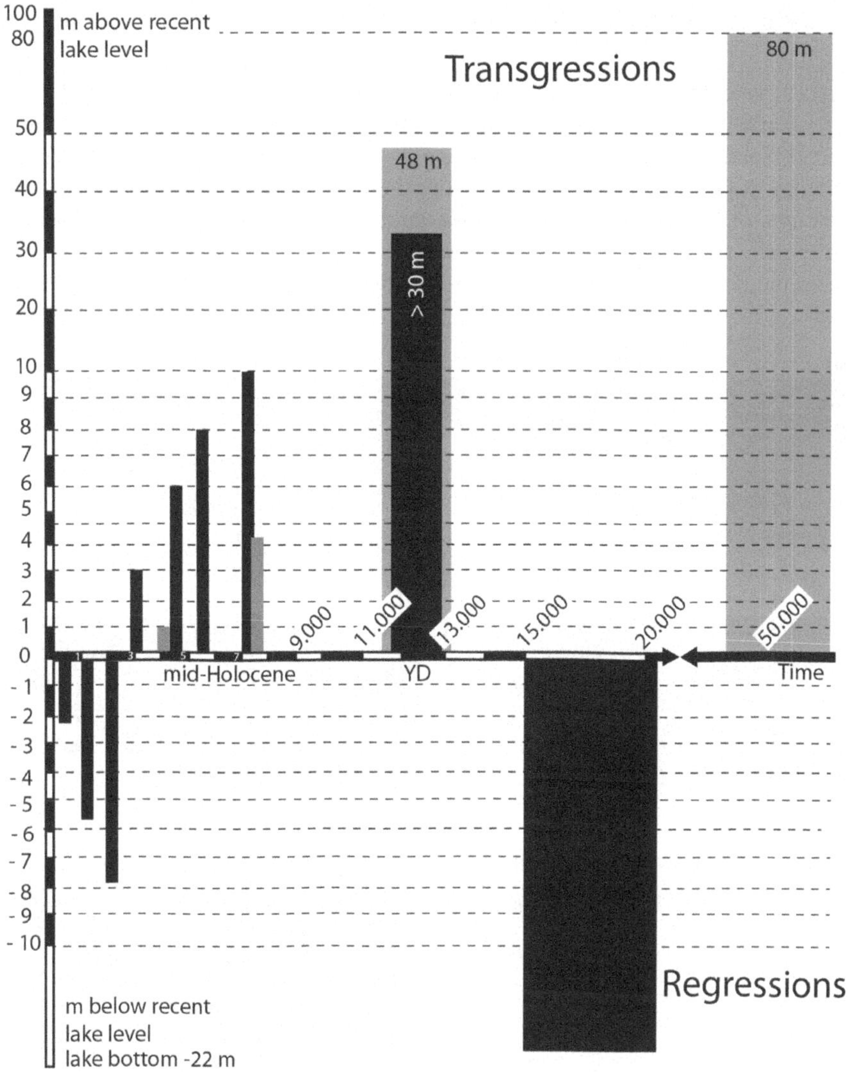

Fig. 8.3 Lake level fluctuations of Uvs Nuur (black) and Bayan Nuur (Zuungovi Sum; gray) after Walther (1999, 2010) and Krengel (2000)

analysis of sediment profiles provide confirming evidence for dry phases of varying length and magnitude characterized by stagnating or falling lake levels during the later Holocene (from the Sub-Boreal onwards and then in the early Sub-Atlantic; Fig. 8.3).

8.4 Recent Lake Conditions

A recent lake-level rise is observed by a series of satellite images in two different regions of Uvs Nuur between 1984 and 2016: at the permafrost-dominated western shores (i) and at the southern Lido-Laguna shoreline (ii; Fig. 8.4). The transgressive tendency of the lake level results from a positive precipitation balance in the catchment area of Uvs Nuur, especially in the large Tes Gol Basin. Satellite images for one specific test plot of the southern shoreline of Uvs Nuur recorded the lake-level rise in the last three decades (Fig. 8.4). The lake level of Uvs Nuur is rising since

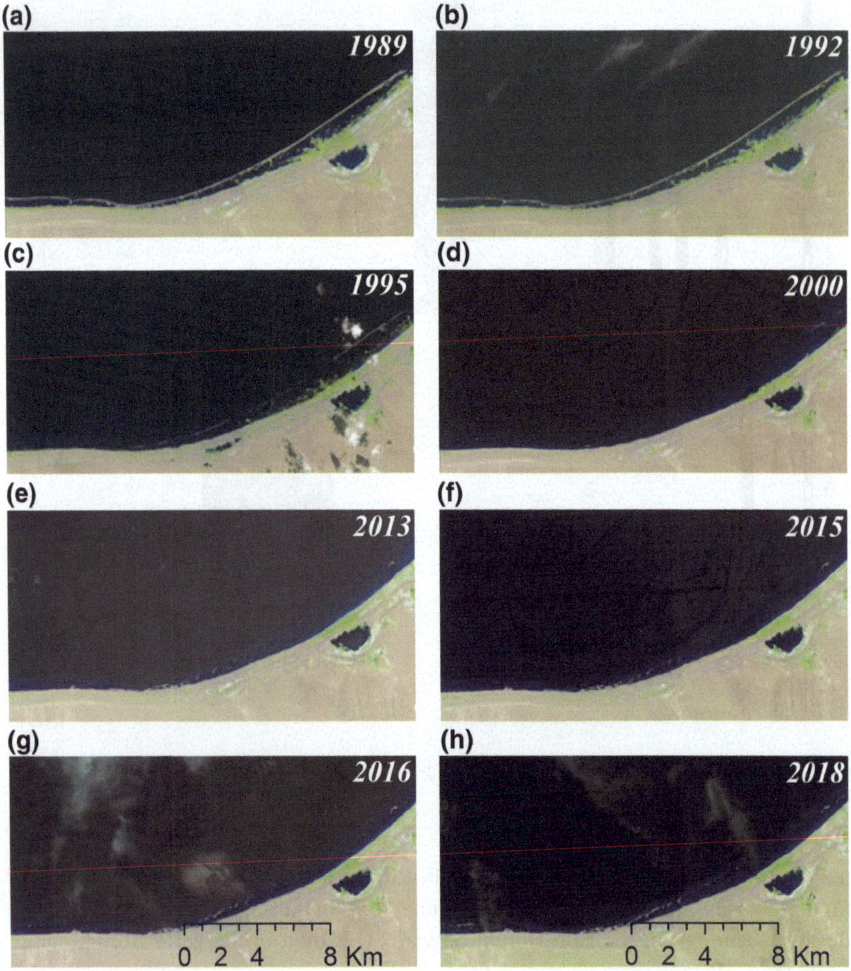

Fig. 8.4 Comparison of the shoreline constellation at the southern side of Uvs Nuur in the years between 1989 and 2018 based on satellite imagery

1984 or possibly a few years earlier. Lake-level data before 1984 are very uncertain and suggest minor lake level-fluctuations.

Batnasan and Batima (2000) pointed out that the sub-recent and recent lake-level rise is caused by increasing precipitation showing a synchronous alternation of lake-level rise and precipitation at the station Ulaangom between 1957 and 1996. The Tes Gol with a river basin of 33,358 km^2 has the second largest catchment area in northwestern Mongolia only outmatched by the larger Khovd Gol River Basin (50,149 km^2) originating in the high mountain area of the Altai south and southwest of the study area. The Tes Gol has a low load of total suspended solids (TSS; annual load 18,420 t) and low catchment erosion (0.552 t/km^2/year; Batnasan and Batima 2000). The reason for the low catchment erosion and low sediment load of the river is the forest cover of the South Siberian Taiga of most of the river basin. Based on a first water-surface calculation by Tserensodnom 1971 (3350 km^2) the lake level first dropped until 1984 and then continuously rose to a maximum in 2011. The Lido-Laguna Shoreline of the southern Uvs Nuur had been flooded rapidly in the first half of the 1990s due to the rising lake level. The lake level slightly fell after 2011 until 2015, and former strand bars became more and more visible although the low level of 1989 was not reached (Fig. 8.5). The lake level is rising again since 2015 as recorded by satellite images (Figs. 8.4 and 8.5).

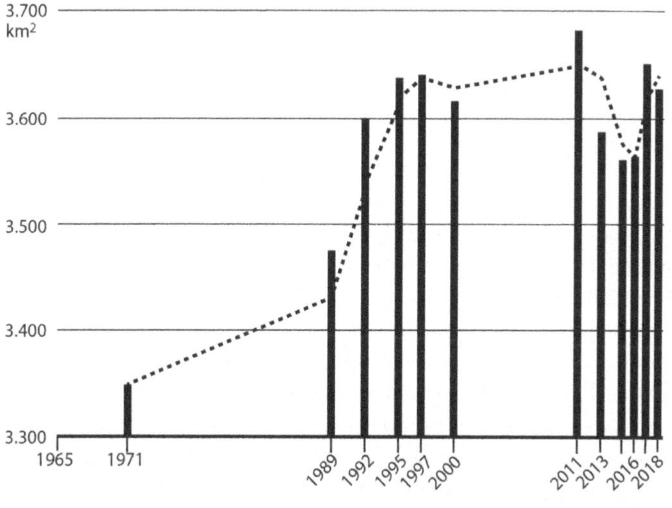

Fig. 8.5 Lake-surface-area changes between 1971 and 2018 (Table 8.2); the black columns show the lake surface, the dotted line shows the moving average of the lake surfaces

Table 8.2 Surface-area changes of Uvs Nuur between 1971 and 2018 based on satellite images (modified after Batnasan and Batima (2000) and own results) and lake levels between 1965 and 2000 [m asl approximately, data source: gauging station Davst, Institute of Meteorology and Hydrology, Ulaanbaatar; provided lake-level data based on an illustration of Horn et al. (2016)]

Sources and year	Lake surface (km^2)	Altitude [m asl (appr.)]
1965	No data	759.50
Tserensodnom (1971, 2000) (data from 1971)	3350.00	760.25
Landsat-4 1984	No data	759.75
Landsat-5 1989	3476.10	No data
NOAA-14 1992	3599.50	760.75
Landsat-5 1995	3637.10	No data
NOAA-14 1997	3641.30	761.50
Landsat-7 2000	3616.40	761.50
Landsat-7 2011	3683.90	No data
Landsat-8 2013	3586.97	No data
Landsat-5 2015	3562.40	No data
Landsat-8 2016	3565.75	No data
Sentinel 2 2018	3628.70	No data

8.4.1 General Conditions of Lakes in the Uvs Nuur Basin

Lakes in large, closed tectonic depressions tend to be terminal salt lakes if the climate is semiarid or arid (Hammer 1986), as it is in the case of the hyposaline (3–20‰) lake Uvs Nuur. Lakes with an outlet are typically fresh (<0.5‰) unless they are fed by saline waters. Two typical examples for such freshwater lakes are the Nogoon Nuur - an alpine lake formed by glacial erosion and a moraine dam, and the probably tectonically originated Bayan Nuur amid the dunes east of Uvs Nuur which is fed by groundwater. In case of the subsaline (0.5–3‰) Doroo Nuur, which is exclusively fed by groundwater from the dunes without any surface outflow, an underground outflow apparently exists. A stronger salt accumulation would be expected, if the inflow is only compensated by evaporation. The morphometric data of these three lakes and of Uvs Nuur are shown in Table 8.3. The provided parameters of Uvs Nuur and Bayan Nuur are partly based on bathymetric maps (Walther 1999; Naumann and Walther 2000). The shape of Uvs Nuur resembles a shallow bowl: the flat central lake bottom with a maximum water depth of 22 m starts to ascend at a distance of ca. 2 to 10 km from the shore. The shores in the southwest and northeast have wide shallow areas with extended reed belts, but major parts of the lake are encircled by surf shores with sand and gravel bars such as the Lido-Laguna shoreline. Bayan Nuur, amid a tectonic Southwest-Northeast oriented fault line, is shaped like a wide graben with its western shore descending steeply to a maximal depth of 29 m. Maximum depths of lakes in the Uvs Nuur Basin is not explicitly related to altitude or lake area. The

Table 8.3 Morphometric parameters of the selected lakes (Horn et al. 2016)

Lake (= Nuur)	A_L (km^2)	A_C (km^2)	z_{max} (m)	V (km^3)	z_r (%)	l_{max} (km)	l_{shore} (km)	IBP	Alt. (m)
Bayan	32	730	29[a]	0.33[f/a]	0.46	12.3	31	11	932
Doroo	72	900	29[e]	0.5*	0.40	15	63		1148
Nogoon	0.09	4.5	35*	0.0014*	10	0.6	1.5	0.9	2650
Uvs	3600	70,220	22[b]	48.4[b/c/d/f]	0.03	85.5	415	116	760

A_L: lake area, A_C: catchment area, z_{max}: maximum depth, V: volume, z_r: relative depth ($z_r = 50\% \, z_{max} \, [\pi/\text{area}]^{1/2}$), l_{max}: maximum length of continuous open-water area, l_{shore}: shoreline length, IBP: Index of Basin Permanence (IBP = 1000 volume (km^3)/l_{shore}); Hutchinson (1957), Keller (1962). [a]Naumann and Walther (2000), [b]Walther (1999), [c]Shil'krot et al. (1993), [d]Dulmaa (1979), [e]Sviridenko et al. (2007), [f]own calculations based on [a], [b] and [c]respectively; *estimation/extrapolation from own field measurements local geomorphology determines the relative lake depth; it is highest in the alpine lake and lowest in mountain-foreland lakes. The Index of Basin Permanence (IBP) is used to assess the importance of littoral areas for the whole lake's volume and can be taken as a rough measure of ageing and "life expectancy" of a lake. The lake Uvs Nuur is the largest and has the highest value (116; Table 8.3).

8.4.2 Hydrological Conditions of Uvs Nuur

Davaa (1996) modelled the water balance of Uvs Nuur for the years 1964 through 1989 based on precipitation, runoff of the important tributaries, calculated evaporation data and lake level. From the data, a mean residence time of 19 years was calculated. More recent calculations of Davaa et al. (2007) yielded 14.7 years. Hebert (2004) estimated a residence time twice as high on the basis of the average tritium content (for details see Horn et al. 2016). The lake-level fluctuations of Uvs Nuur have been recorded since 1964, and they show a mean annual variation of ca. 16 cm with the maximum in July, caused by the periodicity of precipitation (to 2/3) and thermal expansion (to 1/3). The increase of ca. 2 m in the years between 1965 and 1999 reflects a tendency to more humid conditions against the trend of decreasing lake levels of many other large saline lakes (Williams 1993). Salinity measurements from the first half of the twentieth century yielded significantly higher values (up to 19.7 g/l in 1941; Davaa 1996) than those of more recent surveys (mean 1996–1998: 13.6 g/l; Horn et al. 2016). Lower lake levels in the first half of the twentieth century and increasing levels afterwards, sometimes interrupted by phases of lake-level declines, are implied. The lake's hydrograph resembles an ascending sine function with a period of ca. 20 years. Parallels of water-level changes and annual runoff of northwestern Russian rivers with solar activity were found (Zaretskaya et al. 1992; Batjargal et al. 1993).

8.4.3 Physical Water Properties of the Lakes

Water temperature and its temporal and spatial changes are an important factor of the lakes' physical structure, as well as pH, oxygen, nutrients, conductivity, and transparency. The processes of heat exchange of lakes are regulated by local climatic conditions. Decreasing temperatures in autumn and the very cold winters of Mongolia result in a long ice cover. Uvs Nuur has a full ice cover of more than 150 days and an ice thickness of 98 cm on average (data for the years 1980–1997; Institute of Meteorology and Hydrology, Ulaanbaatar). In Bayan Nuur, ice thickness is 150 cm, in Doroo Nuur 125 cm.

The spring is characterized by strong winds and rapidly increasing air temperatures. The consequence is a low thermal stability leading with intensive solar radiation to ice break-up at the end of April, and to high hypolimnion temperatures between 8 and 12 °C. During summer, periods of stagnation are interrupted by stronger winds; epilimnic mixing processes and thus multiple thermoclines can be found in the stratified lakes. A distinct linear temperature profile in Uvs Nuur in August 1999 resulted from strong winds before and during the measurements, when large parts of the lake underwent mixing. Thus, Uvs Nuur is classified as a temperate dimictic lake, but sometimes also as a temperate pleomictic lake, i.e. with periods of full mixing in the summer (Hutchinson and Löffler 1956; Uhlmann and Horn 2001). The dimictic alpine lake Nogoon Nuur has a relatively deep thermocline due to the high altitude; Bayan and Doroo Nuur are dimictic too. In large lakes like Uvs Nuur also significant horizontal temperature differences can be found (Horn et al. 2016).

The vertical distribution of the oxygen content of lake water depends on mixing regime and trophic conditions. The stratified Uvs Nuur shows the profile of an oligotrophic lake; the oxygen saturation is always around 100% in the epilimnion. It dropped to about 70% in the hypolimnion in August 1997 whilst the thermocline reached the lake bottom at the measuring station in the other years. The Nogoon Nuur has an orthograde profile like an oligotrophic lake and a slight oxygen oversaturation, while Bayan Nuur shows a metalimnic oxygen-content maximum and a saturation deficit of 75% in the hypolimnion - both are characteristics of mesotrophic lakes.

The optical characteristics of the water are significantly depending on the lakes' morphometry; shallow lakes with a maximum depth less than 2 m as well as shallow areas of Uvs Nuur are mostly very turbid due to resuspension. Secchi depths less than 0.5 m are common. In the case of Uvs Nuur, the predominant northwestern wind causes a high turbidity along the whole eastern shoreline visible on satellite images. The water is most transparent in the deep central and southern areas. In the deep northwestern bay, Secchi depths of 3.8–5.7 m were measured. Bayan Nuur as a deep lake is better protected from wind and Secchi depths reach values of 4.8–6.0 m although its trophic index is higher than that of Uvs Nuur. In Doroo Nuur, the Secchi depth is only 2 m due to the dominance of picoplanktic cyanobacteria and strong calcite precipitation. Only 3.8 m were measured in Nogoon Nuur, possibly because of the inflow of glacier milk.

8.4.4 Chemical Water Properties of the Lakes

The conductivity range shows large differences between the lakes mainly caused by the vertical climatic gradient and the open or closed lake-basin setting. While the water of the alpine lake Nogoon Nuur has a conductivity of 0.05 mS/cm similar to that of rain water, the salt accumulation increases along the flow to the terminal lake (Bayan Nuur 0.44 mS/cm, Doroo Nuur 0.71 mS/cm, Uvs Nuur 19.3 mS/cm). The pH value of the surface water in all lakes exceeds 8.2 even in the very soft waters of Nogoon Nuur. The high alkalinity is a general property of the surface waters in the Uvs Nuur Basin. The pH of hypolimnic waters of Bayan Nuur is with 7.7 lower, which indicates the accumulation of CO_2 typical for productive lakes. Groundwater inflow can be also significant indicated by the formation of calcareous tufa in Bayan Nuur.

The chemical composition of the lakes' water ranges from Ca-HCO_3 dominance via Mg/Na-HCO_3 to Na–Cl/SO_4 dominance with increasing salinity. While Nogoon Nuur is a typical freshwater lake, Bayan Nuur can be classified as a "hard water" lake of the transition type (no clear dominance of one cation-anion pair) and Uvs Nuur is a hyposaline Na–Cl/SO_4 lake (Paul and Horn 2000; Paul 2012a; Horn et al. 2016). Horizontal differences in the chemical composition of Uvs Nuur are significant, as the most important tributaries flow into the lake in the east where large shallow areas predominate.

The reliability of nutrient analysis is low due to the long transport of samples to the lab. Nevertheless, the available data suggest that there is a lack of dissolved inorganic nitrogen (N) in some of the lakes. An indication for this is the trophic state index [TSI; after Carlson (1977) with a modification for N-limited lakes by Kratzer and Brezonik (1981)]. The concept of TSI assumes that the partial indices for chlorophyll, Secchi depth and total phosphorus (TP) should not differ significantly from each other in normal P-limited lakes. N limitation is indicated if the index for TP is significantly higher than the other two indices (chlorophyll, Secchi depth). Ideally, the smaller one of the indices for TP and total nitrogen (TN) should represent the limiting nutrient. The index for TN was smaller than that for TP in Doroo Nuur and the eastern bay of Uvs Nuur, which possibly indicates N limitation. However, it must be noticed that the index for Secchi depth in these lakes is significantly higher than that of chlorophyll due to the high mineralization of the water, and light might be a limiting factor, at least in parts. The micronutrient Fe^{2+} was not analysed due to the conditions of sample treatment. Growth limitations may occur in alkaline salt lakes with high P content (Evans and Prepas 1997). A further source of uncertainty is the chlorophyll concentration and cyanobacteria dominance in many lakes, which was likely underestimated by Fluorometer measurements. As a result, seemingly more oligotrophic conditions might have been inferred. However, significantly more data for the whole summer are required for a reliable assessment of the lakes' nutrient situation and trophic state. For these reasons, a definite statement about the limiting resources for the growth of phytoplankton in these lakes is not possible.

The trophic state of the lakes shows a distinct dependence on the lakes' depth. Lakes with a depth of <3 m have a strong tendency to be eutrophic. The deeper lakes (Bayan, Doroo, Nogoon, and the deep northwestern part of Uvs Nuur) are oligotrophic, some with the tendency to mesotrophy. The internal nutrient load as resuspended sediments is probably a major source for phytoplankton growth. Shil'krot et al. (1993) identified Uvs Nuur as a meso- to eutrophic lake, whilst the Limnological Catalog of Mongolian Lakes describes the lake as mesotrophic (LcoMl 2018). This is similar to the findings presented here for the southwestern shore and eastern bay of Uvs Nuur and can be seen as a sign of the spatial heterogenity of this large lake. These authors investigated the shallow northeastern part of Uvs Nuur in Russia near the mouth of Tes Gol and other regions of the lake.

8.4.5 Lake Biology

The phytoplankton community is more or less dominated by cyanobacteria, especially in lakes with pH ≥ 9 and possible N limitation. The cyanobacteria dominance can be explained by their high competitive strength for inorganic carbon at high pH and their capability to grow at low N/P proportions. This dominance is apparently independent of the trophic state. *Planktolyngbya contorta* Lemm., a cyanobacterium with coiled filaments, is typical for many sub- and mesosaline lakes of the region. In the less saline, eutrophic lakes, nanoplanktic coccal cyanobacteria forming gelatinous colonies, are dominant and Chlorophyceae (mostly Chlorococcales) are important also. Especially *Ankistrodesmus*, *Lagerheimia* and *Oocystis* spp. are often found. Another important group in the eutrophic, hyposaline lakes are the Dinophyceae, represented by *Ceratium*, different *Gymnodinium* and *Peridinium* species. Diatoms that play a major role in many lakes of the temperate climatic zone are only of secondary importance. This is, however, only true for the phytoplankton biomass; the number of species is highest in diatoms and Chlorococcales.

In zooplankton, that rarely attains high densities (dry weight >1 mg/l), calanoid Copepoda with a few species (mainly *Arctodiaptomus salinus* Daday) are dominant in the large lakes. Cyclopoids of the species *Mesocyclops leuckarti* Claus, *Megacyclops viridis* Jurine and *Cyclops vicinus* Uljanin live in many lakes, but are mostly subdominant. From Nogoon Nuur, a hitherto unknown species - *Cyclops glacialis* (Flößner 2001) is described. Cladocera are found with more species, especially in the shallow lakes. *Moina mongolica* Daday has been found in the salt lakes, whereas *Bosmina longirostris* O.F.M., *Daphnia* spp., *Ceriodaphnia pulchella* Sars and some other species are typical for the less saline lakes. Flößner et al. (2005) gives a complete list of all planktic and benthic Cladocera and Copepoda. In the eutrophic shallow lakes of this area, the crustacean plankton is sparse; here rotifers are the major constituent of the zooplankton. Typical salt lake species are *Brachionus plicatilis* O.F.M., *B. quadridentatus* Hermanns and *Hexarthra* spp. Many other species like *Keratella* spp., *Polyarthra* spp., *Filinia longiseta* Ehrenberg, *Trichocerca* sp., *Lecane luna* O.F.M. and *Asplanchnella* sp. live in the less saline lakes. The crustacean *Leptodora kindtii* Focke found in Bayan Nuur is the only representative of large-sized predatory invertebrates. *Branchinecta orientalis* Sars, an Anostraca, was identified in a small nameless alpine lake (Horn and Paul 2004). Among the few determined planktic protozoa, most remarkable is the newly described choanoflagellate *Acanthocorbis mongolica* (Paul 2012b). It belongs to the mainly marine order Acanthoecida that form delicate, species-specific loricae of siliceous costae, and it is the first of this group found in freshwater.

The pelagic food web is completed by fish (*Oreoleuciscus* spp. in all lakes except Nogoon; in Doroo also *Esox lucius* L.) and water birds: *Phalacrocorax carbo* L., *Larus argentatus* Pont., *Tadorna ferruginea* Pall., *Podiceps cristatus* L., *Ardea alba* L. and many other benthivorous, planktivorous and piscivorous birds. The Altai osman (*Oreoleuciscus* sp.) is a predatory cyprinid especially distributed in the Altai Mountain lakes and river deltas in northwestern Mongolia including deltas of Uvs

Nuur. The fecundity of the genera *Oreoleuciscus* especially in the Uvs Nuur area had been investigated by Dulmaa et al. (2000). *Oreoleuciscus humilis* Warpachowski 1889, *Oreoleuciscus potanini* (Kessler 1879) and *Oreoleuciscus pewzowi* Herzenstein 1883 had been observed in the lakes of the Uvs Nuur Basin. These fishes have only a minor economic importance until today and there is no commercial fishing activity so far.

The littorals of many lakes are colonized by *Potamogeton pectinatus* L., *Chara* sp., *Cladophora* sp. and other filamentous algae; *Hippuris* sp. and *Utricularia* sp. are often found in the shallow water next to stocks of *Phragmites australis* Steudel. The absence of reed belts in some lakes of the lower catchment (Bayan Nuur, eastern bay of Uvs Nuur, etc.) is apparently mainly caused by cattle trampling, but also strong wave action and moving ice are possible reasons. Typical representatives of the bottom fauna in the salt lakes are Gammaridae, Corixidae, Chironomidae, Culicidae, Ephydridae and Stratiomyiidae. Molluscs are virtually absent in the salt lakes; they are numerous in the freshwater lakes, where also Ephemeroptera, Trichoptera, Odonata, Coleoptera, Hirudineae and some other groups are found.

8.4.6 Rivers of Uvs Nuur Basin

The morphological, hydrological, physico-chemical and biological investigations of the rivers reveal four main river types in the Uvs Nuur region: alpine spring streams (i), alpine rivers (ii), mountain (iii) and plain rivers (iv). The distribution of aquatic organisms reflects these four river types.

The main river and tributary of Uvs Nuur is the Tes Gol with a length of 568 km (Great Soviet Encyclopedia, 3rd ed.; 757 km according to the Russian Water Inventory), an average discharge of 56 m^3/s and a river basin of ca. 33,400 km^2. It is the main tributary river besides some local basins of Turgen Gol and Kharkhiraa Gol in the west. Rivers from the south Togtokhiin Shil-Khan Khukhiin Mountain Range and from the north (Russian) Tannu Ola Mountain Range and the Tes Gol deliver 85% of the total amount of surface water into the lake (Davaa et al. 1991).

There are a total of 25,000 km of dry river channels in the Uvs Nuur Basin, corresponding to 0.35 km/km^2 (Dorj et al. 1991). Numerous gullies (Mong. Sair) exist mainly in the foothill regions and at the edge of the low-lying plain. Dry channels are also abundant in the extensive delta of the Tes Gol. It is not permanently filled with water; many river courses are commonly dry. The mountain brooks in the area of runoff formation in the upper and middle reaches are continuously filled with water. They often divide farther downstream into many branches and seep away into the gravelly ground of the debris cones after 10–20 km and reach the terminal lake Uvs Nuur only during periods of flood in spring or following heavy rainfall events. The groundwater flowing towards Uvs Nuur emerges again at places with decreased gradient and/or lower water permeability near the lake forming springs and extended wetlands. Tes Gol and Nariin Gol have permanent discharge all over the year. Tes Gol is perennial because of its large catchment area which covers 47% of the Uvs

Nuur Basin including water-rich tributaries from Siberia. The lowland river Nariin Gol, which is fed by the huge groundwater body of the dunes east of Uvs Nuur, has a relatively steady discharge and no infiltration losses.

Most of the runoff originates in the precipitation-rich mountain areas and the basins surrounding mountain ranges. However, some rivers (e.g. Nariin Gol, Khustay Gol) drain the dune areas of the arid lower part of the basin too. Their springs are mainly fed by groundwater. which originates from mountain rivers blocked by dunes such as the Khustay Gol, which is the continuation of the seeped Baruunturuun Gol (Grunert and Klein 1998).

The annual course of discharge in most of the rivers reflects the precipitation peak in the summer months and the dry, cold winter conditions with low discharge and complete freezing. The time of the snowmelt runoff depends on the catchment altitude. The maximum can be distinct for rivers of the lower mountains (Borshoo, Tes) or superimposed by the rain peak for high alpine rivers such as Kharkhiraa Gol and Turgen Gol (Horn and Paul 2000; Paul 2012a; Horn et al. 2016).

8.5 Relevance of Human Impact

Despite the lake's unique location in the midst of surrounding mountain ranges, with an outstanding importance of flora and fauna, its economic and touristic importance is still very low, with increasing tendency in recent years. Agricultural land use is mainly focused on alluvial plains near settlements such as Ulaangom due to the availability of water for irrigation. Land is mostly cultivated as horticulture. Center-pivot irrigation with water withdrawal especially from Kharkhiraa Gol was used during socialism times before 1992. The areas with sea buckthorn plantations grew in the last ten years near the city of Ulaangom. The number of livestock increased rapidly and overgrazing occurs in the steppe areas of the lower locations near the lake areas as well as in the upper regions of summer pasture in the mountains. Between 1990 and 2017, the number of animals doubled from 1,666,600 to 3,130,000 with two reduction periods between 2000 and 2003, and between 2009 and 2011 due to severe Zud events (i.e., severe winters; Statistical Office Report 2018).

An increasing number of national and international tourists, Ger Camp constructions with insufficient water and sanitation management and an inefficient waste management in the countryside outside the settlements stresses the landscape. The Uvs Nuur Strictly Protected Area (7125.45 km^2) was established in 1994 and combined with an UNESCO Biosphere Reserve in 2000 including the Turgen Mountains in the west, the sand fields of Boorog Deliin Els in the east, and the Khan Khukhiin National Park (2205.5 km^2) and Khyargas Nuur National Park (3320.8 km^2) in the south. These protected areas attracted domestic and international tourists in the past ten years with increasing tendency. Being the water towers for Uvs Nuur, the parks have a high importance for the future water quality and quantity of the lake.

Fishery in the lakes and rivers is so far insignificant because there are no commercial fisheries; even anglers are unimportant.

The central aimag center Ulaangom with its insufficient wastewater management and sanitation is a problem for the water quality of the rivers receiving polluted water. Otherwise, the distance between the discharge of the pollutants and the confluences with the lake Uvs Nuur is relatively large, so that the self purification in the natural water courses eliminates nutrients and other pollutants. Thus, impacts on Uvs Nuur are relatively low. If the pollution pressure will increase due to an increasing population and intensification of industrial and other commercial activities without appropriate wastewater management measures, environmental problems will become relevant. However, present and past pollution levels were not specifically investigated resulting in a lack of reference data.

Human impacts caused by increasing livestock breeding and some mining activities in the upper reaches of tributary rivers (e.g. in the Turgen Gol watershed) are apparently still without impacts on the lake-water quality of Uvs Nuur.

Lake-level oscillations are caused by climate change in the river basins of Tes Gol as the main tributary, and other smaller rivers of the surrounding Turgen, Togtokhiin Shil, Khan Khukhiin and Tannu Ola Mountain ranges. Water withdrawal for drinking-water purposes, industry and irrigation is still of minor significance due to low volumes of diverted water.

8.6 Future Perspective and Challenges

If the present status of Uvs Nuur can be conserved in the future, the lake will remain a remarkable rare example of a large natural lake ecosystem without severe human impact. The present rising lake level suggests that water withdrawal is not a significant issue in the river basins of the main tributaries. However, touristic pressure, water withdrawal for agricultural use and mining activities are a potential threat for the lake and its water quality and quantity. Therefore, the early protection status as Biosphere Reserve (1992) of a shore-near zone around the lake and the Strictly Protected Area Turgen-Kharkhiraa bears a certain warranty for conservation and protection against exploitation of natural/mineral resources with unsustainable techniques. The Uvs Nuur Basin and the adjacent mountain ranges have an outstanding importance as different untouched ecosystems. Nevertheless, the landscape stress is present by overgrazing of livestock, breeding and insufficient wastewater management and sanitation in the aimag and sum centers. Especially the increase of domestic tourism in times of growing economy requires an improved waste management in and outside the settlements.

Lake-level fluctuations in historical and recent times were and are mainly driven by runoff changes of the main tributaries. Hydrological changes in the headwater regions and middle courses of the tributaries as result of human impacts can still be neglected.

Acknowledgements The authors thank the German Research Foundation (DFG) to be members of the Research Group Uvs Nuur between 1995 and 1999 in order to find out most of the presented results and for financial support of the expeditions. We thank the Mongolian Academy of Sciences, Institute of Geography and Geoecology, namely Academician Prof. Dr. D. Dordschgotov and Ms A. Tschimegsaikhan, the Institute of Biology, namely Prof. Dr. A. Dulmaa, and the former Institute of Hydrology, namely Dr. G. Davaa, for their logistic and scientific support.

References

Batjargal Z, Yadamsuren K, Davaa G (1993) K voprosu o svyazi mezhdu urovnem besstochnogo ozera Ubsu-Nur i zentrom tyazhesti solnechnoj sistemy. In: Theses of the international conference. Experiment Uvs Nuur, Kyzyl pp 17–21 (in Russian)

Batnasan N, Batima P (2000) Hydrological regime and water quality of lakes in the Great Lakes Depression, Mongolia. In: Walther M et al. (eds) State and dynamics of geosciences and human geography of Mongolia. extended abstracts of the International Symposium Mongolia 2000, vol 205. Berliner Geowissenschaftliche Abhandlungen, Reihe 4, Berlin, pp 100–106

Carlson R (1977) A trophic state index for lakes. Limnol Ocean 22:361–369

Davaa G (1996) Gidrologo-ekologicheskaya osobennost' osjor Mongolii i tendentsiya ikh razvitiya. Dissertation, Ulaanbaatar (in Mongolian with Russian abstract)

Davaa G, Dashdeleg N, Tsugar T, Tseveendorj N (1991) Dynamics of water balance of Uvs Lake. In: Global Change and Uvs Nuur. State Committee for Nature and Environmental Protection of MPR, National Ecological Center. Thesis of presentations. Ulaanbaatar-Ulaangom, 12–19 August, 1991 pp 18–19

Davaa G, Oyunbaatar D, Sugita M (2007) Surface water of Mongolia. Interpress, Ulaanbaatar, pp 55–68

Delvaux D, Theunissen K, Van der Meer R, Berzin NA (1995) Dynamics and paleostress of the Cenozoic Kurai-Chuya Depression of Gorny Altai (South Siberia): tectonic and climatic control. Russ Geol Geophys 36:31–51

Deviatkin EV (1981) Cenozoic of Inner Asia. Nauka, Moskow (in Russian)

Deviatkin EV, Murzaev EM (1989) The late Cenozoic of Mongolia. The Mongolian Altai. Nauka, pp 42–47 (in Russian)

Dobretsov NL, Bulsov MM, Delvaux D (1996) Meso- and Cenozoic tectonics of the Central Asian Mountain belt: effects of lithospheric plate interaction and mantle plumes. Int Geol Rev 38:430–466

Dorj S, Dashdeleg N, Bayaraa Z, Amarsanaa Y (1991) The resources of surface waters in Uvs lake basin. In: Global Change and Uvs Nuur. Thesis, Intern Scient Conference (12.-19. 08. 1991). State Committee for Nature and Environmental Protection of MPR National Ecological Center, Ulaanbaatar—Ulaangom, pp 27–28

Dulmaa A (1979) Hydrobiological outline of the Mongolian lakes. Int Rev Gesamten Hydrobiol 64:709–736

Dulmaa A, Horn W, Paul M (2000) Fecundity of *Oreoleuciscus* Warpachowski (Pices, Cyprinidae) from the Uvs Nuur hollow. In: Walther M et al. (eds) State and dynamics of geosciences and human geography of Mongolia. Extended abstracts of the International Symposium Mongolia 2000, vol 205. Berliner Geowissenschaftliche Abhandlungen, Reihe 4, pp 139–147

Evans JC, Prepas EE (1997) Relative importance of iron and molybdenum in restricting phytoplankton biomass in high phosphorus saline lakes. Limnol Oceanogr 42(3):461–472

Flößner D (2001) *Cyclops glacialis* n. sp. (Copepoda: Cyclopoida) from a high mountain lake in northwestern Mongolia. Limnologica 31:303–306

Flößner D, Horn W, Paul M (2005) Notes on the cladoceran and copepod fauna of the Uvs Nuur Basin (northwest Mongolia). Int Rev Hydrobiol 90:580–595

Gluchowskaja NB (1989) Late Cenozoic of Mongolia. Ubsunurskaja w padina (in Russian)

Grunert J (2000) Paleoclimatic implications of dunes in the Uvs Nuur-Basin, Western Mongolia. In: Walther M et al. (eds) State and dynamics of geosciences and human geography of Mongolia. Extended abstracts of the International Symposium Mongolia 2000, vol 205. Berliner Geowissenschaftliche Abhandlungen, Reihe 4, pp 2–8

Grunert J, Klein M (1998) Binnendünen im nördlichen Zentralasien (Uws Nuur, westliche Mongolei). Berliner Geographische Abhandlungen 63:45–66

Grunert J, Klein M, Stumböck M, Dash D (1999) Bodenentwicklung auf Altdünen im Uvs Nuur Becken. Die Erde 2:97–115

Grunert J, Lehmkuhl F, Walther M (2000) Paleoclimatic evolution of the Uvs Nuur Basin and adjacent areas (Western Mongolia). Quat Int 65-66:171–192

Hammer UT (1986) Saline lake ecosystems of the world. Monographiae Biologicae, vol 59. Series ed.: Dumont, H.J. Dr W. Junk Publishers, 616pp

Hammer UT, Shamess J, Haynes RC (1983) The distribution and abundance of algae in saline waters of Saskatchewan, Canada. Hydrobiologia 105:1–26

Hebert D (2004) Tritium in Wässern des Uws-Nuur-Beckens (Mongolei). Wissenschaftliche Mitteilungen des Instituts für Geologie, TU Bergakademie Freiberg 27:5–12

Horn W, Paul M (2000) Abiotic and biotic features of typical rivers in the Uvs Nuur Basin. In: Walther M et al. (eds) State and dynamics of geosciences and human geography of Mongolia. In: Extended abstracts of the International Symposium Mongolia 2000, vol 205. Berliner Geowissenschaftliche Abhandlungen, Reihe 4, pp 122–130

Horn W, Paul M (2004) Occurrence and distribution of the Eurasian *Branchinecta orientalis* (Anostraca) in Central Asia (Northwest Mongolia, Uvs Nuur Basin) and in other holarctic areas. Lauterbornia 49:81–91

Horn W, Paul M, Dulmaa A, Davaa G, Tseveendorj N, Uhlmann D (2016) The recent surface and subsurface waters in the endorheic Uvs Nuur Basin (Northwest Mongolia). Abhandlungen der Sächsischen Akademie der Wissenschaften zu Leipzig, Mathematisch-Naturwissenschaftliche Klasse 66(2): 1-32

Hutchinson GE (1957) A Treatise on Limnology, vol 1. Geography, Physics, and Chemistry, 1015pp

Hutchinson GE, Löffler H (1956) The thermal classification of lakes. Proc Natl Acad Sci 42:84–86

Keller R (1962) Gewässer und Wasserhaushalt des Festlandes, Haude & Spener, 520pp

Kratzer CR, Brezonik PL (1981) A Carlson-type trophic state index for nitrogen in Florida Lakes. Water Resour Bull 17(4):713–715

Krengel M (2000) Discourse on history of vegetation and climate in Mongolia—palynological report of sediment core Bayan Nuur I (NW-Mongolia). In: Walther M et al. (eds) State and dynamics of geosciences and human geography of Mongolia. Extended Abstracts of the International Symposium Mongolia 2000, vol 205. Berliner Geowissenschaftliche Abhandlungen, Reihe 4, 80–84

LcoMl (Limnological catalog of Mongolian lakes) (2018). http://oslo.geodata.es/mongolian_lakes/index.php?page=home&lang=en

Lehmkuhl F (1999) Rezente und jungpleistozäne Formungs- und Prozeßregionen im Turgen-Charchira, Mongolischer Altai. Die Erde 130:151–172

Lehmkuhl F (2000) Alluvial fans and pediments in Western Mongolia and their implications for neotectonic events and climatic change. In: Walther M et al. (eds) State and dynamics of geosciences and human geography of Mongolia. Extended abstracts of the International Symposium Mongolia 2000, vol 205. Berliner Geowissenschaftliche Abhandlungen, Reihe 4, pp 14–21

Murzaev EM (1952) Mongolian People's Republic: Physical Geography of Mongolia (in Russian)

Murzaev EM (1954) Mongolian People's Republic: Physical Geography of Mongolia (in Mongolian)

Naumann S (1999) Spät- und postglaziale Landschaftsentwicklung im Bayan Nuur Becken (Nordwestmongolei). Die Erde 130(2):151–172

Naumann S, Walther M (2000) Mid-Holocene Lake-Level Fluctuations of Bayan Nuur (NW Mongolia). Marburger Geographische Schriften 35:15–27

Paul M (2012a) Limnological aspects of the Uvs Nuur Basin in northwest Mongolia. Dissertation, Fakultät Umweltwissenschaften, TU Dresden, 198pp

Paul M (2012b) *Acanthocorbis mongolica* nov. spec. Description of the first freshwater loricate choanoflagellate (Acanthoecida) from a Mongolian lake. Eur J Protistol 48(1):1–8

Paul M, Horn W (2000) Lakes in the Uvs Nuur Basin: ecosystem features and typology. In: Walther M et al. (eds) State and dynamics of geosciences and human geography of Mongolia. Extended abstracts of the International Symposium Mongolia 2000, vol 205. Berliner Geowissenschaftliche Abhandlungen, Reihe 4, pp 131–138

Sevastianov DV, Seliverstov JP, Cernova GM (1993) Landscape development in the Uvs Nuur basin. Vestnik SPbGU Ser 7:71–81 (in Russian)

Sevastianov DV, Schuwalow WF, Neutrujeva IJU (1994) Limnology and Paleolimnology of Mongolia. Nauka, Sankt-Petersburg (in Russian)

Shil'krot GS, Kretova SP, Smirnova EV (1993) Ecosystem of the Ubsu Nuur Lake - Potential and Water Quality. In: Informational problems of biosphere investigations: experiment "Ubsu-Nur". Russian Academy of Sciences, Siberian Division, Pushchino

Sviridenko B, Pyak A, Sviridenko T (2007) Occurrences of Charophycean algae in Mongolia, Tuva and Khakassia. In: Problems of Botany of South Siberia and Mongolia. Proceedings of 6th international scientific-practical conference (Barnaul, 25–28 October 2007), pp 299–302 (in Russian)

Tserensodnom J (1971) Lakes of Mongolia. Ulaanbaatar, Academy of Science, MPR (in Mongolian)

Tserensodnom, J (2000) Catalogue of the Mongolian Lakes. Ulaanbaatar, 69p (in Mongolian)

Uhlmann D, Horn W (2001) Hydrobiologie der Binnengewässer, 528pp

Walther M (1998) Paläoklimatische Untersuchungen zur jungpleistozänen Landschaftsentwicklung im Changai Bergland und in der nördlichen Gobi (Mongolei). Petermanns Geographische Mitteilungen 142:205–215

Walther M (1999) Befunde zur jungquartären Klimaentwicklung rekonstruiert am Beispiel der Seespiegelstände des Uvs Nuur-Beckens (NW-Mongolei). Die Erde 130(2):131–150

Walther M (2010) Paleo-environmental changes in the Uvs Nuur basin (Northwest Mongolia). Erforschung Biologischer Ressourcen in der Mongolei 11:267–279

Walther M, Naumann S (1997) Beobachtungen zur Fußflächenbildung im ariden bis semiariden Bereich der West- und Südmongolei (Nördliches Zentralasien). Stuttgarter Geographische Studien 126:154–171

Walther M, Janzen J, Riedel F, Keupp H (eds) (2000) State and dvnamics of geosciences and human geography of Mongolia. In: Extended Abstracts of the International Symposium Mongolia 2000, vol 205. Berliner Geowissenschaftliche Abhandlungen, Reihe A

Walther M, Enkhjargal V, Gegeensuvd T, Odbaatar E (2016) Environmental changes of Orog Nuur (Bayan Khongor Aimag, South Mongolia). Lake deposits, paleo-shorelines and vegetation history. Erforschung biologischer Ressourcen der Mongolei 13:37–57

Williams WD (1993) The worldwide occurrence and limnological significance of falling water levels in large, permanent saline lakes. Verhandlungen der Internationalen Vereinigung für Theoretische und Angewandte Limnologie 25:980–983

Zaretskaya IP, Lobasenko GV, Ptitsina NG, Rumyantsev VA, Tyasto MI (1992) River runoff changes during solar-activity cycles. Biofizika 37(3):517–523

Index

S. Mischke (ed.), *Large Asian Lakes in a Changing World*, Springer Water,
https://doi.org/10.1007/978-3-030-42254-7

The manufacturer's authorised representative in the EU is Springer Nature Customer Service Centre GmbH, Europaplatz 3, 69115 Heidelberg, Germany. If you have any concerns regarding our products, please contact ProductSafety@springernature.com

Printed and bound by CPI Group (UK) Ltd, Croydon, CR0 4YY
05/10/2026
02228120-0006